Advanced Control of Grid-Integrated Renewable Energy Power Plants

Advanced Control of Grid-Integrated Renewable Energy Power Plants

LMI-based Design in the Takagi-Sugeno Framework

Horst Schulte
Head of Control Engineering Group
HTW Berlin, Germany

WILEY

IEEE Press

This edition first published 2024

Registered Offices
John Wiley & Sons, Inc., 111 River Street, Hoboken, NJ 07030, USA
John Wiley & Sons Ltd, The Atrium, Southern Gate, Chichester, West Sussex, PO19 8SQ, UK

For details of our global editorial offices, customer services, and more information about Wiley products visit us at www.wiley.com.

Wiley also publishes its books in a variety of electronic formats and by print-on-demand. Some content that appears in standard print versions of this book may not be available in other formats.

Library of Congress Cataloging-in-Publication Data:

Names: Schulte, Horst, author.
Title: Advanced control of grid-integrated renewable energy power plants : LMI-based design in the Takagi-Sugeno framework / Horst Schulte.
Description: Hoboken, NJ : Wiley, 2024. | Includes bibliographical references and index.
Identifiers: LCCN 2024017996 (print) | LCCN 2024017997 (ebook) | ISBN 9781119701392 (hardback) | ISBN 9781119701408 (adobe pdf) | ISBN 9781119701286 (epub)
Subjects: LCSH: Power-plants. | Renewable energy sources. | Automatic control. | Convex functions.
Classification: LCC TJ164 .S345 2024 (print) | LCC TJ164 (ebook) | DDC 621.042–dc23/eng/20240517
LC record available at https://lccn.loc.gov/2024017996
LC ebook record available at https://lccn.loc.gov/2024017997

Cover Design: Wiley
Cover Image: © zhihao/Getty Images

Set in 9.5/12.5pt STIXTwoText by Straive, Chennai, India

Contents

Preface

Over the past 20 years, there has been a considerable growth in the installation capacity of renewable energy sources (RES) in power systems all over the world [IRENA, 2020]. In contrast to conventional power plants such as coal-fired or nuclear power plants with large synchronous machines, power plants with RES are much smaller and integrated into the grid via power electronic (PE) interfaces. This leads to a highly distributed structure and heterogeneous integration across all voltage levels into the power grid.

Currently, most renewable energy (RE) systems feed as much power as possible into the grid, but provide little or no grid support (i.e. ancillary services). In the future, this will no longer be sufficient to stabilize the grid due to the decreasing share of synchronous machine-based units. However, it is clear that the decentralized regenerative generators must be able to adapt the power quickly to the demand for frequency and voltage regulation via a suitable mix of grid-following and grid-forming converters, even without or less additional costs arising from battery energy storage systems (BESS) and as few additional control units as possible, such as FACTS and STATCOMS.

There is, still today, no common road map on how to systematically address this problem of grid operation with power-electronically coupled RES in practice. The book aims to contribute to this scientific question using methods of physical modeling, mathematical systems theory, and control engineering. Emphasis is on the development of model-based control concepts for photovoltaic and wind power plants, which are also intended to provide auxiliary services for grid frequency and voltage stabilization.

This book is useful for readers who want to gain knowledge of wind and photovoltaic (PV) power plant modeling and control with a deep understanding of model-based control concepts for optimal resp. flexible energy conversion and grid integration by inverters with grid-following and grid-forming operation. It is suitable for students from the fifth semester onward, experts from industry and academics, and PhD students who want to get an overview and possible ideas on an exciting research topic. The intention was to provide a book with well-understandable description of the design models, the controller functions, and their validation presented in a unified control theory framework. The model and controller validation is done with high fidelity models (used for wind turbine control) or scenarios with measurement uncertainties (considered for PV systems).

This book is based on the results of research projects I have worked on and mostly managed over the past 10 years. The aim was to make the results available to the community in a closed and understandable form, for both professionals and graduate students working on control systems or electrical power systems. I thank my wonderful colleagues and collaborators Prof. S. M. Esmailifar, Prof. J. Fortmann, E. Gauterin, Dr. S. Goerg, N. Goldschmidt, Prof. N. Klaes, Dr. S. Kusche, Dr. Z. R. Labidi, and Dr. F. Pöschke for the many valuable discussions and exchanges without which this book would not have taken the form it has.

Berlin
September 2023

Horst Schulte

Acronyms

AC	alternating current
AGC	automatic generation control
APC	active power control
AS	ancillary services
AVR	automatic voltage regulator
BMI	bilinear matrix inequality
CIG	converter-interfaced generation
DC	direct current
DER	distributed energy resource
DFIG	doubly fed induction generator
DOF	degree of freedom
DPT	demanded power tracking
DSO	distribution system operator
DVPP	dynamic virtual power plant
EMT	electromagnetic transient
EPS	electrical power systems
ESC	extremum-seeking control
FACTS	flexible alternating current transmission system
FAST	Fatigue, Aerodynamics, Structures, and Turbulence model
FLL	frequency-locked loop
HAWT	horizontal-axis wind turbine
HC	hill climbing
HIL	hardware in the loop
HPP	hybrid power plant
HV	high voltage
HVDC	high-voltage direct current
IBR	inverter-based resource
iff	if and only if
IGBT	insulated-gate bipolar transistor

IOS	input-to-output stability
ISS	input-to-state stability
LDI	linear differential inclusion
LHS	left-hand side
LIDAR	light detection and ranging
LMI	linear matrix inequality
LQR	linear quadratic regulator
LPV	linear parameter variable
LTI	linear time invariant
MATLAB®	Matrix Laboratory
MPP	maximum power point
MPPT	maximum power point tracking
MIMO	multiple-input multiple-output
NREL	National Renewable Energy Laboratory
PE	power electronic
PCC	point of common coupling
PDC	parallel distributed compensation
PLL	phase-locked loop
PMSG	permanent magnet synchronous generator
P&O	perturbation and observation
PI	proportional-integral
pu	per unit
PV	photovoltaics
RPP	renewable energy power plant
RES	renewable energy source
RHS	right-hand side
RMS	root mean square
SISO	single-input single output
SL	sector-nonlinearity (approach)
STATCOM	static synchronous compensator
s.t.	such that
STC	standard test condition
TS	Takagi-Sugeno
TSO	transmission system operator
VPP	virtual power plant
VSC	voltage-sourced converters
WPP	wind power plant
WT	wind turbine
WTG	wind turbine generator

Notation

- Italic denotes scalar physical quantities (e.g. R, L, and C) or numerical variables (e.g. x and y).
- Italic boldface denotes a matrix (capital letters) or a vector (small letters), e.g. $\mathbf{A}$, $\mathbf{B}$, $\mathbf{x}$, and $\mathbf{y}$.
- Unit symbols are written using roman type (e.g. Hz, A, and kV).
- Standard mathematical functions are written using roman type (e.g. e, sin, cos, and arctan).
- Lowercase symbols normally denote instantaneous values (e.g. v and i).
- Uppercase symbols normally denote RMS or peak values (e.g. V and I), exception: active and reactive power denoted as P and Q.
- Subscripts d and q refer to the direct- and quadrature-axis components.
- Subscripts α and β refer to α and β components of a three phase system.
- Italic underline type denotes a phasor (e.g. $\underline{I}$, $\underline{V}$, and $\underline{S}$).
- Lowercase symbol with upper arrow denotes a space vector $\vec{\mathbf{x}}$.

1

Introduction

1.1 Energy Transition

Public discussion about the energy transition is dominated by the need to accelerate the planning and installation of wind farms and photovoltaic (PV) plants, as well as the grid expansion by new lines and deferrable loads. That is obvious and quite understandable. Far less known and not a subject of public debate is that the transition to renewable energy also means that the traditional use of synchronous machines, which still regulate frequency and voltage in the grid [Machowski et al., 2008], is being replaced by power electronic (PE)–based converter-interfaced renewable energy sources (RES). The transition to a massive integration of PE-based power plants [Peng et al., 2018a], also called inverter-based (IBG) [Joi, 2018] or converter-interfaced generators (CIG), which began in the decade of the new millennium, enables relatively flexible and efficient power conversion. Such a transition implies all areas within power systems and may be considered a paradigm shift in the sense of Kuhn [1962]. Because power systems fundamentally change in power generation, transmission, and distribution, this has significant implications for the associated engineering and scientific disciplines, as questions of stability, control, and reliability must be addressed in new ways.[1]

Symptomatic of the fundamental change in power systems is the distributed generation by many heterogeneous units feeding power to different grid levels by power electronic interfaces. At the transmission level, for example, offshore wind farms feed in, at the sub-transmission level the onshore wind farms or large PV power plants are integrated, and at the distribution level medium-sized PV power plants. Instead of generators connected to the grid via stator windings, rotating units in renewable energy systems are decoupled from the grid: these cannot

1 Operation and control, stability analysis, computational methods of power systems, protection and automation, power electronics, active distribution systems, and distributed energy resources. The disciplines were chosen partly to follow the names of CIGRE's study committees and their work areas.

Advanced Control of Grid-Integrated Renewable Energy Power Plants: LMI-based Design in the Takagi-Sugeno Framework, First Edition. Horst Schulte.

contribute to system inertia without additional active control. Until now, the frequency and voltage stability characteristics of power systems have been strongly determined by large synchronous generators driven by steam or gas turbines.

Let us first consider the frequency dynamics in conventional power systems. Assuming a common frequency in the AC network,[2] the instantaneous rate of change of frequency (RoCoF) after the disconnection of a generator (generator tripping) or load from a power system before any control becomes active can be computed as follows [ENTSO-E, 2018]

$$\frac{d\Delta f}{dt} = \frac{f_0\,\Delta P}{2\sum_{i=1}^{N_g} H_i\,S_{n,i}}, \qquad \Delta f = f - f_0, \tag{1.1}$$

with $d\Delta f/dt$ as derivative of the difference frequency f related to the nominal system value f_0, where H_i is the inertia constant

$$H_i = \frac{E_{k,i}}{S_{n,i}} = \frac{J_i\,\omega_{sm}^2}{2}\frac{1}{S_{n,i}}, \tag{1.2}$$

of each generator $i = 1, \ldots, N_g$, where $E_{k,i}$ is the kinetic energy of the generator running at synchronous (nominal) speed $\omega_{sm,i}$. Each turbine's total moment of inertia plus the generator rotor is denoted as J_i. The variable $S_{n,i}$ is the rated apparent power of each generator. Here, the inertia constant H_i describes the property of the synchronous machine i to maintain the system state of uniform rotary motion when no external torque is applied. A typical value of H for a synchronous generator can range from two to nine seconds. The active power deviation ΔP in (1.1) is defined by the summation of all mechanical shaft power input to the generator units $P_{m,i}$ for $i = 1, \ldots, N_g$ and the electrical air-gap power $P_{e,i}(\delta_i)$ caused by system loads in the grid and controlled by the power angle or load angle δ, both expressed in watts

$$\Delta P = \sum_{i=1}^{N_g} \left(P_{m,i} - P_{e,i}(\delta_i)\right), \tag{1.3}$$

where N_g denotes the total number of generator units. For the interpretation of (1.1) it must be noted that RoCoF is triggered by unplanned load shedding or generator tripping. In case of generator tripping k (i.e. $P_{m,k} \to 0$) the remaining generators ($\forall i \neq k$) provide a frequency change inversely proportional to $\sum_{i=1,i\neq k}^{N_g} H_i\,S_{n,i}$. This behavior results from the swing equation of each generator represented by two first-order equations [Machowski et al., 2008]

$$\begin{aligned} \frac{2H_iS_{n,i}}{\omega_s}\,\Delta\dot{\omega}_i &= -d_m\,\Delta\omega_i + P_{m,i} - P_{e,i}(\delta_i), \\ \dot{\delta}_i &= \Delta\omega_i + \omega_s, \end{aligned} \tag{1.4}$$

2 The effects of local frequency variations in power systems during transition as deviations from the global nominal frequency, and their modeling and estimation, are studied, i.e. in Milano and Manjavacas [2020].

with $\Delta\omega_i = \omega_i - \omega_s$, where ω_i denotes the electrical rotor speed, $\omega_s = 2\pi f_o$ denotes the electrical synchronous speed resp. synchronous angular frequency, δ_i denotes the power angle, and d_m is the damping coefficient. Introducing an aggregated[3] rotor speed $\Delta\omega_i \approx \Delta\omega$ and $\Delta\dot{\omega} = 2\pi\Delta\dot{f} \approx \Delta\dot{\omega}_i$ in (1.4), and after summation

$$\frac{2\sum_{i=1}^{N_g} H_i S_{n,i}}{f_o} \Delta\dot{f} = -\sum_{i=1}^{N_g} d_m \Delta\omega + \sum_{i=1}^{N_g} \left(P_{m,i} - P_{e,i}(\delta_i)\right), \tag{1.5}$$

and by neglecting the damping term, we obtain the RoCoF index (1.1). It is evident that the RoCoF value increases caused by a decrease of $\sum_{i=1}^{N_g} H_i S_{n,i}$ due to the substitution of synchronous generators by inertia-less (PV power plants) or inertia-decoupled (wind turbines[4]) generator units. In summary, the energy transition with a high penetration of CIGs with over 80% and more in the future has the effect that the RoCoF value of electrical power systems (EPS) significantly increases which leads to a stronger fluctuation of the grid frequency and an increase of the probability of frequency instability.

In addition to frequency regulation, the voltage control of grids with a high percentage of PE-based converter-interfaced generators must also be addressed. In contrast to grid frequency, which can be modeled as a global quantity, the voltage in power grids has to be considered as a local quantity, which can vary strongly in the permitted bounds from node to node. The voltage level could be changing due to factors related to generation, transmission, and distribution. However, the variability of the load is one of the most important factors. The voltage must remain within a specified range because over-voltage implies an increase in active power losses and, as a long-term effect, can increase the probability of insulation failure. Thus control is needed to maintain the grid voltage in a permissible range, and if a short circuit occurs, it must be ensured that voltage drops are limited.

In AC networks with typically overhead transmission line parameters $X \gg R$ (see [Kundur, 1994], Table 6.1) the voltage states are strongly coupled with the reactive power fluxes. For example, the longitudinal voltage drops in the networks are caused by the reactive power currents. For the line element shown in Figure 1.1, the reactive power flow Q of a single-phase system[5] is given by

$$Q = \frac{EV}{X}\cos\delta - \frac{V^2}{X}, \tag{1.6}$$

3 i.e. Based on the center of inertia (COI) frequency ω_{COI} [Kundur, 1994].

4 Note, this is also the case for wind turbines with doubly fed induction generator (DFIG). Although the generator stator is directly connected to the grid, the inverter-based variable frequency excitation in the rotor decouples the speed of the turbine rotor speed from the grid frequency.

5 A balanced three-phase has three times the power of a single-phase system. The more general case with a resistor R and reactance $X = \omega L$ as line impedance is calculated in Chapter 5, with the result (5.58).

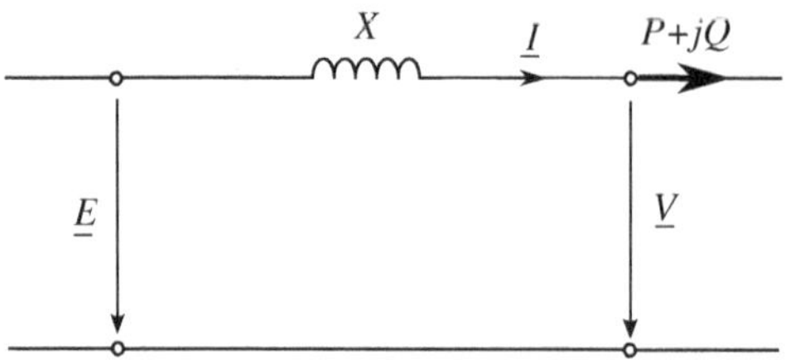

Figure 1.1 Single-phase equivalent circuit of a line with $X = \omega L$

with the impedance $X = \omega L$, where δ denotes the angle between the voltage phasor of the generator resp. converter $\underline{E}$ and $\underline{V}$ at the reference point to the right of the line inductance

$$\underline{E} = E\,e^{j0}, \qquad \underline{V} = V\,e^{j\delta}, \tag{1.7}$$

with the root mean square (RMS) voltage values E and V. The associated active power of a line element of a single-phase system is determined by

$$P(\delta) = \frac{EV}{X} \sin\delta. \tag{1.8}$$

It can be seen that the active power P depends on the product of phase voltages and the sine of the angle δ between their phasors. Because in power systems the node voltages are only within a small percentage of their nominal values, the large changes in active power result from the power angle δ [Machowski et al., 2008]. Note that with the power relations (1.6), (1.8), and the swing equation (1.4) power dynamics of networks with one, two, and multiple generators can be studied in the control loop and under consideration of power plant interactions. In Chapters 5 and 6 such investigations are presented after model-based control concepts of renewable energy (RE) generation (wind and solar) were presented and discussed in Chapters 2–4.

In summary, for reliable power system operation, the provision of ancillary services (AS) will have to be provided to a large extent by renewable energy systems in the future. A possible transformation of how AS will be provided in the future is shown in Figure 1.2 from Komarnicki et al. [2023]. In this process, AS from large synchronous generators are first provided by flexible alternating current transmission system (FACTS), STATCOMS, and high-voltage direct current (HVDC), which are increasingly being replaced by massive distributed inverter-coupled RE generation up to half. Since these are primarily installed in the distribution grid, it will be a significant challenge to make those AS from the distribution also available to the transmission grid. Note that Figure 1.2 from Komarnicki et al. [2023] also includes system recovery and system control services, such as black start capability [DENA, 2014], which are not addressed in this study. The focus of this book is, that in addition to the important regular feed-in control operation, that wind

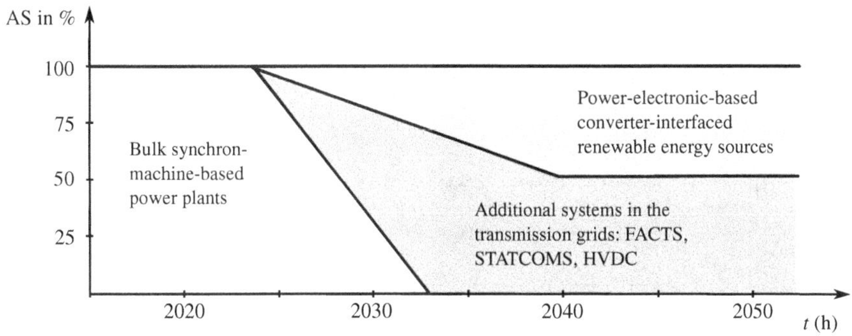

Figure 1.2 Which technology will provide ancillary grid services in the future?

and PV power plants are able to provide flexible services for frequency regulation, such as instantaneous reserve, primary frequency control, and voltage regulation. As shown in Figure 1.2, a share of at least 50% and possibly more should be reliably achieved from 2040. A detailed formal problem description of how wind power and solar PV power plants are able to provide AS is given in Section 1.2.

1.2 Problem Description

Renewable energy systems, such as wind and photovoltaic systems, should either feed as much power as possible into the grid or, if needed but also available, follow a requested power demand. It should be considered that the maximum power that can be fed into the grid obviously depends on the currently available power if no additional storage is used. The power plants are connected to the grid via electronic-based converters and power transformers. The latter is needed to increase the converter's low-voltage level to the grid's medium-voltage level, i.e. between 10 and 35 kV. The grid-side converter is connected to the primary winding of the transformer with an LCL filter. The reference power is generated locally or by an external higher-level controller.

The generic structure of a generation unit to be considered is illustrated in Figure 1.3. The scheme shows the relationships of both wind and photovoltaic (PV) power plants concerning the power flow, measurements, and control signals. Here, the regenerative energy resource as the wind flow in the rotor disc or the irradiation per area is converted into the electrical power P_g of the wind turbine generator or the aggregated DC power P_{dc} of a solar power plant. The wind turbine (WT) generator and PV array, also called PV generator, feed power into the DC link, which is assigned to the "secondary power conversion" block. The secondary

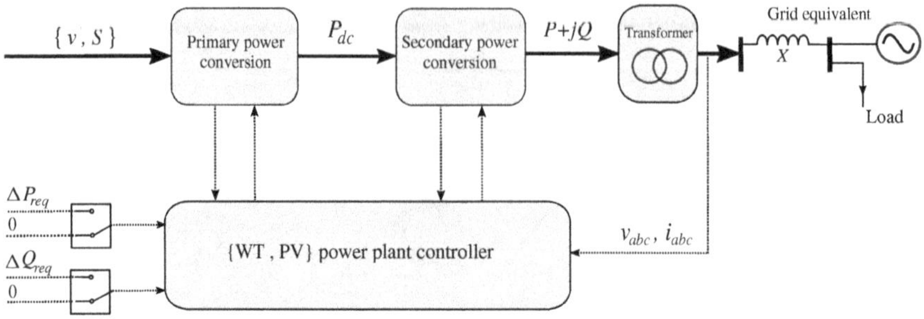

Figure 1.3 Functional diagram of grid-integrated renewable power plant with primary, secondary power conversion, control system, and external reference signals

block contains a voltage-source converter (VSC) that generates a three-phase AC system from the DC link voltage by a suitable switching logic. For this purpose, the standard space vector modulation (SVM) is applied. In combination with an LCL filter to reduce the current harmonics caused by switching, power is fed into the grid.

The two subsystems, the primary and secondary power conversion, are automatically regulated where the controller of the primary conversion subsystem aims to regulate the internal states (rotor speed ω_r or PV voltage v_{pv}) in relation to the states of external resources. The control objective is to either extract the maximum possible power from the renewable resource or to follow a reference power signal, taking into account the achievable limits of the resource. The desired behavior is represented by two modes: (a) generation of the maximum possible electrical power and (b) follow, if possible, the power request ΔP_{req} and ΔQ_{req}.

To clarify the control task of the primary conversion in renewable energy systems, we will use the *P*–*x* curves. Solar PV power plants' primary conversion is characterized by the power–voltage curve (*P*–*V* curve) parameterized via the irradiation S as power per area in W/m^2. Similarly, for wind turbines, the production by the WT generator is given as power over rotor speed (P–ω_r curve) parameterized with the effective wind speed v in front of the rotor. In grid-feeding mode, the task of "primary converter control" is to feed as much power as possible into the grid. In addition, other tasks must be covered by the power plant controller in Figure 1.3. For example, the generated power of wind turbines at high wind speeds should not permanently exceed the rated power of the generator. Therefore, at or above the rated wind speed, the wind turbines are operated at a constant speed (rated speed) with constant torque (rated torque). To achieve this, the pitch angle of the rotor blades is regulated to reduce the rotor's torque and adapt it to the nominal torque of the generator. To provide flexible power control for AS, the strategies described above must be superimposed with a deloading procedure. For a formal

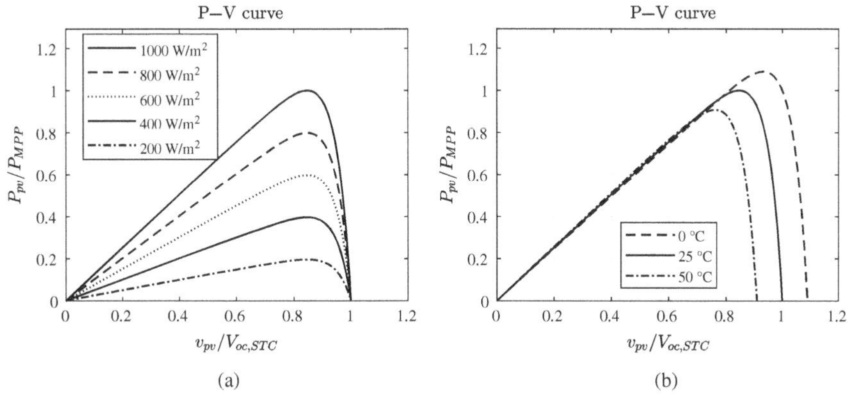

Figure 1.4 *P–V* curve of PV solar modules in pu with variation of irradiation *S* and cell temperature T_c. The power P_{pv} is related to the maximum power by standard test conditions (STC) and the voltage is related to the open-circuit voltage $v_{oc,STC}$ by STC

problem description of the control task of the primary conversion systems, some known characteristic curves for solar PV and wind turbines are now utilized.

First, to characterize the basic control problem of **wind turbines**, the power–rotor speed curve (P_r–ω_r curve) is shown in Figure 1.5. It can be noticed, the point of maximum power changes with varying wind speed. Therefore, to reach the point of maximum power with fluctuating wind speed, the rotor speed must be adjusted according to the P_r–ω_r curve. To achieve this, the rotor speed is changed by the generator torque. Depending on whether the generator torque at the low speed shaft is greater or less than the instantaneous rotor torque, the generator speed increases or decreases. Unless the generator torque reaches the nominal torque limit, the speed can be adjusted within the nominal speed range $\omega_{r,cut\text{-}in} \leq \omega_r \leq \omega_{r,rated}$ 1.2, where $\omega_{r,cut\text{-}in}$ denotes the cut-in and $\omega_{r,rated}$ the nominal rotor speed. If the rotor speed increases to the nominal speed due to the increasing wind speed, the control goal of maximum power tracking is left and the rotor speed is kept on the nominal speed using set point control by increasing generator torque. If the nominal generator torque is reached here as well, the increase in speed is controlled by adjusting the pitch angle of the rotor blades instead.

The scheme described so far is the conventional control scheme without active power control. Additional flexible power change for grid supporting can be achieved by a superimposed generator torque adjustment whereby the operating point related to the rotor torque is not left by adapting the rotor torque by means of pitch angle control. This means that the decoupling of the control problem into

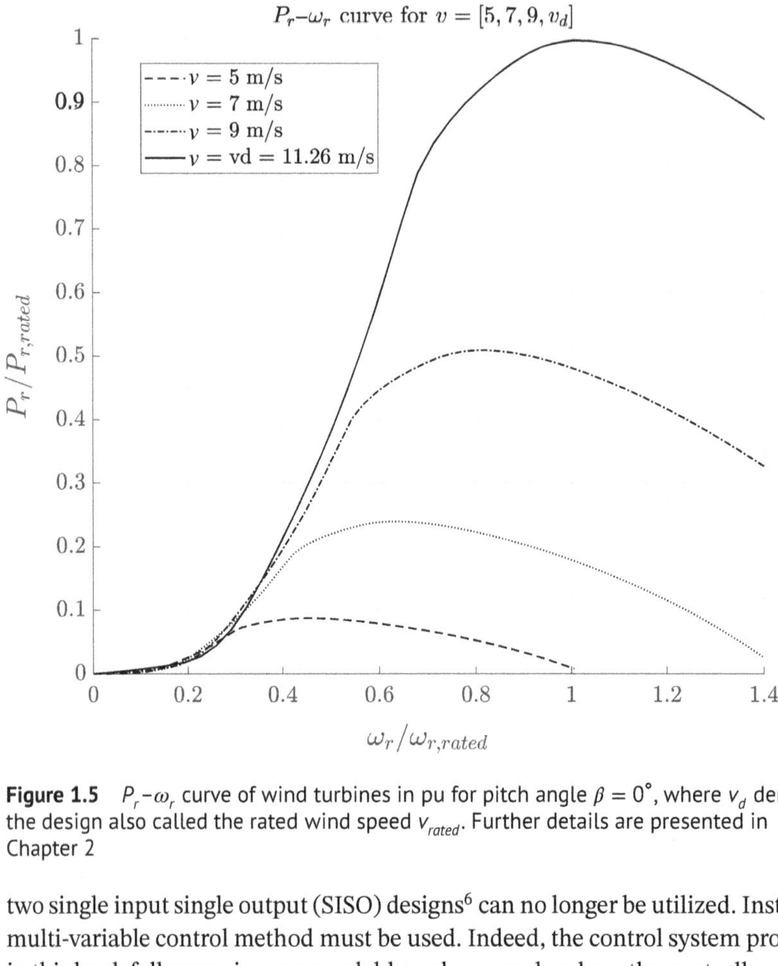

Figure 1.5 P_r–ω_r curve of wind turbines in pu for pitch angle $\beta = 0°$, where v_d denotes the design also called the rated wind speed v_{rated}. Further details are presented in Chapter 2

two single input single output (SISO) designs[6] can no longer be utilized. Instead, a multi-variable control method must be used. Indeed, the control system proposed in this book follows a rigorous model-based approach, where the controller design is performed by solving an optimization problem with linear matrix inequality (LMI) constraints. The modeling of wind turbines for the design is described in Chapter 2. The controller concepts for an optimal and flexible power control and model-based controller synthesis are presented in detail in Chapter 3.

Second, for **solar PV systems**, the essentials of power optimization will be described by the illustration in Figure 1.4. In power-feeding mode the control objective is to track the point of maximum power, also called maximum power point (MPP). Due to the change in radiation $S \in \{200, \ldots, 1000\}\text{W/m}^2$ and the cell temperature $T_c \in \{0, 25, 50\}\,°\text{C}$, see Figure 1.4, the MPP position in the

6 Single-input single-output with generator torque as plant input during partial-load region and the collective pitch-angle as plant input during full-load region with the rotor speed as a common system output.

$P-V$ curve changes. To track the MPP, the voltage v_{pv} is adjusted by a DC–DC converter. For AS, particularly grid support with negative regulating power, the solar PV generator must also be able to reduce the generated power as fast as possible. This is achieved by leaving the MPP. For positive regulating power, the plant can be permanently derated by 10% to 15%. As a result, this percentage is then available without additional storage to increase the generated power. To be seen in the $P-V$ curve of Figure 1.4, increasing or decreasing the voltage v_{pv} set by the DC-DC converter results in a flexible power adjustable output.

Both modeling the controller concepts and synthesis for MPP and flexible power control are presented in detail in Chapter 4.

In addition to the primary conversion presented in PV and wind power systems, the problem of **secondary conversion** is now described **for both systems** together. As shown in Figure 1.3, the secondary system with the DC power, denoted as P_{dc}, is supplied by the primary converter. In wind turbines (Type 4 WT), this is achieved by an AC–DC converter connected in series with the generator. In PV system, the DC power of PV generators, denoted as P_{pv}, is converted into P_{dc} by DC–DC circuits. The main component in the secondary converter block is the VSC. In recent years, a distinction has been made between grid-following (GFL) and grid-forming (GFM) converters, which, in combination with the appropriate primary converter control, determine the overall system behavior of a RE power plant. State-of-the-art control concept for a GFL converter and two concepts for a GFM converter are presented in detail.

1.3 Methodological Framework

The methodological framework in this book is based first on the use of Takagi–Sugeno (TS) models [Takagi and Sugeno, 1985] as a class of nonlinear state-space models, which are similar to linear parameter variable (LPV) models[7] and second the model-based control and observer synthesis formulated as an optimization problem with linear matrix inequality (LMI) constraints. An optimization problem with LMI constraints is usually called a semi-definite programming problem (SDP) which can be considered an essential tool in several branches of control theory and signal processing [Boyd et al., 1994]. The use of TS models for controller design in combination with LMI constraints was first described in the monograph [Tanaka and Wang, 2001]. Since then, the methods have been further developed and applied in areas where classical methods of nonlinear control are either not robust enough or have too many assumptions in the synthesis. To use

7 In Schulte and Pöschke [2016] the differences and similarities are explained using case studies for TS sliding observer-based fault diagnosis of an inverted pendulum and a wind turbine with horizontal axis. The formulation has in common the use of scheduling variables (called premise variables in the fuzzy context) and the weighted combination of LTI models with convex sums.

the TS framework for modeling dynamic systems, either a nonlinear state-space description is either derived from first principle equations or estimated from measured data by suitable system identification methods including model order selection. The latter data-based approach is still an unsolved problem in the general formulation. For certain classes of TS/LPV models, there are solutions which, in their restricted form, are not suitable for the problems dealt with in this book.

In the method applied here, we derive nonlinear state-space models from the first principles and cast them into a suitable TS formulation. The state-space model to be derived has the following form

$$\begin{aligned} \dot{\mathbf{x}} &= \mathbf{f}(\mathbf{x}, \mathbf{u}, \boldsymbol{\theta}), \qquad \mathbf{x}_0 = \mathbf{x}(t_0), \\ \mathbf{y} &= \mathbf{g}(\mathbf{x}), \end{aligned} \tag{1.9}$$

where $\mathbf{f} : \mathbb{R}^n \rightarrow \mathbb{R}^n$ and $\mathbf{g} : \mathbb{R}^n \rightarrow \mathbb{R}^p$ are smooth vector-valued functions, $\mathbf{x} \in \mathbb{R}^n$ denotes the state-space vector, $\mathbf{u} \in \mathbb{R}^m$ is the input vector with controllable and uncontrollable inputs, $\boldsymbol{\theta} \in \mathbb{R}^d$ denotes a time variable parameter vector, and $\mathbf{y} \in \mathbb{R}^p$ denotes the output vector. In this study the Takagi–Sugeno (TS) models are derived by the sector-nonlinearity approach [Ohtake et al., 2001] or by classical linearization at a set of equilibrium points and appropriate choice of functions $h_i : \mathbb{R}^l \rightarrow \mathbb{R}$ with values greater than zero in a local region around the related equilibrium point. In this work two different classes of TS models are used, first weighted combination of linear state-space models

$$\begin{aligned} \dot{\mathbf{x}} &= \sum_{i=1}^{N_r} h_i(\mathbf{z}) \left(\mathbf{A}_i \mathbf{x} + \mathbf{B}_i \mathbf{u}\right), \qquad \mathbf{x}_0 = \mathbf{x}(t_0), \\ \mathbf{y} &= \sum_{i=1}^{N_r} h_i(\mathbf{z}) \mathbf{C}_i \mathbf{x}, \end{aligned} \tag{1.10}$$

or affine state-space models

$$\begin{aligned} \dot{\mathbf{x}} &= \sum_{i=1}^{N_r} h_i(\mathbf{z}) \left(\mathbf{A}_i \mathbf{x} + \mathbf{B}_i \mathbf{u} + \mathbf{a}_i\right), \qquad \mathbf{x}_0 = \mathbf{x}(t_0), \\ \mathbf{y} &= \sum_{i=1}^{N_r} h_i(\mathbf{z}) \left(\mathbf{C}_i \mathbf{x} + \mathbf{c}_i\right), \end{aligned} \tag{1.11}$$

with $\mathbf{A}_i \in \mathbb{R}^{n\times n}$, $\mathbf{B}_i \in \mathbb{R}^{n\times m}$, $\mathbf{C}_i \in \mathbb{R}^{p\times n}$, $\mathbf{a}_i \in \mathbb{R}^n$, and $\mathbf{c}_i \in \mathbb{R}^p$. The functions h_i : $\mathbb{R}^l \rightarrow \mathbb{R}$ are denoted as membership functions and fulfill the convex sum condition

$$0 \leq h_i(\mathbf{z}) \leq 1, \quad i = 1, 2, \dots N_r, \qquad \sum_{i=1}^{N_r} h_i(\mathbf{z}) = 1 \quad \forall \mathbf{z}. \tag{1.12}$$

The vector of the so-called premise variables may directly contain states x_k, inputs u_k, and time variable parameters θ_k or be a function $\mathbf{z} = \mathbf{z}(\mathbf{x}, \mathbf{u}, \boldsymbol{\theta})$. The classic notation of fuzzy rules with linguistic variables is replaced by the sum notation. It has become apparent that the fuzzification and defuzzification from the original work [Takagi and Sugeno, 1985] does not bring any advantages in the system analysis and controller synthesis. The validity of the linear models in the entire system is quantified by the membership function h_i. The name arises from the property that validity resp. membership of the linear or affine models in the overall model (1.10) or (1.11) is determined with h_i.

The steps necessary to construct TS models (1.10) and (1.11) based on (1.9) which are suitable for the controller and observer design are described in Chapter 2 "Modeling of Wind Turbines" and Chapter 4 "Modeling and Control of Photovoltaic Power Plants." It is worth mentioning that the choice of membership functions and sub-models is not unique. When applying the method with sector functions, it must be ensured that the observability and controllability of the linear submodels are preserved. Because, from the clustering of the nonlinearities of the original model (1.9) in the membership functions, it may occur that the structural coupling of the states and inputs in the linear submodels is partially or completely removed. How to solve this problem by a suitable formulation is described in Chapters 2 and 3.

The TS control laws used in this work contain the plant model membership function $h_i(\mathbf{z})$. In the literature, this structure is called a parallel distributed compensator (PDC) [Tanaka and Wang, 2001]. The PDC form used here is a generalization of the integral state feedback controller of linear time-invariant (LTI) systems [Franklin et al., 1994] to the case of nonlinear systems in Takagi–Sugeno form [Schulte and Hahn, 2004]:

$$\mathbf{u} = \sum_{i=1}^{N_r} h_i(\mathbf{z}, \hat{\mathbf{z}}) \left(-\mathbf{K}_{x,i}\,\mathbf{x} + \mathbf{K}_{I,i}\,\mathbf{x}_I\right), \quad \mathbf{x}_I = \int_0^t \left(\mathbf{y}_{ref}(\tau) - \mathbf{y}(\tau)\right) d\tau, \tag{1.13}$$

where $\mathbf{y}_{ref}$ denotes the reference signal. The state feedback gain $\mathbf{K}_x$ and the gain of integral action $\mathbf{K}_I$ are computed by optimization problems with LMI constraints involving the sub-models with $\{\mathbf{A}_i, \mathbf{B}_i, \mathbf{C}_i\}$ resp. $\{\mathbf{A}_i, \mathbf{B}_i, \mathbf{a}_i, \mathbf{C}_i, \mathbf{c}_i\}$ of (1.10) resp. (1.11) and the closed-loop reference dynamics in the form of convex parameterized pole regions (see Appendix B). We do not assume that all premise variables are measurable. Therefore, a distinction is made between measurable $\mathbf{z}$ and estimated $\hat{\mathbf{z}}$ premise variables. For example, a disturbance variable observer is used to estimate the effective wind speed in front of the turbine rotor [Gauterin et al., 2015]. The observer law for state and disturbance variable compensation used in this work reads in its general form

$$\begin{aligned}
\dot{\hat{\mathbf{x}}} &= \sum_{i=1}^{N_r} h_i(\mathbf{z}, \hat{\mathbf{z}}) \left(\tilde{\mathbf{A}}_i\,\hat{\mathbf{x}} + \tilde{\mathbf{B}}_i\,\mathbf{u} + \mathbf{L}_i\,(\mathbf{y} - \hat{\mathbf{y}})\right), \\
\hat{\mathbf{y}} &= \sum_{i=1}^{N_r} h_i(\mathbf{z}, \hat{\mathbf{z}})\,\tilde{\mathbf{C}}_i\,\hat{\mathbf{x}}, \qquad \hat{\mathbf{x}}_0 = \hat{\mathbf{x}}(t = 0),
\end{aligned} \tag{1.14}$$

with $\hat{\hat{\mathbf{x}}} = (\hat{\mathbf{x}}^T, \hat{\mathbf{z}}^T)^T$, where $\mathbf{L}_i$ denotes the observer gain related to each i-th sub-model with $\{\mathbf{A}_i, \mathbf{B}_i, \mathbf{C}_i\}$. The augmented state vector contains the estimated system states $\hat{\mathbf{x}}$ of (1.10) and the disturbance, in this case the unmeasurable premise variables $\hat{\mathbf{z}}$. The observer (1.14) is a generalization of the Luenberger–Observer of LTI systems [Luenberger, 1964] to the case of nonlinear systems in Takagi–Sugeno form.

1.4 Topics of this Book

This book is helpful for readers who want to gain knowledge of mathematical modeling and modern control design of solar photovoltaic and wind power systems. Novel controller structures are presented that enable the RE power plants not only to feed the grid with energy but also to support the grid by fast frequency response and voltage control. The rigorous, model-based approach is presented to encourage readers to understand the paradigm shift in distributed renewable generation and to apply the resulting methods in system development.

The book is suitable for undergraduate and graduate levels with a control system or energy engineering background. In addition to graduate and PhD students, the book may serve as a helpful reference for academic researchers, engineers, and other professionals in the industry. Due to the strict model-based approach, readers with a control system background learn more about the physics and design of wind and solar photovoltaic power generation, including grid integration via power-electronic-based VSC. The control methods are both introduced and explained in the main text, and further technical details are given in the appendixes. This allows readers without a background in control theory to concentrate on the essentials, while others will find further references in the appendix. Most of the chapters contain an introduction to the topic with baseline concepts for modeling and control and advanced concepts to meet modern requirements for grid support. Pictures, illustrations of control schemes, and simulation results support the mathematical and analytical considerations. Furthermore, each chapter contains literature references and problems for self-study. The solutions are provided online.

The book contains six chapters: After the introduction in Chapter 1 the following Chapter 2 proposes mathematical models to describe an entire horizontal axis wind turbine from aerodynamics, structural dynamics, and the actuators for control purposes. The appropriate combination of subsystems results in different design models with varying levels of complexity to achieve applicable optimization criteria. Based on the control objectives for power fed-in and flexible power support for fast frequency control Chapter 3 presents related control schemes and LMI-based design methods in the Takagi–Sugeno framework. The modeling and control of power generation for photovoltaic power plants is addressed in Chapter 4. For the model-based controller design, as with the modeling of wind turbines, all relevant components are modeled and integrated into the TS framework. Methods are presented with which PV systems can flexibly provide instantaneous power for frequency control without battery storage. The primary converters considered for wind and solar energy are connected to the grid via power-electronic-based voltage-source inverters. A generic model consisting of a controlled voltage source and filter elements is presented in Chapter 5. Three different control concepts for inverters of wind and PV systems are proposed in this chapter. A distinction is made between GFL and GFM inverters. The first control concept assumes an existing network and inverter that is controlled in this mode is therefore classified as GFL. The other two presented do not require an existing network and thus enable GFM inverter operation. The two GFM concepts presented vary in terms of control effort and overcurrent protection capability. In the last chapter, the power plant units from the primary converter side proposed in Chapters 2 and 3 for wind turbines and Chapter 4 for PV generators, and the secondary converter side with the proposed voltage-source inverter from Chapter 5 are combined in Chapter 6 to form grid-integrated power plants. Finally, the performance of the wind and PV power plants with all previously designed control loops inside is validated for different scenarios and grid events like active and reactive power request, voltage drops, phase shifts, and changes in the renewable energy sources.

2

Modeling of Wind Turbines

In this chapter, generic models of wind turbines with different degrees of freedom (DOF) are derived and converted into the Takagi–Sugeno (TS) form introduced in Chapter 1. We examine the most common type of turbine, a horizontal-axis wind turbine (HAWT), as illustrated in Figure 2.2. Thereby, the primary power conversion of a turbine is investigated. The secondary power conversion and the integration of both subsystems, including grid connection, will be done in Chapters 5 and 6, respectively.

The generic models are based on the first principles subsequently used in Chapter 3 to develop turbine controllers for the partial-load, full-load region, and the fast power-tracking mode for fast frequency response. In detail, it is explained how dimensionless maps can describe the power generation from a given wind flow, quantifying the aerodynamic rotor torque and the thrust force that causes the blade and tower deflection.

2.1 Introduction

Modern wind turbines are characterized by variable speed operation in the partial-load region and pitch angle control of the rotor blades in the full-load region. Individual wind turbines are usually integrated into a park control system that dispatches the reference power to the individual turbines in an event-driven manner (e.g. in the event of over frequencies or reactive power requirements). Each plant, if there is no override of the reference value from higher-level control, operates as an autonomous power plant in the park grid or as an individual plant with its own grid connection.

The structure of a modern wind turbine with the key hardware components is shown in Figure 2.1. The primary power conversion includes the rotor with pitch

Advanced Control of Grid-Integrated Renewable Energy Power Plants: LMI-based Design in the Takagi-Sugeno Framework, First Edition. Horst Schulte.

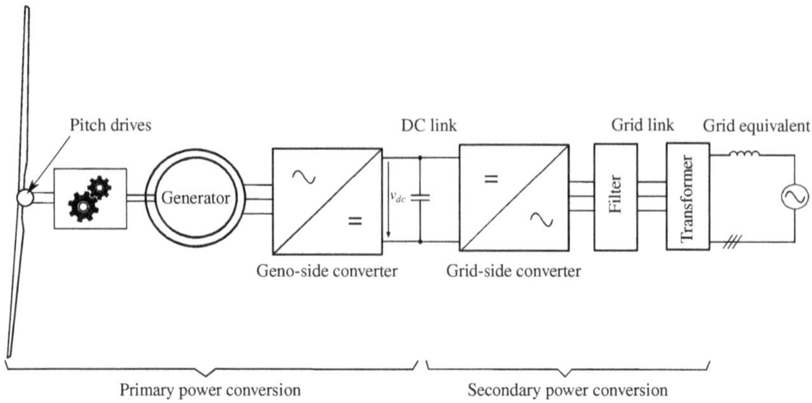

Figure 2.1 Structure of a modern wind turbine with key components

drives for each blade included in the hub, the drive train with low- and high-speed shaft, the gearbox- and generator-side converter. The DC link, the grid-side converter, the LCL filter as well as the transformer are assigned to the secondary power conversion. The wind power plant on the low-voltage (LV) level is connected to the medium-voltage (MV) grid level via a transformer, which is placed either in the nacelle or the turbine's tower base.

2.2 First-Principle Modeling

In the following, generic HAWT models with up to four DOF are proposed to describe the dynamics of primary energy conversion. Depending on the DOFs to be modeled, these contain stationary aerodynamic characteristics, which describe the driving rotor torque and thrust force acting on the turbine and further partial coupled mechanical equivalent models of the rotor blades, the tower, and drive train (Figure 2.2).

2.2.1 Energy Capturing: Actuator Disc Concept

A wind turbine extracts kinetic energy from the wind and converts it into electrical energy. The rather complex aerodynamic processes of energy capturing can be described in a simplified way, but not too simple, by an actuator disc that is placed in a stream tube. A tube, whose boundary surface consists of streamlines, is called a stream tube. That means no fluid exchange takes place through the boundary surface since the velocity vectors of the fluid particles are parallel to the streamlines. Such a stream tube may be treated as if isolated from the adjacent wind field.

Figure 2.2 Example of a horizontal-axis wind turbine (HAWT)

A mass flow through the stream tube illustrated in Figure 2.3 satisfies the continuity law that means the mass flow rate

$$\dot{m} = \rho A v \tag{2.1}$$

with the stream velocity v and density ρ through a control area A must be the same everywhere along the tube. Let us now take a closer look at the flow through three control areas denoted as A_∞, A_D, and A_W. The undisturbed case and the case disturbed by a rotor disk in A_D are considered. Both cases are shown in Figure 2.3,

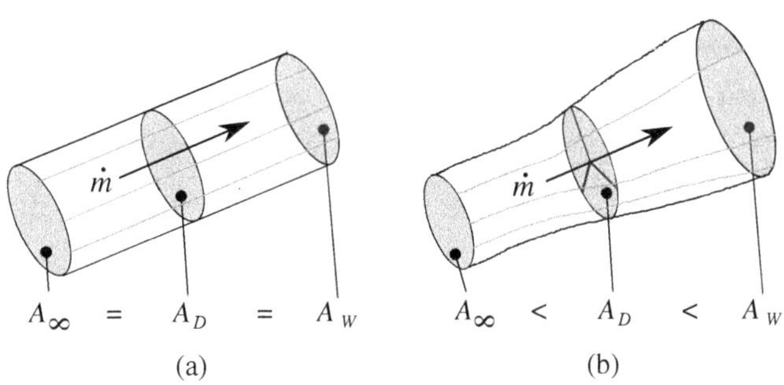

Figure 2.3 Stream-tube without (a) and with (b) actuator disc

where due to the continuity law, the mass flows are the same through all control areas and so

$$\dot{m} = \rho_\infty A_\infty v_\infty = \rho_D A_D v_D = \rho_W A_W v_W, \tag{2.2}$$

where the symbol ∞ refers to conditions far upstream, D refers to conditions in the cross sections at the control area inside the tube (without and with rotor disc), and W refers to conditions in the far wake. For the further consideration the assumptions hold that the flow velocity is quasi-stationary $\dot{v}_{\{\infty,D,W\}} = 0$ and the air is not compressible so that $\rho_\infty = \rho_D = \rho_W = \rho$. The rotor disc induces a velocity reduction in the stream tube which results in $v_\infty > v_D > v_W$ Due to the continuity equation (2.2) the cross sections of the stream tube expand $A_\infty < A_D < A_W$ as shown in Figure 2.3b. This effect of flow influence by a single turbine has to be considered in the wind farm design. To quantify the energy capturing, however, the degree of wind speed reduction ($v_\infty \Rightarrow v_W$) is crucial. In general, the kinetic energy of a moving mass is

$$E = \frac{1}{2} m v^2. \tag{2.3}$$

The time derivative of E considering the assumption $\dot{v}_\infty = 0$ and (2.1) results in the wind power in the absence of an actuator disc

$$P_{wind} = \frac{dE}{dt} = \frac{1}{2}\frac{d}{dt}\left(m v_\infty^2\right) = \frac{1}{2}\left(\dot{m} v_\infty^2\right) + m v_\infty \underbrace{\dot{v}_\infty}_{=0} = \frac{1}{2}\rho A_D v_\infty^3, \tag{2.4}$$

where, however, without limiting the generality, power is related to the control area A_D and the far upstream (undisturbed) wind speed v_∞. If an actuator disc is inserted into the stream tube, the power is divided

$$P_\infty = P_D + P_W, \tag{2.5}$$

where P_∞ refers to wind speed and control area far upstream, P_D denotes the power extraction, and P_W refers to wind speed and control area far upstream. Note that if a rotor disk is placed in the steam-tube it holds $P_\infty \neq P_{wind}$.

2.2.1.1 Power Coefficient

The power extracted from the wind P_D can now be calculated from the wind speed ratio v_W/v_∞ with (2.2), (2.4), and (2.5) as follows

$$P_D = P_\infty - P_W = \frac{1}{2}\rho A_\infty v_\infty^3 - \frac{1}{2}\rho A_W v_W^3.$$

And with $A_W = A_\infty \frac{v_\infty}{v_W}$ we get

$$P_D = \frac{1}{2}\rho A_\infty v_\infty \left(v_\infty^2 - v_W^2\right) = \frac{1}{2}\rho A_\infty v_\infty^3 \left(1 - \frac{v_W^2}{v_\infty^2}\right). \tag{2.6}$$

To obtain the power coefficient defined as

$$c_P = \frac{P_D}{P_{wind}} = \frac{P_D}{\frac{1}{2}\rho A_D v_\infty^3}, \tag{2.7}$$

and after the substitution of P_D by (2.6) we get

$$c_P = \frac{P_D}{P_{wind}} = \frac{A_\infty v_\infty^3 \left(1 - \frac{v_W^2}{v_\infty^2}\right)}{A_D v_\infty^3}.$$

Taking advantage of the fact that the velocity through the disc is equal to the average of velocities far up- and downstream

$$v_D = \frac{v_\infty + v_W}{2}, \tag{2.8}$$

which is proved in the Appendix C based on the Froude–Rankine Theorem and $A_D v_D = A_\infty v_\infty$ holds

$$A_D = 2A_\infty \frac{v_\infty}{v_\infty + v_W}.$$

This results in

$$c_P = \frac{1}{2}\left(1 + \frac{v_W}{v_\infty}\right)\left(1 - \frac{v_W^2}{v_\infty^2}\right), \tag{2.9}$$

where the power coefficient is a function of the ratio $\frac{v_W}{v_\infty}$. In the literature [Burton et al., 2021] the power coefficient is also given as a function of the axial flow induction factor a

$$c_P = 4a(1-a)^2 \tag{2.10}$$

with

$$a = \frac{1}{2}\frac{v_\infty - v_W}{v_\infty}, \tag{2.11}$$

which is mathematically equivalent to (2.9), where

$$v_W = (1-2a)\,v_\infty, \qquad v_D = (1-a)\,v_\infty, \tag{2.12}$$

which fulfill (2.8). The variation of power coefficient with the wind speed ratio $\frac{v_W}{v_\infty}$ and the axial flow induction factor is shown in Figure 2.4. The maximum value of (2.9) and (2.10) thus $c_{P,max}$ is obtained analytically from

$$\frac{dc_P}{dx} = \frac{1}{2}\left(-3x^2 - 2x + 1\right) = 0, \qquad \frac{dc_P}{da} = \left(12a^2 - 16a + 4\right) = 0, \tag{2.13}$$

with $x = \frac{v_W}{v_\infty}$. Hence

$$c_{P,max,Betz} = c_P|_{\frac{v_W}{v_\infty}=1/3} = c_P|_{a=1/3} = \frac{16}{27} = 0.593, \tag{2.14}$$

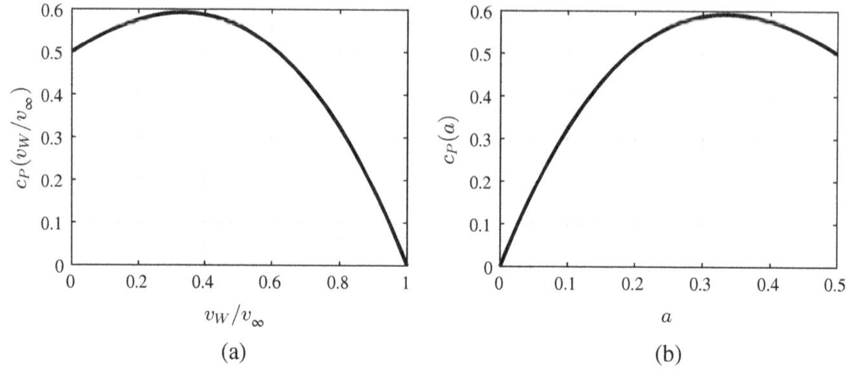

Figure 2.4 Power coefficient c_p as a function of $\frac{v_W}{v_\infty}$ (a) and a (b)

at $\frac{v_W}{v_\infty} = \frac{1}{3}$ or $a = \frac{1}{3}$. The maximum physically achievable value $c_{P,max,Betz} = \frac{16}{27}$ is known as Lanchester–Betz limit [Burton et al., 2021] after [Lanchester, 1915] and [Betz, 1926] or Lanchester–Betz–Joukowsky limit [van Kuik, 2007]. Note that the $c_{P,max,Betz}$ value in (2.14) is only achieved for an ideal wind turbine rotor without losses resulting from

- profile losses caused by drag-forces
- flow around the blade tip, the so-called tip losses
- wake losses due to down stream wake rotation.

2.2.1.2 Thrust Coefficient

Using the simple momentum theory and balance of forces the thrust force F_T on the actuator disc caused by the differential pressure is defined by

$$F_T = \frac{d}{dt}(mv_\infty) - \frac{d}{dt}(mv_W) = \dot{m}(v_\infty - v_W) = (v_\infty - v_W)\rho A_D v_D, \tag{2.15}$$

where $\dot{v}_\infty = \dot{v}_W = 0$ and $\dot{m} = \rho A_D v_D$. The thrust coefficient is given as

$$c_T = \frac{F_T}{\frac{1}{2}\rho v_\infty^2 A_D} \tag{2.16}$$

as the ratio of thrust force (2.15) and reference force caused by the impact pressure $\frac{1}{2}\rho v_\infty^2$ on the control area A_D of the disc. To obtain the thrust coefficient as a function of $\frac{v_W}{v_\infty}$ we substitute F_T by (2.15) and get

$$c_T = 2\frac{v_\infty - v_W}{v_\infty^2} v_D. \tag{2.17}$$

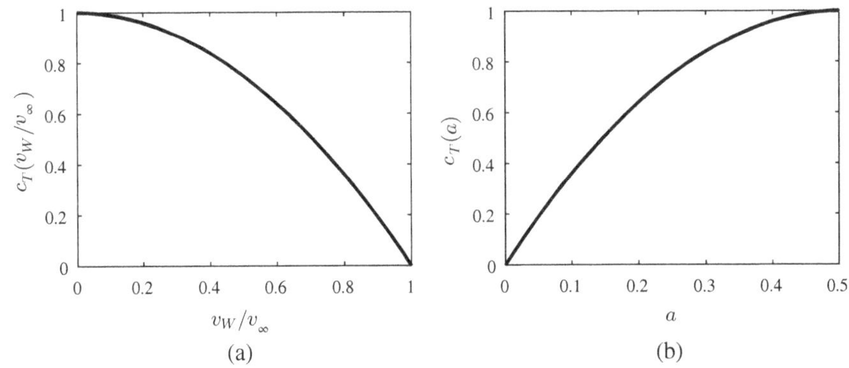

Figure 2.5 Thrust coefficient c_T as a function of $\frac{v_W}{v_\infty}$ (a) and a (b)

Using (2.8) the thrust coefficient is finally given by

$$c_T = \left(1 + \frac{v_W}{v_\infty}\right)\left(1 - \frac{v_W}{v_\infty}\right). \tag{2.18}$$

Like the power coefficient before (2.10), the thrust coefficient can also be given depending on the axial flow induction factor

$$c_T = 4a(1 - a). \tag{2.19}$$

To illustrate the influence of wind speed ratio $\frac{v_W}{v_\infty}$ and axial flow induction factor the graphs of (2.18) and (2.19) are shown in Figure 2.5.

2.2.2 Essentials of Rotor Blade Theory for Control Purpose

Up to now we have investigated the effect of a wind turbine rotor in a flow tube and have derived the maximum power that can be taken from an airflow field with respect to the rotor area A_D. It has been shown that the power coefficient c_P is a function of the stationary flow velocities far upstream and downstream of the rotor disc. However, we have not yet made any statements about how the power can be taken from a wind-stream tube by a rotor. It is obvious that the extracted power P_D is transformed into the mechanical power of the rotor therefore

$$P_r = P_D \qquad \Rightarrow \qquad T_r\,\omega_r = c_p(v_\infty, v_W)\,\frac{1}{2}\,\rho A_D v_\infty^3 \tag{2.20}$$

with (2.4), where P_r denotes the mechanical power of rotor torque, T_r denotes the rotor torque, and ω_r the rotor speed resp. rotor angular velocity. But what causes the rotor torque and how can this torque, and thus the rotor speed, be regulated from a control engineering point of view? For this we consider, as is usual in control engineering, the schematic diagram of the system to be controlled. Here the

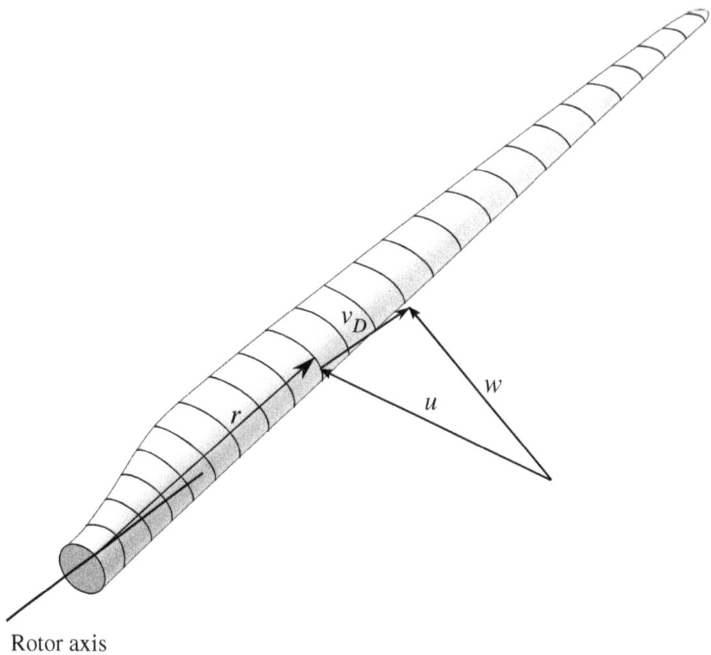

Figure 2.6 Schematic 3D representation of a wind turbine rotor blade comprising airfoil sections (blade elements) with the related relative velocity w resulting from u and v_D for a selected blade element with the radius r from the rotor axis

flow conditions of one rotating rotor blade comprising airfoil sections are considered as illustrated in Figure 2.6. It shows the velocity triangle consists of the axial wind velocity in the rotor area v_D, the circumferential speed $u = \omega_r\, r$ at the airfoil section, where r denotes the distance to the axis of rotation which results in the relative incoming flow velocity

$$w = \sqrt{v_D^2 + (\omega_r\, r)^2} = \sqrt{((1-a)\, v_\infty)^2 + (\omega_r\, r)^2}. \tag{2.21}$$

For better understanding, the velocity vectors, the drag force dD, and lift force dL at a single blade section are illustrated in Figure 2.7. Shown is the velocity vector of the incoming flow w as a vectorial addition of the reduced axial wind velocity $v_D = (1-a)v_\infty$ in the rotor area and the circumferential speed $u = \omega\, r$ related to the radius r as the blade element distance from the rotor axis, see Figure 2.6. The strength of the lift force dL in Figure 2.7 is significantly determined by the angle of attack α_A. It denotes the angle between the incoming flow vector w and the airfoil chord line. An additional design-related angle between the rotor plane and chord

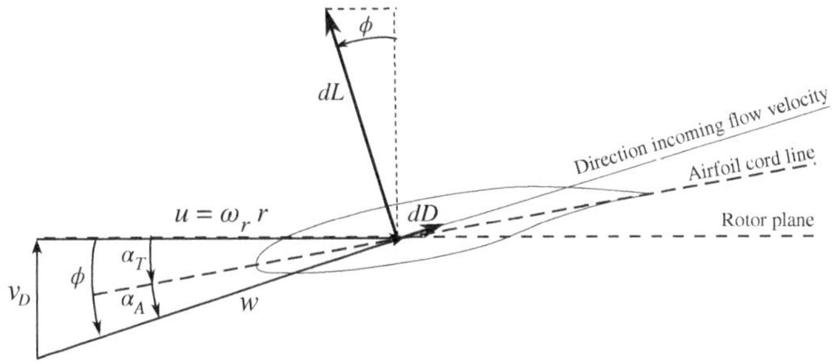

Figure 2.7 Lift force dL and drag force dD at a blade element of length dr caused by pressure disturbance around the 2D airfoil as a function of the angle of attack α_A. Airfoil shape is typical for the blade tip area

is the blade twist angle α_T. Both determine the total angle between the incoming flow vector and rotor plane

$$\phi = \alpha_T + \alpha_A. \tag{2.22}$$

Note that the twisting angle is determined by the design in such a way that α_T of each blade element decreases with increasing distance from the rotor axis. As the velocity triangle (v_D–u–w) becomes sharper toward the blade tip, the additional twist angle to achieve an optimal angle of attack decreases to zero. A typical reduction of alpha with increasing distance from the axis of rotation is shown in Figure 2.8a and is well recognizable in the picture of a real rotor blade (see Figure 2.10).

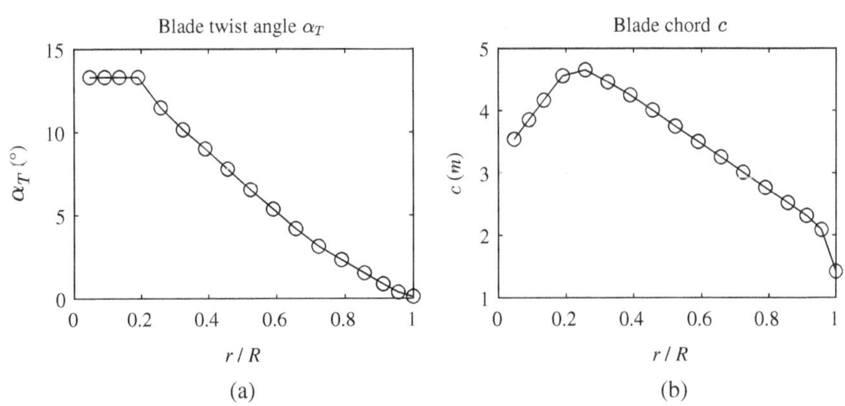

Figure 2.8 Blade twist $\alpha_{T,i}$ (a) and chord c_i (b) parameter of $i = 1,2,\dots,17$ blade elements of the NREL 5-MW turbine illustrated over r_i denoted as the distances of each element to the rotor axis related to the total rotor radius R, see Figure 2.6. Source: Adapted from Jonkman et al. [2009]

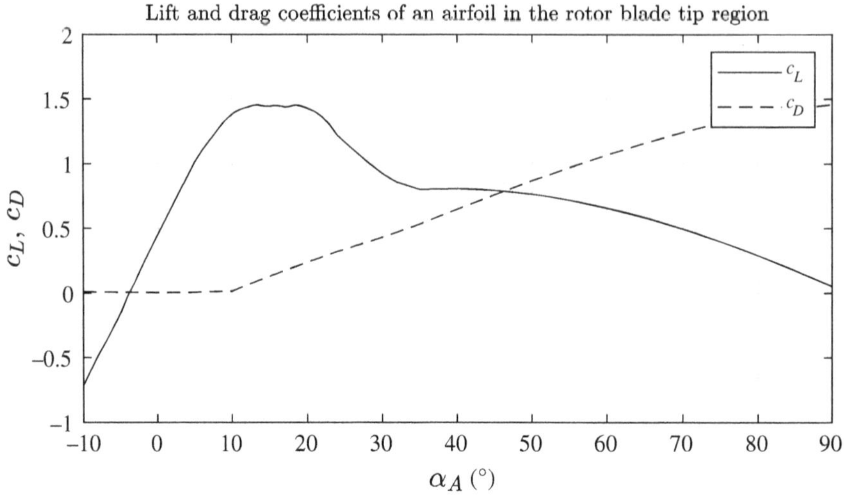

Figure 2.9 Rotor blades of a HAWT with typical twist angle

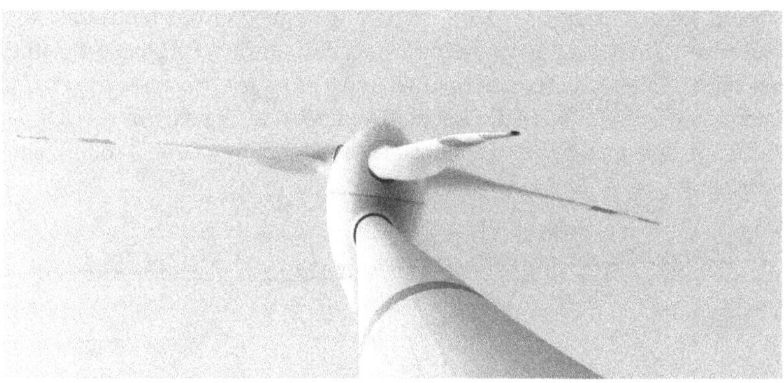

Figure 2.10 Lift coefficient $c_L(\alpha_A)$ and drag coefficient $c_D(\alpha_A)$ of an airfoil (NACA64-A17 airfoil). Source: Adapted from Jonkman et al. [2009]

The positive lift force dL and drag force dD at the blade element of length dr at the radius r with the lift coefficient $c_L(\alpha_A)$ and drag coefficient $c_L(\alpha_A)$ shown in Figure 2.9 are given as

$$dL = \frac{1}{2}\rho w^2 c\,dr\,c_L(\alpha_A), \qquad dD = \frac{1}{2}\rho w^2 c\,dr\,c_D(\alpha_A), \tag{2.23}$$

where c denotes the chord length of the blade element as second design-related blade parameter. Typical characteristic values are shown for all blade elements in Figure 2.8b. However, the crucial coefficients that can be influenced by the control systems are the lift coefficient $c_L(\alpha_A)$ and drag coefficient $c_L(\alpha_A)$. Both coefficients are nonlinear depending on the angle of attack, which are illustrated in Figure 2.9. The lift force dL is normal to the direction of relative velocity w and the drag force is parallel to it (Figure 2.7). Therefore the projection of these forces into the rotor plane with

$$dF_U = dL\sin(\phi) - dD\cos(\phi) \tag{2.24}$$

in the circumferential direction and normal to it in the axial direction

$$dF_T = dL\cos(\phi) + dD\sin(\phi) \tag{2.25}$$

using (2.23) yields

$$\begin{aligned} dF_U &= \frac{\rho}{2} w^2 c\, dr\, \left[c_L(\alpha_A)\sin(\phi) - c_D(\alpha_A)\cos(\phi)\right], \\ dF_T &= \frac{\rho}{2} w^2 c\, dr\, \left[c_L(\alpha_A)\cos(\phi) + c_D(\alpha_A)\sin(\phi)\right]. \end{aligned} \tag{2.26}$$

The thrust force F_T and torque T_r of the entire rotor with z number of blades where each blade is composed of $i = 1,2,\ldots,N_b$ blade elements are given by

$$T_r = z\sum_{i=1}^{N_b} F_{U,i}\cdot r_i, \qquad F_T = z\sum_{i=1}^{N_b} F_{T,i}, \tag{2.27}$$

with

$$F_{U,i} = \int_{r_i}^{r_i+b_i} dF_{U,i}, \qquad F_{T,i} = \int_{r_i}^{r_i+b_i} dF_{T,i},$$

where r_i denotes the distance to the rotor axis and b_i denotes the width of each element. Here the proportional forces on the i blade element are determined individually

$$\begin{aligned} F_{U,i} &= \frac{\rho}{2} w_i^2 c_i b_i \left[c_{L,i}(\alpha_{A,i})\sin(\phi_i) - c_{D_i}(\alpha_{A,i})\cos(\phi_i)\right], \\ F_{T,i} &= \frac{\rho}{2} w_i^2 c_i b_i \left[c_{L,i}(\alpha_{A,i})\cos(\phi_i) + c_{D_i}(\alpha_{A,i})\sin(\phi_i)\right], \end{aligned} \tag{2.28}$$

where $\alpha_{A,i} = \phi_i - \alpha_{T,i}$ and b_i, c_i as well as $\alpha_{T,i}$ are the given design parameters of each blade element, see Figure 2.8. The question is, how can we influence from a control point of the torque and thrust force (2.27). The variable system quantities are ϕ_i and

$$w_i^2 = \sqrt{v_D^2 + (\omega_r r_i)^2},$$

which can be controlled with the rotor speed ω_r, where the wind speed in the rotor plane v_D is a non-controllable variable and is considered as a disturbance.

Remaining parameters $\{\alpha_{T,i}, b_i, c_i, c_{D_i}, c_{L,i}, z\}$ for $i = 1,2,\dots,N_b$ in (2.28) are determined during the blade design. Note that the reasonable choice of these parameters for an optimal blade design is not discussed here. This is treated in Gasch and Twele [2012], for example.

An additional control variable for wind turbines is the blade pitch angle, or in simple terms, pitch angle β. This angle denotes the rotation of the entire rotor blade around the longitudinal axis normal to the rotor rotation axis (see Figure 2.6). To illustrate the influence of an additional common control angle a comparison of the drag and lift forces at an airfoil with different pitch angles and the same wind speed v_D as well as rotor speed ω_r, respectively, circulation speed u is made. Therefore in Figure 2.11 two identical flow profiles (NACA-63-215) are shown. Figure 2.11a with a pitch angle of $\beta = 0°$ and Figure 2.11b with a pitch angle of $\beta = 4°$. Due to the increasing pitch angle from 0° to 4°, dL is significantly reduced and dD increases slightly. This results from a lift coefficient that drops by half and a slightly increasing drag coefficient shown in Figure 2.12 based on the reduced angle of attack α_A caused by the additional pitch angle β with

$$\alpha_A = \phi - \beta - \alpha_T. \tag{2.29}$$

The following values for (2.29) can be seen in Figure 2.11:

$$\text{left airfoil:} \quad \alpha_A = 16° - 0° - 8° = 8°,$$
$$\text{right airfoil:} \quad \alpha_A = 16° - 4° - 8° = 4°,$$

which cause a decreasing lift coefficient from 0.93 to 0.49 and slightly increasing drag coefficient from 0.021 to 0.026 as shown in Figure 2.12.

In summary it can be said that the pitch angle as a control variable for $\beta > 0$ causes a reduction of the rotor torque by reducing the drag coefficient due to the reduction of the angle of attack α_A. This is valid under the assumption that α_A is always set on the left sloping side of $c_L(\alpha_A)$ (see Figures 2.9 and 2.12) by the increase of β which causes a decrease of α_A based on (2.29). The second control

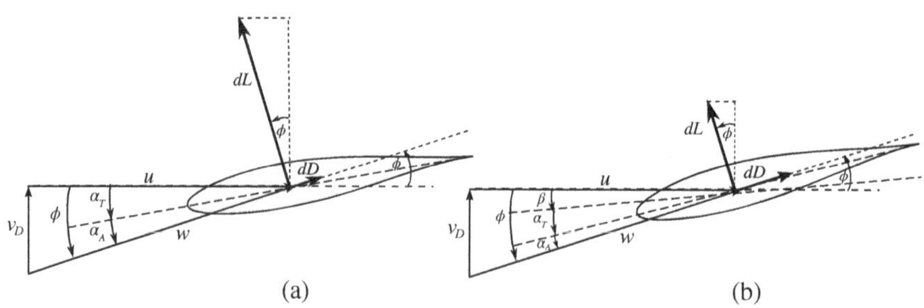

Figure 2.11 Illustration of the influence of an additional control angle β on dL and dD with respect to the lift (a) and drag (b) coefficients in Figure 2.12

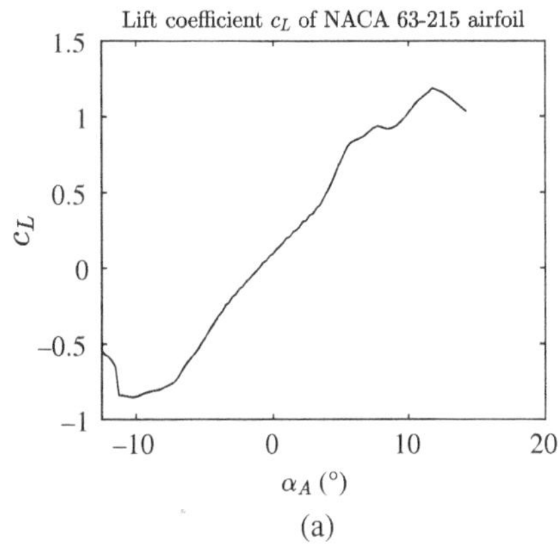

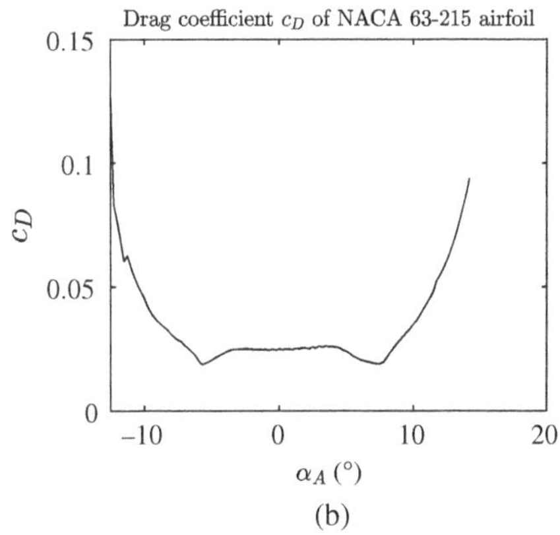

Figure 2.12 Lift coefficient $c_L(\alpha_A)$ (a) and drag coefficient $c_D(\alpha_A)$ (b) of the NACA 63-215 airfoil [Airfoil-Tools, 2020]

variable is the rotor speed, at which α_A is adjusted via the speed vector triangle $w^2 = v_D^2 + (\omega_r r)^2$.

2.2.3 The $c_P-\lambda$, $c_Q-\lambda$, and $c_T-\lambda$ Curves

Based on the results of Sections 2.2.1 and 2.2.2 we are now able to determine the so-called c_x–λ curves $c_P = c_P(\lambda, \beta)$, $c_Q = c_Q(\lambda, \beta)$, and $c_T = c_T(\lambda, \beta)$. Mathematically, these are functions of the pitch angle β and tip-speed ratio determined by

$$\lambda = \frac{\omega_r R}{v_\infty}, \tag{2.30}$$

as the ratio between blade tip-speed $u_R = \omega_r R$, where R denotes the rotor radius and the wind speed v_∞ far upwind the rotor. With these, the rotor power P_r, the rotor torque T_r, and the thrust force F_T acting on a wind turbine can be calculated simply by

$$P_r = \frac{1}{2} \rho \pi R^2 v_\infty^3 c_P(\lambda, \beta) \tag{2.31}$$

$$T_r = \frac{1}{2} \rho \pi R^3 v_\infty^2 c_Q(\lambda, \beta), \tag{2.32}$$

$$F_T = \frac{1}{2} \rho \pi R^2 v_\infty^2 c_T(\lambda, \beta). \tag{2.33}$$

Hence, like already introduced in Section 2.2.2, c_P is often denoted as power coefficient, c_Q as torque coefficient, and c_T as thrust coefficient. Note the relation between c_P and the c_Q curve

$$c_P = \lambda c_Q, \tag{2.34}$$

which results from $P_r = T_r \omega_r$ with (2.30)–(2.32):

$$\begin{aligned} P_r &= T_r \omega_r, \\ \frac{1}{2} \rho \pi R^2 v_\infty^3 c_P &= \frac{1}{2} \rho \pi R^3 v_\infty^2 c_Q \omega_r, \\ v_\infty c_P &= \omega_r R c_Q \quad \Rightarrow \quad c_P = \lambda c_Q. \end{aligned}$$

The relations (2.31)–(2.33) are commonly used for the control-oriented modeling of wind turbines since these depend only on the undisturbed wind speed v_∞ and c–λ curves as a function of the rotor speed ω_r and pitch angle β. The c–λ curves are calculated using the blade element method (BEM), which combines the actuator disc concept from Section 2.2.1 and the 2D theory of a rotating airfoil presented in Section 2.2.2. The basic principles of the calculation for one design tip-speed ratio λ_d are explained in the following. Further calculation methods for the entire λ–β range are described, e.g. in Burton et al. [2021] or Gasch and Twele [2012]. Note, the term design tip-speed ratio determines the ratio between the tip-speed $u = \omega_r R$ and wind speed v_∞ for which the aerodynamically optimal blade chords c_i and blade twist angles $\alpha_{T,i}$ are calculated [Gasch and Twele, 2012].

We start with the rotor power, given by

$$P_D = T_r \omega_r.$$

The left side is replaced by (2.7) and the rotor torque T_r on the right is given by the sum of all forces $F_{U,i}$ for $i = 1,2,\ldots,N_b$ blade elements multiplied with r_i. This gives

$$c_P = \frac{2z \sum_{i=1}^{N_b} F_{U,i} r_i \omega_r}{\rho A_D v_\infty^3}, \qquad A_D = \pi R^2, \tag{2.35}$$

where $F_{U,i}$ is specified by (2.28). The incoming flow velocity (2.21) related to each blade element $i = 1,2,\ldots,N_b$ satisfies the relation

$$w_i^2 = v_D^2 + (\omega_r r_i)^2 = v_D^2 + (\omega_r R)^2 \left(\frac{r_i}{R}\right)^2. \tag{2.36}$$

According to Betz, the wind speed in the rotor plane at maximum c_P value is determined by

$$v_D = \frac{2}{3} v_\infty. \tag{2.37}$$

Substitution of v_D in (2.36) by (2.37) gives

$$w_i^2 = v_\infty^2 R \left(\frac{1}{R}\left(\frac{2}{3}\right)^2 + \lambda_d^2 \frac{r_i^2}{R^3} \right) \qquad \text{with} \qquad \lambda_d = \frac{\omega_r R}{v_\infty}. \tag{2.38}$$

Finally we replace w_i^2 in $F_{U,i}$ (2.28) with (2.38) and insert it into Eq. (2.35). Thus we obtain directly

$$c_P(\lambda_d) = \frac{z\,\lambda_d}{\pi R^2}\sum_{i=1}^{N_b} c_i b_i r_i \left(\frac{1}{R}\left(\frac{2}{3}\right)^2 + \lambda_d^2 \frac{r_i^2}{R^3}\right)\left(c_{L,i}(\alpha_{A,i})\sin(\phi_i) - c_{D_i}(\alpha_{A,i})\cos(\phi_i)\right), \tag{2.39}$$

where $c_P(\lambda_d) = c_{P,max}$ with

$$\phi_i = \arctan\left(\frac{2}{3\lambda_d}\frac{R}{r_i}\right), \qquad \alpha_{A,i} = \phi_i - \alpha_{T,i} - \beta, \tag{2.40}$$

for $\beta = 0$. This means the result (2.39) is just an expression of the parameters of blade and rotor geometry $\{b_i, c_i, r_i, R, z\}$ for all $i = 1,2,\dots,N_b$ blade elements (see Figure 2.8) combined with the function values $c_{D_i}(\alpha_{A,i}), c_{L_i}(\alpha_{A,i})$ of the lift and drag coefficients with regard to the optimal angle of attack for a given design tip-speed ratio λ_d. This means that the blade geometry was designed according to Schmitz–Glauert [Gasch and Twele, 2012] for λ_d.

From (2.34) follows directly the expression for the torque coefficient

$$c_Q(\lambda_d) = \frac{1}{\lambda_d} c_P(\lambda_d). \tag{2.41}$$

The value of c_T–λ curve (2.33) at $\lambda = \lambda_d$ and $\beta = 0$ is derived analogous to the previous calculation of c_P. The total thrust force F_T results from the sum of proportional forces $F_{T,i}$ (2.28). This gives for (2.16)

$$c_T = \frac{z\sum_{i=1}^{N_b} F_{T,i}}{\frac{1}{2}\rho v_\infty^2 A_D}, \qquad A_D = \pi R^2$$

with (2.28)

$$F_{T,i} = \frac{\rho}{2} w_i^2\, c_i\, b_i \left(c_{L,i}(\alpha_{A,i})\cos(\phi_i) + c_{D_i}(\alpha_{A,i})\sin(\phi_i)\right).$$

By replacing w_i^2 with (2.36) we get

$$c_T(\lambda_d) = \frac{z}{\pi R^2}\sum_{i=1}^{N_b} c_i b_i \left(\frac{1}{R}\left(\frac{2}{3}\right)^2 + \lambda_d^2 \frac{r_i^2}{R^3}\right)\left(c_{L,i}(\alpha_{A,i})\sin(\phi_i) - c_{D_i}(\alpha_{A,i})\cos(\phi_i)\right), \tag{2.42}$$

which is obviously an expression of $\{b_i, c_i, r_i, R, z\}$ for $i = 1,2,\dots,N_b$ blade elements.

From the viewpoint of modeling, we have got a first impression of the determination of the c–λ curves in (2.31)–(2.33). But only for the optimal operating point, where $\lambda = \lambda_d$ and $\beta = 0$ is valid whereby ϕ_i in (2.39), (2.42) are determined by (2.40). For $\lambda \neq \lambda_d$ this is no longer the case and therefore ϕ_i has to be determined

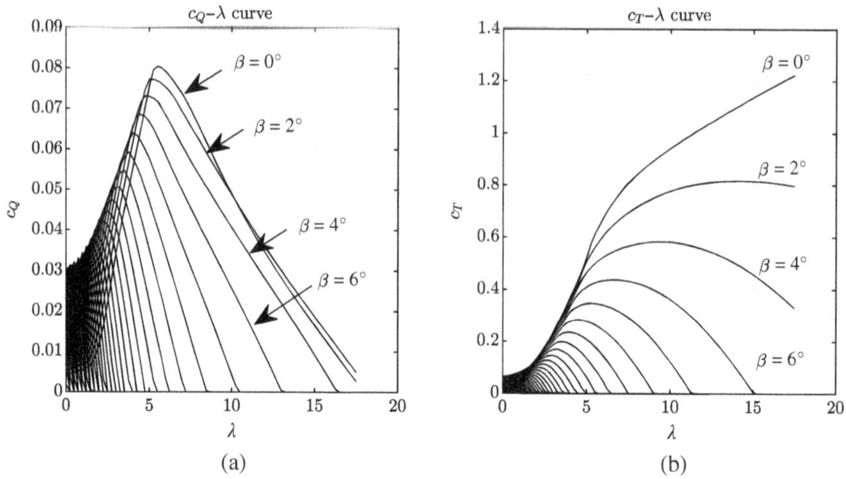

Figure 2.13 c_Q–λ (a) and c_T–λ (b) curves of NREL 5-MW wind turbine related to rotor pitch angles $\beta \in \{0^\circ, 2^\circ, \ldots, 88^\circ, 90^\circ\}$

iteratively for all N_b blade elements for a grid of λ and β values. The full iterative algorithms based on the BEM method are described in Gasch and Twele [2012], but the results are presented below.

The full c_Q–λ and c_T–λ curves based on to the NREL 5-MW wind turbine airfoil parameter including the lift and drag coefficients $c_{L,i}$ and $c_{D,i}$ for $i = 1,2,\ldots,17$ blade elements [Jonkman et al., 2009] calculated by the BEM method are illustrated in Figure 2.13. The c_P–λ curve in Figure 2.14 was calculated from the c_Q–λ (Figure 2.13) with the relation (2.34). Here we can see that the maximum value

$$c_{P,max} = c_P(\lambda_{opt}, \beta = 0) = 0.5279 \tag{2.43}$$

is reached by a tip-speed ratio

$$\lambda = \lambda_{opt} = 7.75, \tag{2.44}$$

which is called the optimal tip-speed ratio, where $c_{P,max} < c_{P,max,Betz}$ (2.14).

For the purpose of model based design, especially when using TS systems composed of Taylor linearized local models (see Appendix A.2.2), it is necessary, that the c_x–λ curves are available in analytical form. Since the BEM code only delivers look-up tables on a given λ–β-grid, the c_Q and c_T values must be approximated by analytical functions. In Georg et al. [2012b] it was shown that the c_Q–λ and c_T–λ curves can be approximated using a modified and expanded function

Figure 2.14 c_P–λ curve of NREL 5-MW wind turbine related to rotor pitch angles $\beta \in \{0°, 5°, \ldots, 85°, 90°\}$

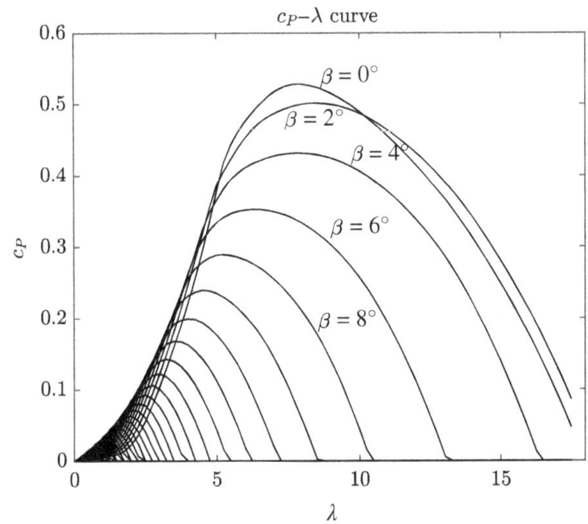

given in Heier [2006] which is itself a modified version of the expressions given in Wasynczuk et al. [1981] and Amlang et al. [1992]:

$$\tilde{c}_Q(\lambda, \beta) = c_1 \left(1 + c_2 (\beta + c_3)^{\frac{1}{2}}\right) + \frac{c_4}{\lambda} \left(c_5 \lambda_i(\lambda, \beta) - c_6\beta - c_7\beta^{c_8} - c_9\right) e^{(-c_{10}\lambda_i(\lambda,\beta))}, \tag{2.45}$$

where

$$\lambda_i(\lambda, \beta) = \frac{1}{\lambda + 0.08\beta} - \frac{0.035}{c_{11} + c_{12}\beta^3} \tag{2.46}$$

valid for $\lambda > 0$. The function $\tilde{c}_Q$ may assume negative values [Georg, 2015]. If the analytical form of c_Q shall be used in simulations, negative coefficients must be excluded. In case of $\tilde{c}_Q < 0$ the power flow reversed, the generator must be operated as a motor to further increase λ, i.e. increase the rotor speed and/or reduce the wind speed. For this purpose, the function (2.45) is multiplied by an additional term, which yields the final approximation for c_Q:

$$c_Q(\lambda, \beta) = \tilde{c}_Q(\lambda, \beta) \frac{1 + \operatorname{sgn} \tilde{c}_Q(\lambda, \beta)}{2}. \tag{2.47}$$

The coefficients were published in Georg [2015] and determined by fitting (2.47) to the values of c_Q obtained from FAST/AeroDyn simulations of the 5-MW turbine. The following values were obtained:

$$\begin{array}{llll} c_1 = 0.005 & c_2 = 1.53 & c_3 = 0.5 & c_4 = 0.18 \\ c_5 = 121 & c_6 = 27.9 & c_7 = 198 & c_8 = 2.36\,. \\ c_9 = 5.74 & c_{10} = 11.35 & c_{11} = 16.1 & c_{12} = 201 \end{array} \tag{2.48}$$

Note that these coefficients correspond to a calculation using the SI unit [rad] for the pitch angle.

The map for c_T was approximated using the following polynomial–exponential function:

$$\tilde{c}_T(\lambda, \beta) = a_1 + a_2\,(\lambda - a_3\beta)\,e^{-a_4\beta} + a_5\,\lambda^2\,e^{-a_6\beta} + a_7\,\lambda^3\,e^{-a_8\beta}. \tag{2.49}$$

Again, to exclude nonphysical negative coefficients, function (2.49) is multiplied by an additional term, which yields the final approximation for c_T:

$$c_T(\lambda, \beta) = \tilde{c}_T(\lambda, \beta)\,\frac{1 + \operatorname{sgn}\tilde{c}_T(\lambda, \beta)}{2}. \tag{2.50}$$

The coefficients were determined as

$$\begin{aligned} &a_1 = 0.006 \qquad a_2 = 0.095 \qquad a_3 = -4.15 \qquad a_4 = 2.75 \\ &a_5 = 0.001 \qquad a_6 = 7.8 \qquad a_7 = -0.00016 \qquad a_8 = -8.88. \end{aligned} \tag{2.51}$$

2.2.4 Wind Turbine Models with 1- and 4-DOF

First, a 1-DOF generic wind turbine model including a first-order pitch dynamic is presented. This provides the relationship between the far upstream wind speed v_∞, instantaneous rotor speed ω_r, and the demanded collective pitch angle β_{ref} and generator torque T_g as control variables. In the following to simplify the notation and readability, the undisturbed upstream wind speed is denoted by v instead of v_∞.

$$\boxed{v := v_\infty} \tag{2.52}$$

Further, a 4-DOF wind turbine model is presented as an extension of the previous one (only rotor rotation), which contains the elastic drive train with rotor and generator rotation, the fore–aft bending of the tower, and flap-wise bending of the collective blades as DOFs.

Using the first principles and the aerodynamic rotor torque (2.32), one obtains the 1-DOF equation of motion of the rotor rotation

$$\begin{aligned} \dot{\omega}_r &= \frac{1}{J}\left(T_r - n_g\,T_g\right), \\ T_r &= \frac{1}{2}\,\rho\,\pi\,R^3\,v^2\,c_Q(\lambda, \beta), \quad \lambda = \frac{\omega_r R}{v}, \end{aligned} \tag{2.53}$$

where n_g denotes the gear ratio and J the inertia of the entire drive train

$$J = J_r + n_g^2\,J_g. \tag{2.54}$$

The mass inertia J_r summarizes the mass inertia of the low-speed shaft, the gearbox inertia assigned to the low-speed shaft, the hub and rotor blades around the rotor axis. Similarly, J_g includes the mass inertia of the high-speed shaft, the gearbox inertia assigned to the high-speed shaft, and the generator inertia. The generator speed is determined by the rotor speed

$$\omega_g = n_g \, \omega_r. \tag{2.55}$$

Rotor blades in wind turbines of the megawatt class are individually pitched either with hydraulic or electric drives. For hydraulic pitch systems, the dynamics can be modeled as a second-order model [Odgaard et al., 2009], which is able to display oscillatory behavior. For pitch systems with electric drives, which are more widespread in use, a first-order model is sufficient and used in this work for control design purposes:

$$\dot{\beta} = -\frac{1}{\tau_\beta}\beta + \frac{1}{\tau_\beta}\beta_{ref}, \tag{2.56}$$

where β denotes the collective pitch angle, β_{ref} the demanded pitch angle, respectively, and τ_β denotes the time constant of the pitch model. Additionally for controller validation a rate limitation is still taken into account. The rate is typically limited to 8°/s to avoid excessive loads in the pitch bearing and blade root. For illustration purposes, an electrical pitch system with a blade bearing for a 1.5 MW turbine is illustrated in Figure 2.15. The angle sensor for detecting the pitch angle , the electric motor, and the power electronics in the background are clearly shown.

Figure 2.15 Pitch drive testbed system at HTW Berlin, Control Engineering Group

Finally, the 1-DOF model in the form of a state-space model results from Eqs. (2.53) and (2.56)

$$\dot{\mathbf{x}} = \begin{pmatrix} \frac{1}{2}\frac{\rho\,\pi\,R^3}{J}\,v^2\,c_Q(\lambda(x_1, v), x_2) \\ -\frac{1}{\tau_\beta}x_2 \end{pmatrix} + \underbrace{\begin{pmatrix} -\frac{n_g}{J} & 0 \\ 0 & \frac{1}{\tau_\beta} \end{pmatrix}}_{\mathbf{B}}\mathbf{u},$$
$$\mathbf{y} = \underbrace{\begin{pmatrix} n_g & 0 \\ 0 & 1 \end{pmatrix}}_{\mathbf{C}}\mathbf{x} \tag{2.57}$$

with c_Q (2.47), the tip-speed ratio $\lambda = (x_1 R)/v$, the state vector $\mathbf{x} = \left(x_1, x_2\right)^T = \left(\omega_r, \beta\right)^T$, input vector $\mathbf{u} = \left(u_1, u_2\right)^T = \left(T_g, \beta_{ref}\right)^T$, measurement output vector $\mathbf{y} = \left(y_1, y_2\right)^T = \left(\omega_g, \beta_{ref}\right)^T$, and the upstream wind speed as disturbances $d = v$ that affects the system and cannot be directly influenced by the controller. Note, the measurement equation in (2.57) takes into account that in real wind turbines the generator speed ω_g is used for control instead of the rotor speed ω_r.

To derive the 4-DOF model we start with the structural mechanics of a generic HAWT. Here, the collective flap-wise bending of the blades (conic mode), the fore–aft bending of the tower, and the elastic drive train with a gearbox are taken into account. The bending and torsional elasticity are modeled through equations of motion, each describing the dynamics of the deflection at a specific position. For the blades, it is the flap-wise tip deflection related to the body-fixed coordinates at the blade root denoted by y_B. The fore–aft tower deflection is described by tower top deflection denoted by y_T. The DOF are shown in Figure 2.16. Note that the blade tip deflection related to absolute coordinates is given by

$$y_{B,a} = y_B + y_T. \tag{2.58}$$

The motion of the drive train illustrated in the schematic diagram in Figure 2.17 is determined by the rotor speed $\omega_r = \dot{\theta}_r$, generator speed $\omega_g = \dot{\theta}_g$, and torsion angle related to the high-speed shaft determined by

$$\Delta\theta_S = \theta_r n_g - \theta_g. \tag{2.59}$$

The equations of motion are obtained using Lagrangian equation of second order including dissipation [Landau and Lifschitz, 1990]

$$\frac{d}{dt}\left(\frac{\partial L}{\partial \dot{q}_i}\right) - \frac{\partial L}{\partial q_i} = f_i - \frac{\partial P_D}{\partial \dot{q}_i}, \tag{2.60}$$

where q_i denotes the generalized coordinates, f_i denotes the generalized forces, and P_D denotes the dissipation function first introduced by Rayleigh [1877].

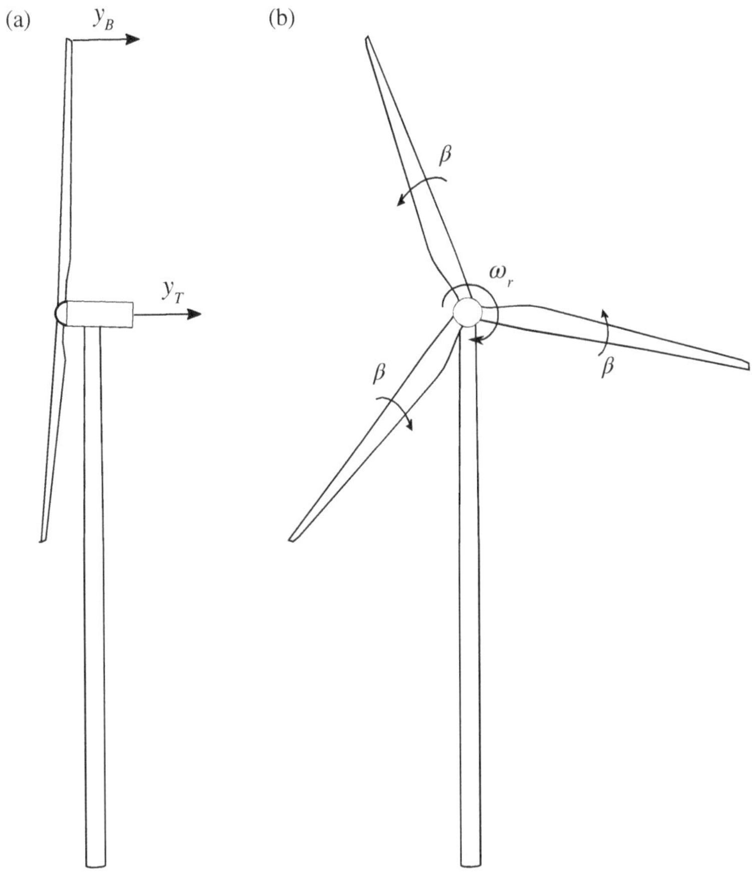

Figure 2.16 Schematic diagram of a horizontal-axis wind turbine with collective pitch angle β

The Lagrangian function L is determined by the difference between kinetic energy E_K and potential energy E_P

$$L = E_K - E_P. \tag{2.61}$$

For the 4-DOF wind turbine model related to the specific generalized coordinates and generalized forces

$$\mathbf{q} = \begin{pmatrix} q_1 \\ q_2 \\ q_3 \\ q_4 \end{pmatrix} = \begin{pmatrix} y_T \\ y_B \\ \theta_r \\ \theta_g \end{pmatrix}, \qquad \mathbf{f} = \begin{pmatrix} f_1 \\ f_2 \\ f_3 \\ f_4 \end{pmatrix} = \begin{pmatrix} F_T \\ F_T \\ T_r \\ -T_g \end{pmatrix}, \tag{2.62}$$

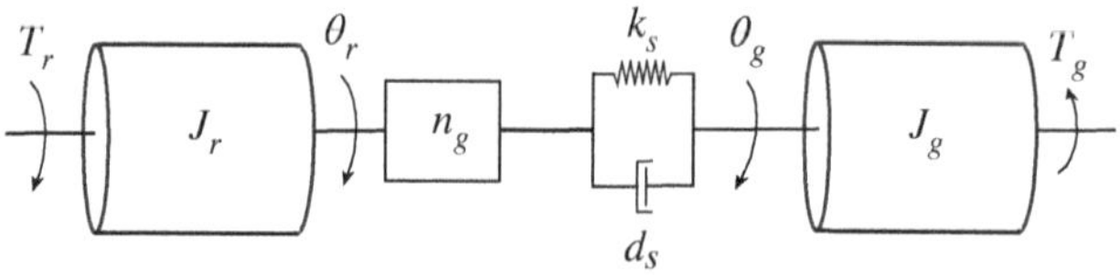

Figure 2.17 Schematic diagram of a wind turbine drive train with gear box

the kinetic energy expression in the Lagrangian function (2.60) is given by

$$E_K = \underbrace{\frac{1}{2} m_T \dot{y}_T^2}_{tower} + \underbrace{\frac{1}{2} N m_B \left(\dot{y}_{B,a}\right)^2}_{blades} + \underbrace{\frac{1}{2} J_r \dot{\theta}_r^2}_{rotor} + \underbrace{\frac{1}{2} J_g \dot{\theta}_g^2}_{generator} \tag{2.63}$$

with $\omega_r = \dot{\theta}_r$ and $\omega_g = \dot{\theta}_g$, where $y_{B,a}$ is defined by (2.58) and m_T denotes the effective mass of nacelle–tower motion, which can be calculated according to Gasch and Twele [2012] with the real rotor, nacelle, and tower mass by

$$m_T = m_{rotor} + m_{nacelle} + 0.25\, m_{tower}. \tag{2.64}$$

Further, m_B denotes the effective mass of the blade tip motion, which can also be estimated by

$$m_B = 0.25\, m_{blade} \tag{2.65}$$

with m_{blade} as the real blade mass. The potential energy can be obtained as

$$E_P = \underbrace{\frac{1}{2} k_T y_T^2}_{tower} + \underbrace{\frac{1}{2} N k_B \underbrace{\left(y_{B,a} - y_T\right)^2}_{y_B^2}}_{blades} + \underbrace{\frac{1}{2} k_s \left(\theta_r n_g - \theta_g\right)^2}_{drivetrain} \tag{2.66}$$

and the dissipation function P_D is given by

$$P_D = \underbrace{\frac{1}{2} d_T \dot{y}_T^2}_{tower} + \underbrace{\frac{1}{2} N d_B \underbrace{\left(\dot{y}_{B,a} - \dot{y}_T\right)^2}_{\dot{y}_B^2}}_{blades} + \underbrace{\frac{1}{2} d_s \left(\dot{\theta}_r n_g - \dot{\theta}_g\right)^2}_{drivetrain}. \tag{2.67}$$

By applying the Lagrange equation (2.61), (2.60) related to the q_i coordinates and generalized forces f_i for $i = 1, \ldots, 4$ (2.62), we obtain the motion equation for the 4-DOF mechanical model. Thereby for preparation of the state-space representation it is given in matrix form:

$$\mathbf{M}\ddot{\mathbf{q}} + \mathbf{D}\dot{\mathbf{q}} + \mathbf{K}\mathbf{q} = \mathbf{Q}\,\mathbf{u}_m \tag{2.68}$$

with the input vector

$$\mathbf{u}_m = \left(F_T, T_r, T_g\right)^T, \tag{2.69}$$

the mass **M** and damping matrix **D**

$$\mathbf{M} = \begin{pmatrix} m_T + N\,m_B & N\,m_B & 0 & 0 \\ N\,m_B & N\,m_B & 0 & 0 \\ 0 & 0 & J_r & 0 \\ 0 & 0 & 0 & J_g \end{pmatrix}, \qquad \mathbf{D} = \begin{pmatrix} d_T & 0 & 0 & 0 \\ 0 & N\,d_B & 0 & 0 \\ 0 & 0 & d_S\,n_g^2 & -d_S\,n_g \\ 0 & 0 & -d_S\,n_g & d_S \end{pmatrix}, \tag{2.70}$$

and the stiffness **K** and input distribution matrix **Q**

$$\mathbf{K} = \begin{pmatrix} k_T & 0 & 0 & 0 \\ 0 & N\,k_B & 0 & 0 \\ 0 & 0 & k_S\,n_g^2 & -k_S\,n_g \\ 0 & 0 & -k_S\,n_g & k_S \end{pmatrix}, \qquad \mathbf{Q} = \begin{pmatrix} 1 & 0 & 0 \\ 1 & 0 & 0 \\ 0 & 1 & 0 \\ 0 & 0 & -1 \end{pmatrix}. \tag{2.71}$$

This results in the state-space model

$$\dot{\mathbf{x}}_m = \underbrace{\begin{pmatrix} \mathbf{0}_{4\times4} & \mathbf{I}_4 \\ -\mathbf{M}^{-1}\mathbf{K} & -\mathbf{M}^{-1}\mathbf{D} \end{pmatrix}}_{\mathbf{A}_m} \mathbf{x}_m + \underbrace{\begin{pmatrix} \mathbf{0}_{4\times3} \\ \mathbf{M}^{-1}\mathbf{Q} \end{pmatrix}}_{\mathbf{B}_m} \mathbf{u}_m \tag{2.72}$$

with the states $\mathbf{x}_m = \left(\mathbf{q}^T, \dot{\mathbf{q}}^T\right)^T \in \mathbb{R}^{8\times1}$. The 4-DOF model (2.72) may be reduced by one coordinate if instead of two absolute angles θ_r and θ_r the torsion angle (2.59) is used, see Figure 2.17. Hence, we obtain, based on the new state vector,

$$\tilde{\mathbf{x}}_m = \left(q_1, q_2, \Delta\theta_S, \dot{\mathbf{q}}^T\right)^T \in \mathbb{R}^{7\times1} \tag{2.73}$$

and (2.69) the final form of the mechanical part of the wind turbine model

$$\dot{\tilde{\mathbf{x}}}_m = \underbrace{\begin{pmatrix} \mathbf{0}_{3\times3} & \tilde{\mathbf{A}}_{m_{12}} \\ -\mathbf{M}^{-1}\tilde{\mathbf{K}} & -\mathbf{M}^{-1}\mathbf{D} \end{pmatrix}}_{\tilde{\mathbf{A}}_m} \tilde{\mathbf{x}}_m + \underbrace{\begin{pmatrix} \mathbf{0}_{3\times3} \\ \mathbf{M}^{-1}\mathbf{Q} \end{pmatrix}}_{\tilde{\mathbf{B}}_m} \begin{pmatrix} F_T \\ T_r \\ T_g \end{pmatrix} \tag{2.74}$$

with

$$\tilde{\mathbf{A}}_{m_{12}} = \begin{pmatrix} 1 & 0 & 0 & 0 \\ 0 & 1 & 0 & 0 \\ 0 & 0 & n_g & -1 \end{pmatrix}, \qquad \mathbf{M}^{-1}\mathbf{Q} = \begin{pmatrix} 0 & 0 & 0 \\ \frac{1}{N\,m_B} & 0 & 0 \\ 0 & \frac{1}{J_r} & 0 \\ 0 & 0 & \frac{1}{J_g} \end{pmatrix}, \tag{2.75}$$

$$\tilde{\mathbf{K}} = \begin{pmatrix} d_T & 0 & 0 \\ 0 & N\,k_B & 0 \\ 0 & 0 & k_S\,n_g \\ 0 & 0 & -k_S \end{pmatrix}.$$

Finally, substitution of the rotor-speed-dependent thrust force F_T and rotor torque T_r by the aerodynamic equations (2.32) and (2.33) and the consideration of the pitch dynamics as first-order model (2.56) results in a state-space form separated into a linear and nonlinear part

$$\dot{\mathbf{x}} = \underbrace{\begin{pmatrix} \mathbf{0}_{3\times3} & \tilde{\mathbf{A}}_{m_{12}} & \mathbf{0}_{3\times1} \\ -\mathbf{M}^{-1}\tilde{\mathbf{K}} & -\mathbf{M}^{-1}\mathbf{D} & \mathbf{0}_{4\times1} \\ \mathbf{0}_{1\times3} & \mathbf{0}_{1\times4} & -\frac{1}{\tau_\beta} \end{pmatrix}}_{\mathbf{A}} \mathbf{x} + \underbrace{\begin{pmatrix} \mathbf{0}_{4\times1} \\ \frac{1}{N\, m_B} F_T(\mathbf{x}, v) \\ \frac{1}{J_r} T_r(\mathbf{x}, v) \\ \mathbf{0}_{2\times1} \end{pmatrix}}_{\mathbf{f}(\mathbf{x},v)} + \underbrace{\begin{pmatrix} \mathbf{0}_{6\times1} & \mathbf{0}_{6\times1} \\ -\frac{1}{J_g} & 0 \\ 0 & \frac{1}{\tau_\beta} \end{pmatrix}}_{\mathbf{B}} \mathbf{u} \tag{2.76}$$

where

$$\mathbf{u} = \begin{pmatrix} T_g \\ \beta_{ref} \end{pmatrix}, \qquad \mathbf{x} = \left(y_T,\, y_B,\, \Delta\theta_S,\, \dot{y}_T,\, \dot{y}_B,\, \omega_r,\, \omega_g,\, \beta\right)^T \tag{2.77}$$

with

$$F_T(\mathbf{x}, v) = \frac{1}{2}\,\rho\,\pi\,R^2\,v^2\,c_T\left(\lambda(x_6, v), x_8\right),$$

$$T_r(\mathbf{x}, v) = \frac{1}{2}\,\rho\,\pi\,R^3\,v^2\,c_Q\left(\lambda(x_6, v), x_8\right),$$

where $x_6 = \omega_r$, $x_8 = \beta$, see (2.77).

2.2.5 Simulation-Based Model Validation

A wind turbine benchmark, the NREL 5-MW reference turbine [Jonkman et al., 2009] of the NREL's aeroelastic simulation tool FAST[1] v8.16, is used to evaluate the accuracy of the 1-DOF (2.57) and 4-DOF (2.76) model with c–λ curves calculated by the BEM. The structural parameters of (2.57) and (2.76) related to the NREL 5-MW wind turbine are given in Appendix D based on the consideration in Georg et al. [2012b] and Georg [2015].

The model validation is carried out with an automatic control system to prevent overspeeding of the turbine at wind speeds above the nominal wind speed. For this, the so-called basic controller is used which is presented in detail in Section 3.2. To illustrate the validation procedure of the 1-DOF model using FAST, a simulation scenario is shown as an example in Figure 2.18. Starting from a rotor speed of $\omega_r(t=0) = 0.765 \cdot \omega_{r,rated}$ below the nominal speed, the turbine is first excited by a wind speed jump of 16 m/s. From $t = 40$ seconds a gust of wind excites the turbine

1 FAST Fatigue, Aerodynamics, Structures, and Turbulence model.

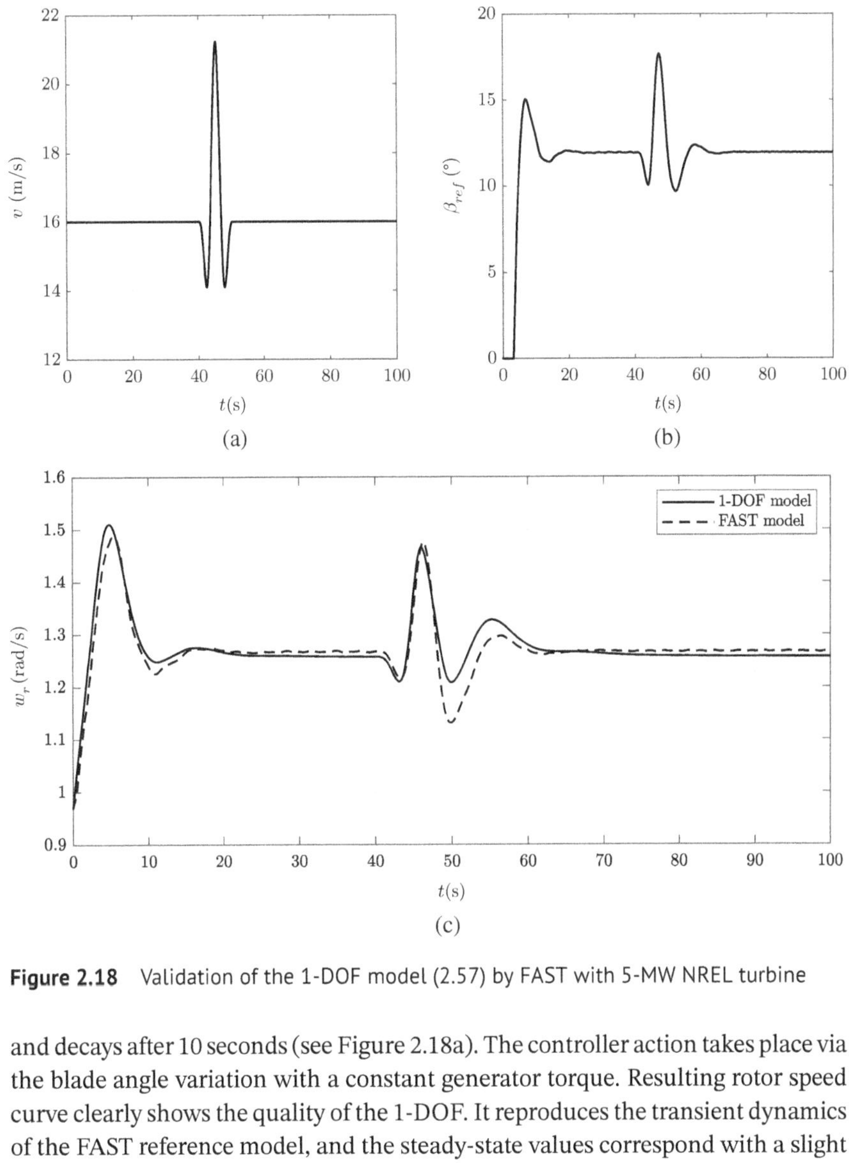

Figure 2.18 Validation of the 1-DOF model (2.57) by FAST with 5-MW NREL turbine

and decays after 10 seconds (see Figure 2.18a). The controller action takes place via the blade angle variation with a constant generator torque. Resulting rotor speed curve clearly shows the quality of the 1-DOF. It reproduces the transient dynamics of the FAST reference model, and the steady-state values correspond with a slight deviation.

The validation of the 4-DOF model (2.76) is done by the same wind speed curve as previously used for the 1-DOF model, see Figure 2.18. The operation of the FAST model in the closed-loop results in a blade pitch angle curve which is given as input u_2 to (2.76). The generator torque as input u_1 is thereby kept constant at the rated value with $u_1 = T_{g,rated}$. Rotor speed ω_r from the 4-DOF model calculated

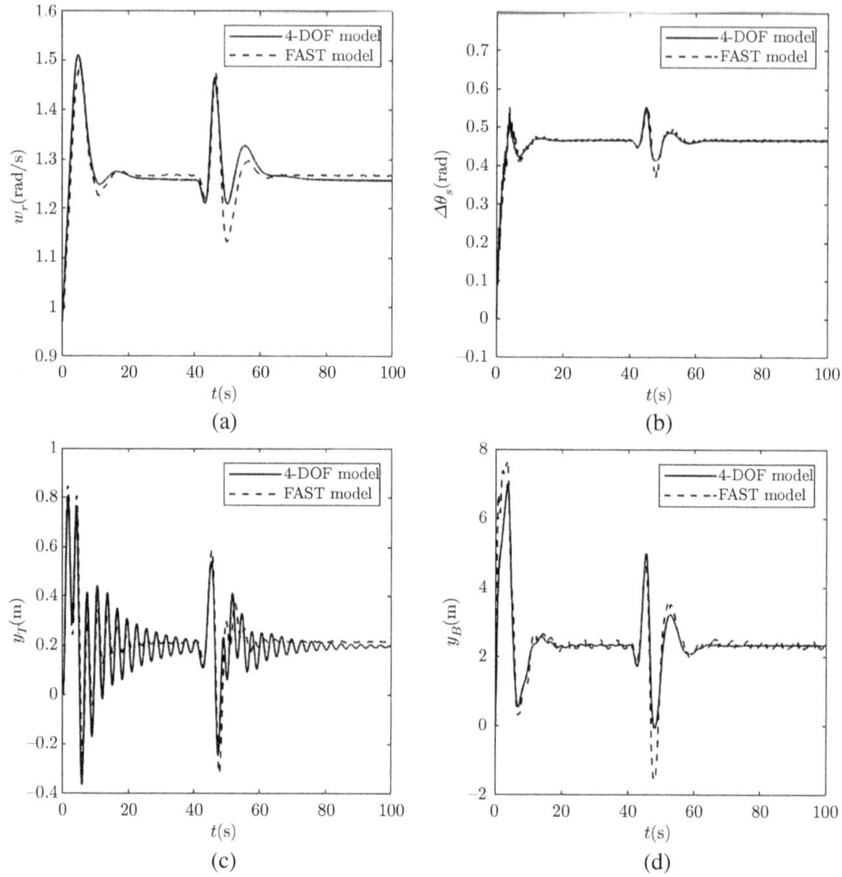

Figure 2.19 Validation of the 4-DOF model (2.57) by FAST using 5-MW NREL Turbine with (a) rotor speed, (b) torsional angle, (c) tower deflection and (d) blade deflection

with (2.76) corresponds to the 1-DOF model, compare Figure 2.18 with Figure 2.19. The curves of the three additional DOFs of 4-DOF model are plotted in Figure 2.19b with the torsion angle $\Delta\theta_s$, and Figure 2.19c with the fore–after tower displacement y_T related to the tower top at the nacelle junction point and Figure 2.19d with the collective blade tip displacement y_B. Comparison of $\Delta\theta_s$ and y_B curves shows a good fit with the FAST model. A larger difference can be seen in the tower movement. Butthe large deflections between t = [0,10] seconds and t = [40,50] seconds are reproduced well. However, the decay behavior that then follows is different. However, suppose the model is used for the controller design. In that case, it must be decided based on the controller performance whether the quality of the model is sufficient or not.

Finally, we consider the eigenvalues of $\mathbf{A} \in \mathbb{R}^{n\times n}$ as the linear part of the 4-DOF wind turbine model (2.76) without feedback of the state-dependent aerodynamic forces F_T and T_r. The following eigenvalues $\text{eig}(\mathbf{A}) = \{EV_1, \dots, EV_n\}$ are obtained for the NREL wind turbine parameters given in Tables D.2 and D.3:

$$\begin{aligned} EV_{1,2} &= -2.3187 \pm 2.7819i, \\ EV_{3,4} &= -0.0849 \pm 2.0882i, \\ EV_{5,6} &= -0.6985 \pm 13.9481i, \\ EV_7 &= 0, \quad EV_8 = -10. \end{aligned} \tag{2.78}$$

Here, the conjugate complex eigenvalues $EV_{1,2}$ are related to the collective flap-wise blade motion, $EV_{3,4}$ to the tower motion and $EV_{5,6}$ to the torsional dynamics of the drive train. The natural frequencies specified in Jonkman et al. [2009][2] of the rotor blade motion $f_{n,R} = 0.6993$ Hz and the tower fore–aft motion with $f_{n,T} = 0.3240$ Hz agree well in a first approximation with the values $|EV_{1,2}|/2\pi = 0.5764$ and $|EV_{3,4}|/2\pi = 0.3326$ given here. The remaining eigenvalue EV_7 refers to the integrating behavior of the rotary motion of the drive train with $\omega_{r,g} \propto \int_0^t (T_r - n_g T_g)d\tau$ and EV_8 follows directly from the time constant τ_β of the first-order model of the blade pitch system, see (2.56) with $EV_8 = -1/\tau_\beta$.

2.3 Control-Oriented Models in TS Form

To prepare the controller development in Section 3.3 different design-oriented models in TS form (see Appendix A) based on the generic wind turbine models (2.57) and (2.76) are now derived. First, after a short treatment of the control ranges of a wind turbine, two TS models are presented for a separate design, each specialized for the partial-load and full-load region of wind turbines. Subsequently, a model augmented by a disturbance component is derived to estimate the effective wind speed. Finally, a TS model for the integrated controller design for both regions (partial and full load) is presented.

2.3.1 Partial- and Full-Load Operating Region

It is common to divide the control task of a wind turbine into a partial- and full-load operating region. However, this does not mean that the controllers have to be designed separately. A model approach that utilizes the design DOF of the TS framework to achieve an integrated multiple-input multiple-output (MIMO)

2 see the Table 9-1 "Full-system natural frequencies in hertz" in Jonkman et al. [2009].

controller design is presented in Section 2.3.5. However, let's first start with the usual separate controller design, which is currently standard in industrial practice. Depending on the operating region, the control objectives are different. Whereas in the partial-load region, the optimization of the energy extraction from the wind is the main objective, in the full-load region, the energy extraction is limited to the turbine's rated power. For the achievement of these purposes, different actuators are employed. The energy optimization in the partial-load region is conducted by a variation of the applied generator torque as a function of the current generator speed, i.e. $T_g = f(\omega_g)$, whereas for the power limitation in region 3 the pitch actuator of the wind turbine is used. The region-specific control objectives are now the initial point for the following design-oriented modeling.

2.3.2 Single-Region 1-DOF Models

(I) Full-load operating region: We start with a model to describe the dynamics in the full-load operating region. For this purpose, the nonlinear model (2.57) is converted into a TS model (A.2) using the sector nonlinearity (SL) approach (see Appendix A.2.1). As the control only operates in the full load range, u_1 in (2.57) is set to the constant nominal generator torque $u_1 = T_{g,rated}$ and $u_2 = \beta_{ref}$ remains as a control input. With the common input matrix $\mathbf{B} = \mathbf{B}_i \in \mathbb{R}^{2\times 1}$ and output matrix $\mathbf{C} = \mathbf{C}_i \in \mathbb{R}^{2\times 2}$ $\forall i$ according to (2.57) one is given

$$\dot{\mathbf{x}} = \sum_{i=1}^{N_r} h_i(\mathbf{z})\,\mathbf{A}_i\,\mathbf{x} + \mathbf{B}\,\mathbf{u},$$
$$\mathbf{y} = \underbrace{\sum_{i=1}^{N_r} h_i(\mathbf{z})}_{=1}\,\mathbf{C}\,\mathbf{x} \Rightarrow \mathbf{y} = \mathbf{C}\,\mathbf{x}. \tag{2.79}$$

In the following it will be shown how to obtain the TS form (2.79) as a controller design model for the full-load region. In Georg [2015], it could be demonstrated that a direct substitution of the single nonlinearity $f_1(v, \mathbf{x}) = v^2 c_Q(\lambda(x_1, c), x_2)$ in (2.57) by a sector function (A.6) leads to a TS model with non-controllable sub-models. For this reason, according to the approach in Pöschke et al. [2017] the analytical function of the c_Q–λ curve (2.47) is divided into two functions

$$c_Q(\lambda, \beta) = c_{Q,1}(\lambda, \beta) + c_{Q,2}(\lambda, \beta), \quad \lambda = \frac{\omega_r R}{v}, \tag{2.80}$$

where

$$c_{Q,i}(\lambda, \beta) = \tilde{c}_{Q,i}(\lambda, \beta)\,\frac{1 + \operatorname{sgn}\tilde{c}_Q(\lambda, \beta)}{2}, \quad i = 1,2 \tag{2.81}$$

with $\tilde{c}_Q$, see Eq. (2.45) and

$$\begin{aligned}\tilde{c}_{Q,1}(\lambda,\beta) &= c_1\left(1+c_2\left(\beta+c_3\right)^{\frac{1}{2}}\right)\\ &\quad+\frac{c_4}{\lambda}\left(c_5\,\lambda_i(\lambda,\beta)-c_7\beta^{c_8}-c_9\right)e^{\left(-c_{10}\,\lambda_i(\lambda,\beta)\right)},\\ \tilde{c}_{Q,2}(\lambda,\beta) &= -\frac{c_4}{\lambda}\,c_6\,e^{\left(-c_{10}\,\lambda_i(\lambda,\beta)\right)}\cdot\beta\end{aligned}\tag{2.82}$$

so that the TS model remains controllable. Hence, the full-load region model follows directly from (2.57) with (2.80)–(2.82)

$$\dot{\mathbf{x}} = \begin{pmatrix} f_1(\mathbf{x},v) & f_2(\mathbf{x},v) \\ 0 & -\frac{1}{\tau_\beta} \end{pmatrix}\mathbf{x} + \begin{pmatrix} 0 \\ \frac{1}{\tau_\beta} \end{pmatrix}u, \tag{2.83}$$

where

$$u = \beta_{ref}, \qquad \mathbf{x} = \left(x_1, x_2\right)^T = \left(\omega_r, \beta\right)^T$$

with

$$\begin{aligned}f_1(\mathbf{x},v) &= \frac{1}{2}\left(\frac{\rho\,\pi\,R^3\,v^2\,c_{Q,1}(\mathbf{x},v)}{J} - \frac{n_g\,T_{g,rated}}{J}\right)\frac{1}{x_1},\\ f_2(\mathbf{x},v) &= \frac{\rho\,\pi\,R^3\,v^2\,c_{Q,2}(\mathbf{x},v)}{2J}\,\frac{1}{x_2}.\end{aligned}\tag{2.84}$$

Note that in (2.84) the

- generator torque is set to the nominal value $T_{g,rated}$ = const. in the valid model range (full-load region)
- function $f_2(\mathbf{x}, v)$ although it looks like it is not singular for $x_2 = \beta = 0$ because with (2.81) and (2.82) we get

$$f_2(\mathbf{x},v) = -\frac{\rho\,\pi\,R^3\,v^2}{2J}\,\frac{c_4\,c_6}{\lambda(x_1,v)}\,e^{\left(-c_{10}\,\lambda_i(\lambda(x_1,v),x_2)\right)}. \tag{2.85}$$

To convert (2.83) into a TS form (A.2), the two nonlinear functions $f_j(\mathbf{x}, v)$ are substituted by the sector functions (A.6)

$$f_j(\mathbf{x},v) = \underbrace{\frac{\bar{f}_j - f_j(\mathbf{x},v)}{\bar{f}_j - \underline{f}_j}}_{=w_{1,j}}\,\underline{f}_j + \underbrace{\frac{f_j(\mathbf{x},v) - \underline{f}_j}{\bar{f}_j - \underline{f}_j}}_{=w_{2,j}}\,\bar{f}_j, \tag{2.86}$$

where the weighting functions $w_{k,j}$ depending on states $\mathbf{x}$ and v are given as

$$w_{1,j} = \frac{\bar{f}_j - f_j(\mathbf{x},v)}{\bar{f}_j - \underline{f}_j}, \qquad w_{2,j} = \frac{f_j(\mathbf{x},v) - \underline{f}_j}{\bar{f}_j - \underline{f}_j}. \tag{2.87}$$

The definition of min/max values in (2.86) is given as

$$\overline{f}_j = \max_{\substack{\mathbf{x}\in[\underline{\mathbf{x}},\overline{\mathbf{x}}]\\ v\in[\underline{v},\overline{v}]}} f_j, \qquad \underline{f}_j = \min_{\substack{\mathbf{x}\in[\underline{\mathbf{x}},\overline{\mathbf{x}}]\\ v\in[\underline{v},\overline{v}]}} f_j. \tag{2.88}$$

Thereby the valid range of the function $f_i \in [\underline{f}_j, \overline{f}_j]$ follows from the state-space region to be considered $\mathbf{x} \in [\underline{\mathbf{x}}, \overline{\mathbf{x}}]$ and the wind speed $v \in [\underline{v}, \overline{v}]$ that occurs in the full-load region. An appropriate specification for standard wind turbines in the megawatt power scale is

$$\begin{aligned} x_1 &= \omega_r \in [0.8\,\omega_{r,rated},\ 1.4\,\omega_{r,rated}]\ \text{rad/s},\\ x_2 &= \beta \in [0,\ 0.45]\ \text{rad},\\ v &\in [10,25]\ \text{m/s}, \end{aligned} \tag{2.89}$$

where $\omega_{r,rated}$ denotes the rated rotor speed. To yield the general TS system (2.79), the variables of f_i are aggregated into the vector of premise variables

$$\mathbf{z} = (z_1, z_2, z_3)^T = (x_1, x_2, v)^T. \tag{2.90}$$

Thus, the weighting functions from (2.86) can be noted as follows

$$w_{1,j} = w_{1,j}(\mathbf{z}), \qquad w_{2,j} = w_{2,j}(\mathbf{z}). \tag{2.91}$$

The next step toward the TS model form defines the membership functions $h_i(\mathbf{z})$ using the generic relation

$$\sum_{k=1}^{N_r} h_i(\mathbf{z}) = \prod_{j=1}^{N_l} \sum_{k=1}^{2} w_{k,j}(\mathbf{z}), \quad N_r = 2^{N_l}. \tag{2.92}$$

Due to two nonlinear functions in (2.83) with $N_l = 2$ we obtain $N_r = 2^{N_l} = 4$ membership functions

$$\begin{aligned} h_1(\mathbf{z}) &= w_{1,1}(\mathbf{z})\,w_{1,2}(\mathbf{z}), & h_2(\mathbf{z}) &= w_{1,1}(\mathbf{z})\,w_{2,2}(\mathbf{z}),\\ h_3(\mathbf{z}) &= w_{2,1}(\mathbf{z})\,w_{1,2}(\mathbf{z}), & h_4(\mathbf{z}) &= w_{2,1}(\mathbf{z})\,w_{2,2}(\mathbf{z}). \end{aligned} \tag{2.93}$$

The matrices $\{\mathbf{A}_i, \mathbf{B}_i\}$ of the associated linear sub-models in (2.79) are obtained by substituting $f_j(\mathbf{x}, v) = f_j(\mathbf{z})$ in (2.83) by the sector function

$$f_j(\mathbf{z}) = w_{1,j}(\mathbf{z})\,\underline{f}_j + w_{2,j}(\mathbf{z})\overline{f}_j \tag{2.94}$$

as follows

$$\dot{\mathbf{x}} = \begin{pmatrix} w_{1,1}(\mathbf{z})\,\underline{f}_1 + w_{2,1}(\mathbf{z})\overline{f}_1 & w_{1,2}(\mathbf{z})\,\underline{f}_2 + w_{2,2}(\mathbf{z})\overline{f}_2 \\ 0 & -\frac{1}{\tau_\beta} \end{pmatrix}\mathbf{x} + \begin{pmatrix} 0 \\ \frac{1}{\tau_\beta} \end{pmatrix} u. \tag{2.95}$$

Extending the system matrix elements in (2.95) by exploiting the convex sum condition $w_{1,j}(\mathbf{z}) + w_{2,j}(\mathbf{z}) = 1$ for $j = 1,2$

$$\begin{aligned}
a_{11} &= \left(w_{1,1}(\mathbf{z})\,\underline{f}_1 + w_{2,1}(\mathbf{z})\overline{f}_1\right)\left(w_{1,2}(\mathbf{z}) + w_{2,2}(\mathbf{z})\right)\\
a_{12} &= \left(w_{1,2}(\mathbf{z})\,\underline{f}_2 + w_{2,2}(\mathbf{z})\overline{f}_2\right)\left(w_{1,1}(\mathbf{z}) + w_{2,1}(\mathbf{z})\right)\\
a_{22} &= -\frac{1}{\tau_\beta}\left(w_{1,1}(\mathbf{z}) + w_{2,1}(\mathbf{z})\right)\left(w_{1,2}(\mathbf{z}) + w_{2,2}(\mathbf{z})\right)
\end{aligned}$$

and using the definition of h_i (2.93), we obtain the design-oriented TS model for the full-load region

$$\begin{aligned}
\dot{\mathbf{x}} = \Bigg(&\underbrace{w_{1,1}(\mathbf{z})\,w_{1,2}(\mathbf{z})}_{h_1(\mathbf{z})}\underbrace{\begin{pmatrix}\underline{f}_1 & \underline{f}_2\\ 0 & -\frac{1}{\tau_\beta}\end{pmatrix}}_{\mathbf{A}_1} + \underbrace{w_{1,1}(\mathbf{z})\,w_{2,2}(\mathbf{z})}_{h_2(\mathbf{z})}\underbrace{\begin{pmatrix}\underline{f}_1 & \overline{f}_2\\ 0 & -\frac{1}{\tau_\beta}\end{pmatrix}}_{\mathbf{A}_2}\\
&+ \underbrace{w_{2,1}(\mathbf{z})\,w_{1,2}(\mathbf{z})}_{h_3(\mathbf{z})}\underbrace{\begin{pmatrix}\overline{f}_1 & \underline{f}_2\\ 0 & -\frac{1}{\tau_\beta}\end{pmatrix}}_{\mathbf{A}_3} + \underbrace{w_{2,1}(\mathbf{z})\,w_{2,2}(\mathbf{z})}_{h_4(\mathbf{z})}\underbrace{\begin{pmatrix}\overline{f}_1 & \overline{f}_2\\ 0 & -\frac{1}{\tau_\beta}\end{pmatrix}}_{\mathbf{A}_4}\Bigg)\mathbf{x} + \underbrace{\begin{pmatrix}0\\ \frac{1}{\tau_\beta}\end{pmatrix}}_{\mathbf{B}}u.
\end{aligned} \tag{2.96}$$

This results in the standard sum notation (2.79) for systems with common **B** and **C**

$$\dot{\mathbf{x}} = \sum_{i=1}^{N_r=4} h_i(\mathbf{z})\,\mathbf{A}_i\,\mathbf{x} + \mathbf{B}\,u, \qquad \mathbf{y} = \mathbf{C}\,\mathbf{x}, \tag{2.97}$$

where $\mathbf{x} = (\omega_r, \beta)^T$, $u = \beta_{ref}$, $\mathbf{y} = (\omega_g, \beta)^T$ with the above-presented h_i-functions (2.93), related $\{\mathbf{A}_i, \mathbf{B}\}$ (2.96) and

$$\mathbf{C} = \begin{pmatrix} n_g & 0\\ 0 & 1\end{pmatrix}. \tag{2.98}$$

(II) Partial-load operating region: Like in the previous full-load region, the reduced design model of the partial-load operating region also has a single input. However, the torque of the generator and not the pitch angle is the actuating input to be adjusted.

$$u = T_g, \qquad \beta = 0, \tag{2.99}$$

where the pitch angle β is set to zero for the entire partial-load region. Consequently, the system state $x_2 = \beta$ vanishes, and the 1-DOF WT model is described by

$$\dot{x} = f(x, v)x - \frac{n_g}{J}u, \quad f(x, v) = \frac{1}{2}\frac{\rho \pi R^3}{J} v^2 c_Q(\lambda(x, v), \beta = 0)\frac{1}{x}, \tag{2.100}$$

where $x = \omega_r$ with $c_Q = c_{Q,1}$ since the second function values for the overall definition range are set to zero $c_{Q,2} = 0$ due to $\beta = 0$, see (2.80). As in the derivation of the previous TS model, $f(x, v)$ is replaced by a single sector function (2.86)

$$f(x, v) = w_1 \underline{f} + w_2 \overline{f} \tag{2.101}$$

with

$$w_1 = \frac{\overline{f} - f(x, v)}{\overline{f} - \underline{f}}, \qquad w_2 = \frac{f(x, v) - \underline{f}}{\overline{f} - \underline{f}}. \tag{2.102}$$

Note that we have to consider only one nonlinearity; thus, $N_l = 1, j = 1$, and $N_r = 2^{N_l} = 2$. Hence it follows from (2.92) that

$$h_i(\mathbf{z}) = w_i(\mathbf{z}), \quad i = 1,2, \tag{2.103}$$

where $x = \omega_r$ and v are aggregated in

$$\mathbf{z} = (z_1, z_2)^T = (x, v)^T = (\omega_r, v)^T \tag{2.104}$$

and the function (2.101) is denoted as $f_1(\mathbf{z})$. The validity range of $f_1(\mathbf{z}) \in [\overline{f}_1, \underline{f}_1]$ determined by

$$\overline{f}_j = \max_{\substack{\omega_r \in [\underline{\omega}_r, \overline{\omega}_r] \\ v \in [\underline{v}, \overline{v}]}} f_j, \qquad \underline{f}_j = \min_{\substack{\omega_r \in [\underline{\omega}_r, \overline{\omega}_r] \\ v \in [\underline{v}, \overline{v}]}} f_j. \tag{2.105}$$

follows from the partial-load specification, where an overlapping of around 10% into the full-load range is still taken into account:

$$\begin{aligned} \omega_r &\in [0.5\,\omega_{r,rated}, 1.1\,\omega_{r,rated}]\ \text{rad/s}, \\ v &\in [0.25\,v_d, 1.1\,v_d]\ \text{m/s}. \end{aligned} \tag{2.106}$$

For the NREL 5-MW wind turbine, the numerical values of the design wind speed v_d related to design tip-speed ratio λ_d (see Section 2.2.3) and the rated rotor speed $\omega_{r,rated}$ are given in Appendix D. The substitution of $f_1(\mathbf{z})$ in (2.100) by (2.101) yields

$$\dot{x} = \underbrace{w_{1,1}(\mathbf{z})}_{h_1(\mathbf{z})} \underbrace{\underline{f}_1}_{a_1} x + \underbrace{w_{2,1}(\mathbf{z})}_{h_2(\mathbf{z})} \underbrace{\overline{f}_1}_{a_2} x + \underbrace{\left(-\frac{n_g}{J}\right)}_{b} u. \tag{2.107}$$

It follows directly the control-oriented model for the partial-load in TS form

$$\dot{x} = \sum_{i=1}^{N_r=2} h_i(\mathbf{z}) a_i x + b u, \qquad y = x, \tag{2.108}$$

where $x = \omega_r$ and $u = T_g$ with the above-introduced h_i-functions (2.103) and the $\{a_i, b\}$ coefficients $a_1 = \underline{f}_1$, $a_2 = \overline{f}_1$ and $b = -n_g/J$. Related to the parameters of the NREL 5 MW plant, the lower and upper bounds result in

$$\underline{f}_1 = 1.0486 \times 10^{-4}, \qquad \overline{f}_1 = 0.1700. \tag{2.109}$$

2.3.3 Single-Region 2- and 3-DOF Models

Until now, in the control-oriented TS modeling task, we have considered the 1-DOF model. It was shown how the nonlinear aerodynamic rotor torque is integrable into the TS framework. Based on the 1-DOF model, in Chapter 3, all necessary steps of controller design using linear matrix inequality (LMI) criteria are proposed for power optimizing and limiting. However, by achieving these primary control objectives alone, the control algorithms can also significantly affect the mechanical loads experienced by the wind turbine [Bossanyi, 2003]. Clearly, the controller must be tuned[3] after to prevent excessive loading, but one can also explicitly consider the objective of load reduction by representing it in the design model. This aspect of controller design has become increasingly important in recent years as the size of wind turbines increases, and the cost of energy must be reduced through cost-effective mechanical structure design (e.g. reduction of steel tower thicknesses). 2-DOF and 3-DOF models with additional DOF are introduced as design models to address this aspect. In order not to exceed the scope of the book, approaches to the derivation of TS models employing the previous techniques (1-DOF TS model in Section 2.3.2) are given. However, a complete derivation with all intermediate steps is omitted. But at the end of this chapter in Section 2.5 a problem is given and the derivation is provided as a solution.

2.3.3.1 Model with Tower and Rotor DOF

Let's start with the 2-DOF model, which represents the rotor rotation and tower deflection. Analogous to Section 2.2.4, the energy method of Lagrange is used for modeling. Here, the generalized coordinates and generalized forces of the rotor and tower DOF can be specified

$$\mathbf{q} = \begin{pmatrix} q_1 \\ q_2 \end{pmatrix} = \begin{pmatrix} y_T \\ \theta_r \end{pmatrix}, \qquad \mathbf{f} = \begin{pmatrix} F_1 \\ F_2 \end{pmatrix} = \begin{pmatrix} F_T \\ T_r - n_g T_g \end{pmatrix}. \tag{2.110}$$

It should be noted that the generalized force f_2 that causes the rotation results from the difference of the rotor torque T_r and the generator torque related to the

3 Controller tuning in the sense of adjusting the closed-loop pole region parameter of the LMI criteria, see Section 3.3.3.

slow shaft, i.e. $n_g\,T_g$. The kinetic energy E_K, potential energy E_P, and dissipation function P_D of the 2-DOF model are given by

$$E_K = \underbrace{\frac{1}{2} m_T \dot{y}_T^2}_{tower} + \underbrace{\frac{1}{2} J\,\omega_r^2}_{drivetrain}, \qquad E_P = \underbrace{\frac{1}{2} k_T y_T^2}_{tower}, \qquad P_D = \underbrace{\frac{1}{2} d_T \dot{y}_T^2}_{tower}. \tag{2.111}$$

The parameters contained here are already introduced in (2.63), (2.66), with the exception of J as the inertia of the entire drive train specified in (2.54). Using the Lagrangian equation of motion (2.60) for $i = 1{,}2$, we obtain

$$\begin{aligned} m_T \ddot{y}_T &= -k_T y_T - d_T \dot{y}_T + F_T, \\ \dot{\omega}_r &= \frac{1}{J}\left(T_r - n_g T_g\right). \end{aligned} \tag{2.112}$$

Together with the dynamics model of the blade pitch drive (2.56) and the state vector and input definition

$$\mathbf{x} = \left(y_T, \dot{y}_T, \omega_r, \beta\right)^T, \quad \mathbf{u} = \left(T_g, \beta_{ref}\right)^T \tag{2.113}$$

the state-space model of the 2-DOF wind turbine is given

$$\dot{\mathbf{x}} = \underbrace{\begin{pmatrix} 0 & 1 & 0 & 0 \\ -\frac{k_T}{m_T} & -\frac{d_T}{m_T} & 0 & 0 \\ 0 & 0 & 0 & 0 \\ 0 & 0 & 0 & -\frac{1}{\tau_\beta} \end{pmatrix}}_{\mathbf{A}} \mathbf{x} + \underbrace{\begin{pmatrix} 0 & 0 \\ 0 & 0 \\ -\frac{n_g}{J} & 0 \\ 0 & \frac{1}{\tau_\beta} \end{pmatrix}}_{\mathbf{B}} \mathbf{u} + \underbrace{\begin{pmatrix} 0 \\ \frac{1}{m_T} F_T(\mathbf{x}, v) \\ \frac{1}{J} T_r(\mathbf{x}, v) \\ 0 \end{pmatrix}}_{\mathbf{f}(\mathbf{x},v)} \tag{2.114}$$

with the aerodynamic thrust force and rotor torque (generalized forces)

$$\begin{aligned} F_T(\mathbf{x}, v) &= \frac{1}{2}\rho\,\pi\,R^2\,v^2\,c_T\left(\lambda(x_3, v), x_4\right), \\ T_r(\mathbf{x}, v) &= \frac{1}{2}\rho\,\pi\,R^3\,v^2\,c_Q\left(\lambda(x_3, v), x_4\right), \end{aligned} \tag{2.115}$$

where $x_3 = \omega_r, x_4 = \beta$, see (2.113). This model has the same structure as the 4-DOF model (2.76). The term $\mathbf{Ax}$ represents the linear mechanics of the tower and rotor and $\mathbf{f}(\mathbf{x}, v)$ the nonlinear internal feedback due to the state-dependent generalized aerodynamic forces. Accordingly, the eigenvalues of $\mathbf{A}$ of the 2-DOF model with $\mathrm{eig}(\mathbf{A}) = \{-0.0849 \pm 2.0882i,\ 0,\ 10\}$ are a subset of the eigenvalues of the 4-DOF model, see (2.78). In practice, the tower movement is detected by an acceleration sensor in the nacelle. The reconstructed tower velocity is obtained by integration of acceleration signal and subsequent high-pass filtering to eliminate the offset

$$\dot{y}_{Tf} = \mathcal{HP}\left\{\int_0^\infty \ddot{y}_T(t)\,dt\right\}, \tag{2.116}$$

where $\mathcal{HP}\{\cdot\}$ denotes the high-pass filter. This yields the output vector with the measurable quantities

$$\mathbf{y} = \left(\dot{y}_{Tf},\, \omega_g,\, \beta\right)^T, \tag{2.117}$$

where $\omega_g = n_g\,\omega_r$.

2.3.3.2 Model with Tower, Elastic Drive Train, and Rotor DOF

For the application of the Lagrange formalism, we will first specify the generalized coordinates $\mathbf{q}$ and force vector $\mathbf{f}$ of the considered 3-DOF model:

$$\mathbf{q} = \begin{pmatrix} q_1 \\ q_2 \\ q_3 \end{pmatrix} = \begin{pmatrix} y_T \\ \theta_r \\ \theta_g \end{pmatrix}, \qquad \mathbf{f} = \begin{pmatrix} F_1 \\ F_2 \\ F_3 \end{pmatrix} = \begin{pmatrix} F_T \\ T_r \\ -T_g \end{pmatrix}. \tag{2.118}$$

The kinetic energy E_K, potential energy E_P, and dissipation function P_D for this are as follows

$$\begin{aligned} E_K &= \underbrace{\frac{1}{2} m_T \dot{y}_T^2}_{tower} + \underbrace{\frac{1}{2} J_r \dot{\theta}_r^2}_{rotor} + \underbrace{\frac{1}{2} J_g \dot{\theta}_g^2}_{generator}, \\ E_P &= \underbrace{\frac{1}{2} k_T y_T^2}_{tower} + \underbrace{\frac{1}{2} k_s \left(\theta_r n_g - \theta_g\right)^2}_{drivetrain}, \\ P_D &= \underbrace{\frac{1}{2} d_T \dot{y}_T^2}_{tower} + \underbrace{\frac{1}{2} d_s \left(\dot{\theta}_r n_g - \dot{\theta}_g\right)^2}_{drivetrain}. \end{aligned} \tag{2.119}$$

Similar to the 2-DOF model, the use of the Lagrangian equation of motion (2.60) for $i = 1,2,3$ yields

$$\begin{aligned} m_T \ddot{y}_T &= -k_T y_T - d_T \dot{y}_T + F_T, \\ J_r \dot{\omega}_r &= -k_s n_g \underbrace{(n_g \theta_r - \theta_g)}_{\Delta\theta_s} - d_s n_g \underbrace{(n_g \omega_r - \omega_g)}_{\Delta\dot{\theta}_s} + T_r, \\ J_g \dot{\omega}_g &= k_s \underbrace{(n_g \theta_r - \theta_g)}_{\Delta\theta_s} + d_s \underbrace{(n_g \omega_r - \omega_g)}_{\Delta\dot{\theta}_s} - T_g, \end{aligned} \tag{2.120}$$

where $\Delta\theta_s$ denotes the torsion angle (2.59) related to the high-speed shaft. With the introduction of the state vector $\mathbf{x} = \left(y_T,\, \Delta\theta_s,\, \dot{y}_T,\, \omega_r,\, \omega_g,\, \beta\right)^T$, since only the torsion angle and not the absolute angles $\theta_{r,g}$ are included in the modeling, the state-space model results from (2.120)

$$\dot{\mathbf{x}} = \underbrace{\begin{pmatrix} 0 & 0 & 1 & 0 & 0 & 0 \\ 0 & 0 & 0 & n_g & -1 & 0 \\ -\frac{k_T}{m_T} & 0 & -\frac{d_T}{m_T} & 0 & 0 & 0 \\ 0 & -\frac{k_s}{J_r} n_g & 0 & -\frac{d_s}{J_r} n_g^2 & \frac{d_s}{J_r} n_g & 0 \\ 0 & \frac{k_s}{J_g} & 0 & \frac{d_s}{J_g} n_g & -\frac{d_s}{J_g} & 0 \\ 0 & 0 & 0 & 0 & 0 & -\frac{1}{\tau_\beta} \end{pmatrix}}_{\mathbf{A}} \mathbf{x} + \underbrace{\begin{pmatrix} 0 & 0 \\ 0 & 0 \\ 0 & 0 \\ 0 & 0 \\ -\frac{1}{J_g} & 0 \\ 0 & \frac{1}{\tau_\beta} \end{pmatrix}}_{\mathbf{B}} \mathbf{u} + \underbrace{\begin{pmatrix} 0 \\ 0 \\ \frac{1}{m_T} F_T(\mathbf{x}, v) \\ \frac{1}{J} T_r(\mathbf{x}, v) \\ 0 \\ 0 \end{pmatrix}}_{\mathbf{f}(\mathbf{x}, v)} \tag{2.121}$$

with the input $\mathbf{u} = \left(T_g, \beta_{ref}\right)^T$ and the same F_T and T_r expressions as in (2.115) (note that the numbering state components x_i is different). Also as in the 2-DOF model, the eigenvalues of

$$\text{eig}(\mathbf{A}) = \{-0.0849 \pm 2.0882i, -0.6985 \pm 13.9481i, 0, -10\} \tag{2.122}$$

are a subset of the eigenvalues of the 4-DOF model (2.78) related to the tower, elastic drive train, integrating behavior of the rotor, and pitch drive dynamics. The outputs of the model (2.121) are

$$\mathbf{y} = \left(\dot{y}_{Tf}, \omega_g, \beta\right)^T, \tag{2.123}$$

where $\dot{y}_{Tf}$ is given in (2.116).

2.3.3.3 Model with Tower, Blades, and Rotor DOF

The generalized coordinates $\mathbf{q}$ and force vector $\mathbf{f}$ representing these DOF are:

$$\mathbf{q} = \begin{pmatrix} q_1 \\ q_2 \\ q_3 \end{pmatrix} = \begin{pmatrix} y_T \\ y_B \\ \theta_r \end{pmatrix}, \qquad \mathbf{f} = \begin{pmatrix} F_1 \\ F_2 \\ F_3 \end{pmatrix} = \begin{pmatrix} F_T \\ F_T \\ T_r - n_g T_g \end{pmatrix}. \tag{2.124}$$

The kinetic energy, potential energy, and the dissipation function are given as

$$\begin{aligned}
E_K &= \underbrace{\frac{1}{2} m_T \dot{y}_T^2}_{tower} + \underbrace{\frac{1}{2} N m_B \left(\dot{y}_{B,a}\right)^2}_{blades} + \underbrace{\frac{1}{2} J \omega_r^2}_{rotor}, \\
E_P &= \underbrace{\frac{1}{2} k_T y_T^2}_{tower} + \underbrace{\frac{1}{2} N k_B y_B^2}_{blades}, \\
P_D &= \underbrace{\frac{1}{2} d_T \dot{y}_T^2}_{tower} + \underbrace{\frac{1}{2} N d_B \dot{y}_B^2}_{blades}
\end{aligned} \tag{2.125}$$

with the blade flap-wise deflection related to the blade root coordinates (body-fixed) as $y_B = y_{B,a} - y_T$, where $y_{B,a}$ denotes the blade deflection related to absolute coordinates. Utilizing the Lagrangian equation of motion (2.60) with (2.125) yields in

$$\begin{aligned}
\ddot{y}_T &= \frac{1}{m_T} F_T - \frac{d_T}{m_T} \dot{y}_T - \frac{k_T}{m_T} y_T + \frac{N d_B}{m_T} \dot{y}_B + \frac{N k_B}{m_T} y_B, \\
\ddot{y}_B &= \frac{1}{N m_B} F_T + \frac{d_T}{m_T} \dot{y}_T - \frac{k_B}{m_B} \frac{m_T + N m_B}{m_T} y_B + \frac{k_T}{m_T} y_T - \frac{d_B}{m_B} \frac{m_T + N m_B}{m_T} \dot{y}_B, \\
\dot{\omega}_r &= \frac{1}{J} T_r - \frac{n_g}{J} T_g.
\end{aligned} \tag{2.126}$$

Using the state and input vector

$$\mathbf{x} = \left(y_T,\, y_B,\, \dot{y}_T,\, \dot{y}_B,\, \omega_r,\, \beta\right)^T, \quad \mathbf{u} = \left(T_g,\, \beta_{ref}\right)^T \tag{2.127}$$

the state-space model results from (2.126) and (2.56)

$$\begin{aligned}
\dot{\mathbf{x}} &= \begin{pmatrix}
0 & 0 & 1 & 0 & 0 & 0 \\
0 & 0 & 0 & 1 & 0 & 0 \\
-\frac{k_T}{m_T} & \frac{N k_B}{m_T} & -\frac{d_T}{m_T} & \frac{N d_B}{m_T} & 0 & 0 \\
\frac{k_T}{m_T} & -\frac{k_B}{m_B} \frac{m_T + N m_B}{m_T} & \frac{d_T}{m_T} & -\frac{d_B}{m_M} \frac{m_T + N m_B}{m_T} & 0 & 0 \\
0 & 0 & 0 & 0 & 0 & 0 \\
0 & 0 & 0 & 0 & 0 & -\frac{1}{\tau_\beta}
\end{pmatrix} \mathbf{x} \\
&+ \begin{pmatrix}
0 & 0 \\
0 & 0 \\
0 & 0 \\
0 & 0 \\
-\frac{n_g}{J} & 0 \\
0 & \frac{1}{\tau_\beta}
\end{pmatrix} \mathbf{u} + \begin{pmatrix}
0 \\
0 \\
\frac{1}{m_T} F_T(\mathbf{x}, v) \\
\frac{1}{N m_B} F_T(\mathbf{x}, v) \\
\frac{1}{J} T_r(\mathbf{x}, v) \\
0
\end{pmatrix}.
\end{aligned} \tag{2.128}$$

Again, as is the case of the previous models with 2-DOF (2.114) and 3-DOF (2.121), the eigenvalues of **A** are included in those of the 4-DOF model with

$$\mathrm{eig}(\mathbf{A}) = \{-0.0849 \pm 2.0882i, -2.3187 \pm 2.7819i, 0, -10\} \tag{2.129}$$

related to blade (flap-wise motion), tower (fore–aft motion), rotor, and pitch drive dynamics. The systems outputs are the measurable quantities

$$\mathbf{y} = \left(\dot{y}_{T,f}, y_B, \omega_g, \beta\right)^T. \tag{2.130}$$

It should be noted that the y_B can be reconstructed from the blade root bending moment in the flap-wise direction measured strain gauges.

The last derived 2-DOF model (2.114) and (2.121), (2.128) contain the thrust force and rotor torque in comparison with 1-DOF model (2.83) that contain the rotor torque only. It is illustrated how the nonlinear function $F_T(\mathbf{x}, v)$ can be reformulated by two sector functions in preparation to get a TS design model, which is full controllable. First, however, in (2.114), the rotor and constant rated generator torque for the full-load region are combined and replaced with two separate functions f_1 and f_2 (2.84) as before for the 1-DOF model (2.83):

$$\dot{\mathbf{x}} = \begin{pmatrix} 0 & 1 & 0 & 0 \\ -\frac{k_T}{m_T} & -\frac{d_T}{m_T} & 0 & 0 \\ 0 & 0 & f_1(\mathbf{x}, v) & f_2(\mathbf{x}, v) \\ 0 & 0 & 0 & -\frac{1}{\tau_\beta} \end{pmatrix} \mathbf{x} + \begin{pmatrix} 0 \\ 0 \\ 0 \\ \frac{1}{\tau_\beta} \end{pmatrix} u + \begin{pmatrix} 0 \\ \frac{1}{m_T} F_T(\mathbf{x}, v) \\ 0 \\ 0 \end{pmatrix}$$

with $u = \beta_{ref}$ since in the full-load range only the pitch angle is used as a command variable. Analogous to the splitting of the c_Q function in (2.80) C_T is also split in

$$c_T(\lambda, \beta) = c_{T,1}(\lambda, \beta) + c_{T,2}(\lambda, \beta) \tag{2.131}$$

where

$$c_{T,i}(\lambda, \beta) = \tilde{c}_{T,i}(\lambda, \beta)\, \frac{1 + \operatorname{sgn} \tilde{c}_T(\lambda, \beta)}{2}, \quad i = 1,2. \tag{2.132}$$

The polynomial (2.49) is thereby divided into $\tilde{c}_{T,1}$ and $\tilde{c}_{T,2}$

$$\begin{aligned} \tilde{c}_{T,1} &= a_1 + a_2\, \lambda\, e^{-a_4\beta} + a_5\, \lambda^2\, e^{-a_6\beta} + a_7\, \lambda^3\, e^{-a_8\beta}, \\ \tilde{c}_{T,2} &= -a_2\, a_3\, e^{-a_4\beta}\, \beta. \end{aligned} \tag{2.133}$$

From this, the functions $f_{3,4}$ can now be introduced as

$$
\begin{aligned}
f_3 &= \frac{1}{2}\frac{\rho\,\pi\,R^3\,v^2\,C_{T,1}}{m_T}\frac{1}{x_3},\\
f_4 &= \frac{1}{2}\frac{\rho\,\pi\,R^3\,v^2\,C_{T,2}}{m_T}\frac{1}{x_4}.
\end{aligned}
\tag{2.134}
$$

Finally, this results in the state-space model

$$
\dot{\mathbf{x}} = \begin{pmatrix} 0 & 1 & 0 & 0 \\ -\frac{k_T}{m_T} & -\frac{d_T}{m_T} & f_3(\mathbf{x},v) & f_4(\mathbf{x},v) \\ 0 & 0 & f_1(\mathbf{x},v) & f_2(\mathbf{x},v) \\ 0 & 0 & 0 & -\frac{1}{\tau_\beta} \end{pmatrix} \mathbf{x} + \begin{pmatrix} 0 \\ 0 \\ 0 \\ \frac{1}{\tau_\beta} \end{pmatrix} u, \tag{2.135}
$$

which can be transformed similarly to (2.83) into a TS model with $N_r = 2^4$ submodels using sector functions (A.6). It is worth mentioning that this model is always fully controllable as long as f_i for $i = 1, \dots, 4$ do not become zero. Here, this is fulfilled.

2.3.4 Multi-region Model for Disturbance Observer Design

Previously derived control-oriented models in TS form[4] (2.97) and (2.108) include the upstream wind speed v $(= v_\infty)$ in the premise variable vector $\mathbf{z}$ and therefore in the h_i-function. The TS framework used in this work utilizes a controller structure also known as parallel distributed compensation (PDC) framework that contains the same h_i-functions as the design-oriented models. This has the consequence that the upstream wind speed must be available to the controller.

However, the wind speed in front of the rotor cannot be measured with standard sensors. For the certification of wind turbine prototypes, it is common to use wind speed measurement masts, see Figure 2.20. In addition, light detection and ranging (LIDAR) systems are used for control purposes in research wind turbines for wind speed measurements, either placed next to the turbine or mounted in the spinner of the turbine as illustrated in Figure 2.21 [Simley et al., 2014]. Alternatively, a disturbance observer can be used for wind speed reconstruction. In Gauterin et al. [2015] observer-based methods were compared how to estimate the rotor effective wind speed with standard sensors such as speed and position sensor alone. The rotor effective wind speed is the concept of a virtual single point wind speed v that yields the same variations in aerodynamic torque T_r as the corresponding

4 Additionally the renewable power plant models in TS form, which will be further presented in this and the following Section 2.3.5.

Figure 2.20 Wind turbine with wind measurement mast in front

(a) (b) (c)

Figure 2.21 An upwind scanning spinner LIDAR installed on a turbine axis in the spinner tip of a 2.3-MW turbine

3D turbulent wind field [van der Hooft and van Engelen, 2004]. In the following, a TS model as an augmentation of the 1-DOF model (2.57) is presented for the disturbance observer design.

For the observer design we start with the 1-DOF[5] model (2.57) already introduced in Section 2.2.4

$$\begin{aligned} \dot{\mathbf{x}} &= \begin{pmatrix} f(\mathbf{x}, v) \\ -\frac{1}{\tau_\beta} x_2 \end{pmatrix} + \underbrace{\begin{pmatrix} -\frac{n_g}{J} & 0 \\ 0 & \frac{1}{\tau_\beta} \end{pmatrix}}_{\mathbf{B}} \mathbf{u}, \\ \mathbf{y} &= \underbrace{\begin{pmatrix} n_g & 0 \\ 0 & 1 \end{pmatrix}}_{\mathbf{C}} \mathbf{x} \end{aligned} \tag{2.136}$$

with

$$f(\mathbf{x}, v) = \frac{1}{2} \frac{\rho \pi R^3}{J} v^2 c_Q(\lambda(x_1, v), x_2), \tag{2.137}$$

where $\mathbf{x} = (\omega_r, \beta)^T$, $\mathbf{u} = (T_g, \beta_{ref})^T$, and $\mathbf{y} = (\omega_g, \beta)^T$. Here, the output equation takes into account that the generator speed ω_g and the pitch angle β are measurable. When converting model (2.136) to a TS form with the purpose of observer design, it must be considered that full observability is satisfied. A necessary condition here is that the sub-models of the TS form are each observable, i.e. the pairs $\{\mathbf{A}_i, \mathbf{C}_i\}$, $i = 1,2, \dots, N_r$ are observable. To fulfill this, the function $f(\mathbf{x}, v)$ (2.137) is decomposed into

$$f(\mathbf{x}, v) = f_1(\mathbf{x}, v) x_2 + f_2(\mathbf{x}, v) v \tag{2.138}$$

with

$$\begin{aligned} f_1(\mathbf{x}, v) &= \frac{1}{2} \frac{\rho \pi R^3}{J} v^2 c_{Q,2}(\lambda(x_1, v), x_2) \frac{1}{x_2}, \\ f_2(\mathbf{x}, v) &= \frac{1}{2} \frac{\rho \pi R^3}{J} v\, c_{Q,1}(\lambda(x_1, v), x_2). \end{aligned} \tag{2.139}$$

Note that the TS model (2.96) and (2.97) cannot be used for the observer design, because the pairs $\{\mathbf{A}_i, \mathbf{C}_i\}$ are not full observable. As a next step, a disturbance

5 The advantage of this approach is that the observer design with only the rotational DOF does not require additional sensors. Here, existing standard sensor system is sufficient to measure the rotor speed and pitch angles. In Section 3.3.2, it will be shown that the 1-DOF model-based design provides sufficiently good estimation results. However, an improvement can be obtained by adding the collective flap-wise bending of the blades as an extra DOF in the observer model [Gauterin et al., 2023].

model for the effective wind speed has to be integrated into (2.136). For this purpose, the model

$$\dot{v} = -\frac{1}{\tau_v} v + \frac{1}{\tau_v} v_0 \tag{2.140}$$

proposed in Ekelund [1997] is used, where τ_v denotes the time constant of disturbance model and v_0 represents the mean wind speed. The choice of a suitable value of τ_v is still explained by the observer design in Section 3.3.2. With the assumption that the mean values as reference wind speed v_0 changes slowly, as suggested in Georg [2015] it can be calculated over an appropriate time period T_p (e.g. 10 minutes) from the anemometer sampled wind measurement

$$v_0 = v_\Sigma = \sum_{i=1}^{N_s} v(t - i\,T_s), \quad T_b = N_s\,T_s, \tag{2.141}$$

where T_s denotes the sampling time and N_s the number of samples in T_s. Furthermore, the mean wind speed can be obtained by additional prior knowledge by means of static maps which contain relations of the equilibrium points

$$v_0 = v_0(\beta, \omega_g)|_{\{\beta_i^e, v_i^e\}}. \tag{2.142}$$

The use of (2.142) to calculate v_0 in (2.140) is applied in Chapter 3. Let's remind once again what is the purpose of the previous consideration. The objective is to design an observer model for the estimation of the wind speed far in front of the rotor. Therefore, the state of the wind model must be integrated into (2.136) to estimate the current wind speed from the known inputs $\mathbf{u}$ and outputs $\mathbf{y}$. Formally, an extended state vector is introduced for this purpose

$$\tilde{\mathbf{x}} = (x_1, x_2, x_3)^T = (\omega_r, \beta, v)^T. \tag{2.143}$$

Hence, from (2.136), (2.138), and (2.140), the related augmented state-space model results in

$$\begin{aligned} \dot{\tilde{\mathbf{x}}} &= \begin{pmatrix} 0 & f_1(\tilde{\mathbf{x}}) & f_2(\tilde{\mathbf{x}}) \\ 0 & -\frac{1}{\tau_\beta} & 0 \\ 0 & 0 & -\frac{1}{\tau_v} \end{pmatrix} \tilde{\mathbf{x}} + \underbrace{\begin{pmatrix} -\frac{n_g}{J} & 0 & 0 \\ 0 & \frac{1}{\tau_\beta} & 0 \\ 0 & 0 & \frac{1}{\tau_v} \end{pmatrix}}_{\tilde{\mathbf{B}}} \tilde{\mathbf{u}}, \\ \mathbf{y} &= \underbrace{\begin{pmatrix} n_g & 0 & 0 \\ 0 & 1 & 0 \end{pmatrix}}_{\tilde{\mathbf{C}}} \tilde{\mathbf{x}}, \end{aligned} \tag{2.144}$$

where $\tilde{\mathbf{u}} = (T_g, \beta_{ref}, v_0)^T$. In accordance with the previous procedure in Section 2.3.2 the nonlinear functions $f_1(\tilde{\mathbf{x}})$ and $f_2(\tilde{\mathbf{x}})$ (2.139) are replaced by sector functions (2.86) with the related weights (2.87). The subsequent appropriate extension by convex sum conditions $w_{1,j}(\mathbf{z}) + w_{2,j}(\mathbf{z}) = 1$ for $j = 1{,}2$ yields an equivalent TS form of (2.144) valid for the defined state-space and input range. Typical values of a wind turbine operating range are

$$
\begin{aligned}
x_1 &= \omega_r \in [0.5708\,\omega_{r,rated},\ 1.4\,\omega_{r,rated}]\ \mathrm{rad/s},\\
x_2 &= \beta \in [0,\ 0.45]\ \mathrm{rad},\\
x_3 &= v \in [3,\ 25]\ \mathrm{m/s},
\end{aligned}
\tag{2.145}
$$

which cover the partial- and full-load region. With the definition of the physical working range, the lower and upper bounds of the sector functions (2.88) are also defined: $f_i \in [\underline{f}_j, \overline{f}_j]$.

To summarize, we obtain

$$
\begin{aligned}
\dot{\tilde{\mathbf{x}}} &= \left(h_1(\mathbf{z})\,\tilde{\mathbf{A}}_1 + h_2(\mathbf{z})\,\tilde{\mathbf{A}}_2 + h_3(\mathbf{z})\,\tilde{\mathbf{A}}_3 + h_4(\mathbf{z})\,\tilde{\mathbf{A}}_4\right)\tilde{\mathbf{x}} + \tilde{\mathbf{B}}\,\tilde{\mathbf{u}},\\
\mathbf{y} &= \tilde{\mathbf{C}}\,\tilde{\mathbf{x}},
\end{aligned}
\tag{2.146}
$$

where the membership functions h_i are determined by (2.93) and the system and input matrices are given as

$$
\begin{aligned}
\tilde{\mathbf{A}}_1 &= \begin{pmatrix} 0 & \underline{f}_1 & \underline{f}_2 \\ 0 & -\frac{1}{\tau_\beta} & 0 \\ 0 & 0 & -\frac{1}{\tau_v} \end{pmatrix}, \quad
\tilde{\mathbf{A}}_2 = \begin{pmatrix} 0 & \underline{f}_1 & \overline{f}_2 \\ 0 & -\frac{1}{\tau_\beta} & 0 \\ 0 & 0 & -\frac{1}{\tau_v} \end{pmatrix},\\
\tilde{\mathbf{A}}_3 &= \begin{pmatrix} 0 & \overline{f}_1 & \underline{f}_2 \\ 0 & -\frac{1}{\tau_\beta} & 0 \\ 0 & 0 & -\frac{1}{\tau_v} \end{pmatrix}, \quad
\tilde{\mathbf{A}}_4 = \begin{pmatrix} 0 & \overline{f}_1 & \overline{f}_2 \\ 0 & -\frac{1}{\tau_\beta} & 0 \\ 0 & 0 & -\frac{1}{\tau_v} \end{pmatrix},\\
\tilde{\mathbf{B}} &= \begin{pmatrix} -\frac{n_g}{J} & 0 & 0 \\ 0 & \frac{1}{\tau_\beta} & 0 \\ 0 & 0 & \frac{1}{\tau_v} \end{pmatrix} \quad \text{and} \quad \tilde{\mathbf{C}} = \begin{pmatrix} n_g & 0 & 0 \\ 0 & 1 & 0 \end{pmatrix}.
\end{aligned}
\tag{2.147}
$$

Thus, the final TS form with (2.146) and (2.147) results in

$$
\dot{\tilde{\mathbf{x}}} = \sum_{i=1}^{N_r=4} h_i(\mathbf{z})\,\tilde{\mathbf{A}}_i\,\tilde{\mathbf{x}} + \tilde{\mathbf{B}}\,\tilde{\mathbf{u}}, \qquad \mathbf{y} = \tilde{\mathbf{C}}\,\tilde{\mathbf{x}},
\tag{2.148}
$$

where $\mathbf{z} = (\omega_r,\ \beta,\ v)^T$, $\tilde{\mathbf{x}} = (\omega_r,\ \beta,\ v)^T$, $\tilde{\mathbf{u}} = (T_g,\ \beta_{ref},\ v_0)^T$, and $\mathbf{y} = (\omega_g,\ \beta)^T$.

2.3.5 Multi-region Model for Controller Design

So far, dedicated models for the part-load and full-load regions have been derived for use in the controller design. The separated modeling corresponds, for example, to the approach proposed in Díaz de Corcuera et al. [2015]. For implementation in practice, these subsystems must then be controlled by a higher-level unit that decides which controller is active. Alternatively, the region changes are integrable into the TS framework to avoid the superordinate layer. For this purpose, a mixed-fuzzy SL approach is proposed. The TS region models are connected via convex fuzzy membership functions to form an overall TS region model. Note that instead of SLs, the region models can also be obtained by the Taylor linearization at selected stationary operating points and fuzzy interpolation (method description see A.2.2), proposed in Pöschke et al. [2020].

In order to demonstrate the mixed-fuzzy SL approach, we now form the overall model from the two region-specific systems with (2.108) in the partial-load and (2.97) in the full-load region. The partial-load region model (2.108) can be mathematically equivalent reformulated as follows:

$$\dot{\mathbf{x}} = \mathbf{A}_I(\mathbf{x}, v)\,\mathbf{x} + \mathbf{B}_I\mathbf{u} \tag{2.149}$$

with

$$\mathbf{A}_I(\mathbf{x}, v) = \begin{pmatrix} f_I(\mathbf{x}, v) & 0 \\ 0 & 0 \end{pmatrix}, \quad \mathbf{B}_I = \begin{pmatrix} -\frac{n_g}{J} & 0 \\ 0 & 0 \end{pmatrix} \tag{2.150}$$

and

$$f_I(\mathbf{x}, v) = \frac{1}{2}\,\rho\,\frac{\pi R^3}{J}\,v^2\,c_Q(\lambda, \beta)\,\frac{1}{x_1}.$$

The zero entries take into account that the blade pitch system, where $x_{2,0} = \beta_0 = \beta(t = 0)$ is not active. Similarly, the full-load region can be reformulated

$$\dot{\mathbf{x}} = \mathbf{A}_{II}(\mathbf{x}, v)\mathbf{x} + \mathbf{B}_{II}\mathbf{u} \tag{2.151}$$

with

$$\mathbf{A}_{II}(\mathbf{x}, v) = \begin{pmatrix} f_{II,1}(\mathbf{x}, v) & f_{II,2}(\mathbf{x}, v) \\ 0 & -\frac{1}{\tau_\beta} \end{pmatrix}, \quad \mathbf{B}_{II} = \begin{pmatrix} -0 & 0 \\ 0 & \frac{1}{\tau_\beta} \end{pmatrix}, \tag{2.152}$$

where $f_{II,1} = f_1$ and $f_{II,2} = f_1$ are defined in (2.84). The common state vector and input vector in (2.149) and (2.151) are

$$\mathbf{x} = (\omega_r, \beta)^T, \quad \mathbf{u} = (T_g, \beta_{ref})^T. \tag{2.153}$$

Both models can be transformed into a TS form using the SL approach:

$$\begin{aligned} \dot{\mathbf{x}} &= \sum_{i=1}^{2} h_{I,i}(\mathbf{z})\,\mathbf{A}_{I,i}\,\mathbf{x} + \mathbf{B}_I\,\mathbf{u}, \\ \dot{\mathbf{x}} &= \sum_{j=1}^{4} h_{II,j}(\mathbf{z})\,\mathbf{A}_{II,j}\,\mathbf{x} + \mathbf{B}_{II}\,\mathbf{u} \end{aligned} \tag{2.154}$$

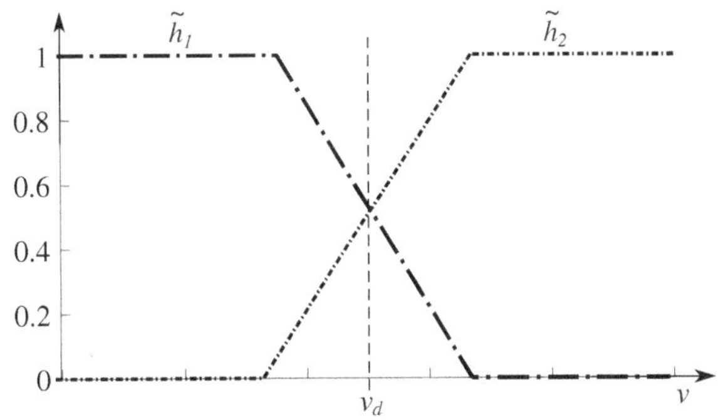

Figure 2.22 Membership function to join two single region models given in (2.154) to one multi-region model

with a common premise variable vector $\mathbf{z} = (\omega_r, \beta, v)^T$. Multiplying the first TS equation by $\sum_{j=1}^{4} h_{II,j}(\mathbf{z}) = 1$ and the second by $\sum_{i=1}^{2} h_{II,i}(\mathbf{z}) = 1$ we get

$$\dot{\mathbf{x}} = \sum_{i=1}^{2}\sum_{j=1}^{4} h_{I,i}(\mathbf{z})\, h_{II,j}(\mathbf{z})\, \mathbf{A}_{I,i}\, \mathbf{x} + \mathbf{B}_I\, \mathbf{u},$$

$$\dot{\mathbf{x}} = \sum_{i=1}^{2}\sum_{j=1}^{4} h_{I,i}(\mathbf{z})\, h_{II,j}(\mathbf{z})\, \mathbf{A}_{II,j}\, \mathbf{x} + \mathbf{B}_{II}\, \mathbf{u}.$$

This two TS models of the individual regions are joined by a convex combination via the membership function $\tilde{h}_i$ shown in Figure 2.22 and gives an overall control-oriented model:

$$\begin{aligned}\dot{\mathbf{x}} =& \tilde{h}_1(v) \sum_{i=1}^{2}\sum_{j=1}^{4} h_{I,i}(\mathbf{z})\, h_{II,j}(\mathbf{z})\, \mathbf{A}_{I,i}\, \mathbf{x} + \mathbf{B}_I\, \mathbf{u} \\ &+ \tilde{h}_2(v) \sum_{i=1}^{2}\sum_{j=1}^{4} h_{I,i}(\mathbf{z})\, h_{II,j}(\mathbf{z})\, \mathbf{A}_{II,i}\, \mathbf{x} + \mathbf{B}_{II}\, \mathbf{u}.\end{aligned} \tag{2.155}$$

Finally, (2.155) is written into a standard TS form. This is done by the combination of all h_x-functions $(\tilde{h}_k, h_{I,i}, h_{II,j})$ to a single h_m-function with $m = 1, \dots, N_r = 2 \cdot 4 \cdot 2 = 16$. The appropriate $\mathbf{A}_m$ and $\mathbf{B}_m$ matrices are associated with these membership functions h_m for $m = 1, \dots, 16$:

$$h_m \in \{h_{I,1}, h_{I,2}\} \times \{h_{II,1}, h_{II,2}, h_{II,3}, h_{II,4}\} \times \{\tilde{h}_1, \tilde{h}_2\} \tag{2.156}$$

All necessary details are listed in the Appendix C.1.3. With this, the multi-region control-oriented TS model

$$\begin{aligned}\dot{\mathbf{x}} &= \sum_{m=1}^{16} h_m(\mathbf{z})\, \mathbf{A}_m\, \mathbf{x} + \mathbf{B}_m\, \mathbf{u} \\ \mathbf{z} &= (\omega_r, \beta, v)^T, \ \mathbf{x} = (\omega_r, \beta)^T, \ \mathbf{u} = (T_g, \beta_{ref})^T\end{aligned} \tag{2.157}$$

can be created for the readers interested in it.

2.4 Summary

The models derived in this chapter will be used for the controller and observer design in the following. First-principle modeling was utilized to derive generic state-space models of the HAWT (primary power conversion). Models with different complexity related to the structural dynamics with 1- to 4-DOF were proposed. DOFs considered here are the rotor rotation, the flap-wise tip reflection of the blades (conic mode), tower top deflection, and generator rotation for elastic drive train. The aerodynamic c_Q–λ and c_T–λ curves were introduced to calculate the driving rotor torque T_r and the thrust force F_T on the tower and rotor blades. Essentials of the rotor blade theory and BEM method for determining the c_Q–λ and c_T–λ curves using the blade profile data were briefly presented. It was shown, that generic state-space models with different DOF have the common structure $\dot{\mathbf{x}} = \mathbf{A}\mathbf{x} + \mathbf{f}(\omega_r, \beta, v) + \mathbf{B}\mathbf{u}$ with $\mathbf{x} = (\omega_r, \beta, \dots)^T$, where the state vector $\mathbf{x}$ contains at least the rotor speed ω_r and pitch angle β. Further states result from the DOF to be described, i.e. y_T, $\dot{y}_T$, y_B, $\dot{y}_B$, and $\Delta\theta_s$. The structure shows clearly that the nonlinearities result from the internal state feedback given by the rotor speed and pitch angle, and the effective wind speed as an external disturbance. The linear part with the $\mathbf{A}$ matrix, on the other hand, contains only the concentrated spring-mass-damper coefficients of the mechanical equivalent models. The magnitudes of the eigenvalues correspond to the natural frequencies of the mechanical structure. The quality of the models was verified by simulation with the aeroelastic program FAST and comparison of the calculated natural frequencies (2.78) with those published in Jonkman et al. [2009].

Based on these generic state-space models, control-oriented models in TS form were derived for different design objectives. Thereby, the desired objective determines the model depth related to the dimension of the system and input matrix. For example, if the load mitigation is to be addressed in the control design by specifying the pole regions, the tower and/or rotor-blades must be integrated. Furthermore, a tailored decomposition of the functions containing $c_{Q,T}$–λ was applied; thus that for the integral state feedback, the augmented sub-models are controllable, and for the observer design, all sub-models are observable.

2.5 Problems

2.1 Show that the augmented TS model as an extension of (2.96)–(2.98) with $\tilde{\mathbf{x}} = (\mathbf{x}^T, v)^T$ contains sub-models that are not fully observable.

2.2 In contrast to the first problem, the sub-models in the TS form (2.148) are observable. Prove the structural observability of the related pairs $\{\mathbf{A}_i, \mathbf{C}\}$ (2.147) for all i.

2.3 Derive a TS model from the 2-DOF state-space model (2.135). Use the sector functions f_i given in (2.84) and (2.134).

2.4 Build an open-loop simulation for the model derived in third. Plot the h_i-functions by the wind gust excitation given in Figure 2.18.

2.6 Bibliography

For a more detailed study of the aerodynamics of wind turbines as presented in Section 2.2, the textbooks [Hansen, 2008, Gasch and Twele, 2012] are recommended. Both describe the effects of dynamics and how this can be modeled in an aeroelastic code as implemented in e.g. FLEX, FAST, or QBlade [Marten et al., 2021], which are widely used in the design and verification of modern wind turbines. An important reference book is [Burton et al., 2021], emphasizing the aerodynamics of HAWTs, the design loads, and the derived concept and component design of HAWTs.

Control-oriented models of wind turbines, which are derived in Section 2.2.4, and related TS fuzzy/linear parameter variable (LPV) forms in Section 2.3 are proposed in Bianchi et al. [2007], Georg et al. [2012a,2012b], Georg [2015], and Pöschke et al. [2017] for example. Instead of using first-principles models, a method for computing TS fuzzy models for MIMO controller design using numerical linearization from FAST was presented in Pöschke et al. [2020]. A special class of wind turbines, the Type 5 WT with hydrostatic continuous variable drive-trains, was modeled in Schulte [2014] and Aschemann and Kersten [2017] with the goal of controller design. The Type 5 turbines are connected directly to the grid via synchronous generators. Here, the rotor speed fluctuations of the turbines are compensated by a dedicated nonlinear control of a hydrostatic transmission.

Instead of generating design models from existing models represented by differential equations or numerical aeroelastic models, data-driven methods based on system identification technique can also be used [Ljung, 1999, Schoukens and Pintelon, 2001]. It is still an open research question how to use this technique for the data-based modeling of the entire working space of wind turbines. So far, only certain aspects are identified from measurement data, such as structure dynamic models using experimental modal analysis in Carcangiu et al. [2011], or subspace identification is used in Rinker and Gavin [2015] to create a simple gain-scheduled controller. The approach proposed in van der Veen et al. [2011] assumes that all nonlinearities over the operating range result from a static aerodynamic mapping associated with linear, time-invariant dynamics of the turbine's mechanical structure with only the rotor torque as input. This simplification allows the description of the wind turbine via a Hammerstein model whereby appropriate system identification techniques can be applied to this system class.

3

Control of Wind Turbines

3.1 Introduction

In this chapter, with the models of the previous one, different control systems of the primary conversion of wind turbine generators are now designed. In the first section, a model-free controller scheme is introduced that has been standard in the industry so far. The scheme contains a generator torque adjustment by preset characteristic curve and proportional-integral (PI) gain-scheduling set-point control of the generator speed at rated power through collective pitch variation. In this chapter, advanced model-based approaches are presented to achieve better system performance as standard baseline controller and to formally take into account predefined control objectives by linear matrix inequality (LMI) constraints. It will be shown how the choice of suitable design models in the TS framework can meet the desired control requirements. Furthermore, a new controller structure with power-tracking capability is presented which enables services such as fast frequency reserve for grid stabilization. All designed controllers are evaluated by realistic simulation runs using NREL's FAST, a load calculation program approved for wind turbine certification.

3.2 Baseline Generator-Torque and Blade-Pitch Controller

Before starting with the model-based systematic design, let's consider the almost model-free controller still in use in many commercial turbines. Almost model-free means that no dynamic model is used for the controller design. Based on static considerations, heuristic switching between operating regions and a gain-scheduling PI controller in the full-load region is applied without formal proof of stability and systematic formulation of control requirements. As already explained in Section 2.3.1 the operation of wind turbines is broadly divided into

Advanced Control of Grid-Integrated Renewable Energy Power Plants: LMI-based Design in the Takagi-Sugeno Framework, First Edition. Horst Schulte.

two main regions. Below the design wind speed v_d also known as rated wind speed v_{rated}, the rotor speed ω_r is less than the rated rotor speed. This region is called the partial-load region because the turbine generator power is less than the rated power. Therefore in the partial-load region, the main control objective is power optimization. If the wind speed exceeds the rated wind speed, the generator power is actively limited to the rated power. Therefore, this region is called the full-load region.

For the model-free baseline controller, the partial-load operating region is sub-divided into regions 1.5, 2, and 2.5.[1] The full-load operating region is related to Region 3 only. The automated decision of which region is currently active is made by measuring the generator speed and torque. The individual regions and the corresponding characteristic of the generator torque as a piece-wise linear/quadratic positive function of the generator speed

$$T_g = T_g(\omega_g), \tag{3.1}$$

are shown in Figure 3.1. For better comparability, the generator torque and speed quantities are related to the nominal parameters $\omega_{g,rated}$ and $T_{g,rated}$. The ranges of the regions may vary and depend on the plant design and the choice of converter/generator type. Shown in Figure 3.1 is the generator characteristic curve $T_g = T_g(\omega_g)$ of the NREL 5-MW plant as proposed in Jonkman et al. [2009]. Note that in Region 3 two different strategies (dashed lines 1 and 2) are presented. Due to the limited control variable action of the pitch actuators (limited pitch rate and

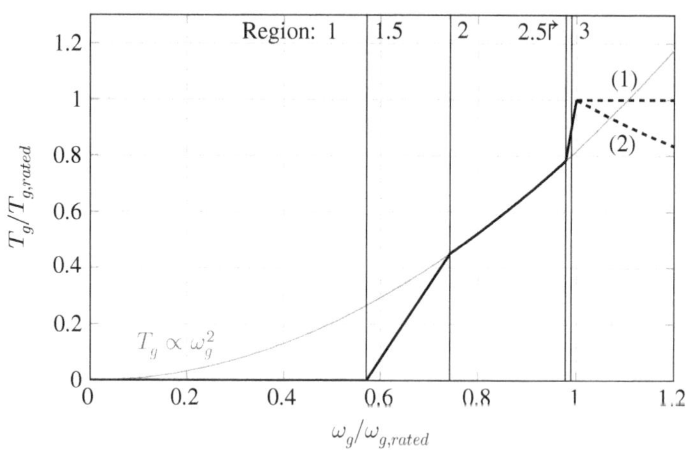

Figure 3.1 T_g as a function of the generator speed ω_g.

1 Note that the numbering of these operating regions is not uniform. This study uses the frequently used numbering from Jonkman et al. [2009].

the inflow delay) and the desired mitigation of mechanical loads, over speeds of $\omega_g \leq 1.15\ \omega_{g,rated}$ are usually allowed. To reduce power fluctuations, the generator torque can be reduced with increasing speed so that the power is constant, see curve (2). The disadvantage is that this leads to an increase in rotor speed before the slower pitch adjustment compensates it. For this reason, because power fluctuations of the individual turbines in a wind farm are smoothed by summation at the grid connection point, the generator torque is often kept constant in Region 3 represented by the curve (1) in Figure 3.1. In the following, the generic control concept is described, i.e. related to the rated quantities of a turbine.

For the baseline concept, but also for the model-based control scheme presented here, wind turbines (related to primary conversion) are controlled with two control inputs: (1) reference collective pitch angle β_{ref} and (2) reference generator torque T_g. Due to the quickly adjustable generator torque by the lower-level control in the frequency converter, no distinction is made between the reference torque and the actual torque ($T_{g,ref} \approx T_g$). In the partial-load region with the related regions 2 and 2.5 is the generated power lower as the rated power ($P_g < P_{g,rated}$). The control objective in the partial-load with the related regions 2 and 2.5 is the optimization of the generator power, whereby as much power as possible should be extracted from the wind. Thereby the pitch angle is kept constant at zero, since a value greater than zero always results in a reduction of the power efficiency, see the c_P-λ curves in Section 2.2.3, Figure 2.14. For power optimization, a control strategy is followed which is based on the steady-state assumptions that

$$P_g \overset{!}{=} P_r = T_r\,\omega_r \quad \text{valid for} \quad \frac{d\omega_r}{dt} = 0 \quad \text{with the condition} \quad \lambda \overset{!}{=} \lambda_{opt}, \tag{3.2}$$

where P_g denotes the generator power, P_r the power of the turbine rotor. With $P_g = T_g\,\omega_g$, $P_r = T_r\,\omega_r$ and the description of T_r by (2.32) and the relation (2.34) one obtains

$$T_g\,\omega_g = \frac{1}{2}\,\rho\,\pi\,R^3\,v^2\,\frac{c_P}{\lambda}\,\omega_r. \tag{3.3}$$

Using $\omega_g = n_g\,\omega_r$ (2.55) and the extension with $\frac{\omega_r^2}{\omega_r^2}\,\frac{R^2}{R^2}$ results in

$$T_g\,n_g\,\omega_r = \frac{1}{2}\,\rho\,\pi\,R^5\,\underbrace{\frac{v^2}{\omega_r^2\,R^2}}_{1/\lambda^2\,:\,(2.30)}\,\frac{c_P}{\lambda}\,\omega_r^3. \tag{3.4}$$

Finally, by substituting $\omega_r = \omega_g/n_g$ by (2.55) and $\lambda = \lambda_{opt}$ we obtain the quadratic function

$$T_g = f(\omega_g) = \underbrace{\frac{1}{2}\,\rho\,\pi\,R^5\,\frac{c_{P,max}}{n_g^3\,\lambda_{opt}^3}}_{k_{opt}}\,\omega_g^2 = k_{opt}\,\omega_g^2, \tag{3.5}$$

Table 3.1 Lower and upper bounds of the operating regions of baseline controller (can shift by ±10% if a switching hysteresis is implemented).

Region	$\underline{\omega}_{g,x}/\omega_{g,rated}$	$\overline{\omega}_{g,x}/\omega_{g,rated}$	$\underline{T}_{g,x}/T_{g,rated}$	$\overline{T}_{g,x}/T_{g,rated}$
1	0.000	0.5708	0	0
1.5	0.5708	0.7421	0	0.4503
2	0.7421	0.9782	0.4503	0.7823
2.5	0.9782	0.99	0.7823	0.9
3	0.99	1.2	0.9	1.2

with $c_{P,max} = c_P(\lambda_{opt}, \beta = 0)$, where k_{opt} is a constant defined by the optimal tip-speed ratio λ_{opt}, gear box ratio n_g, the rotor radius R and $c_{P,max}$ of the turbine to be controlled. The function (3.5) is denoted in Figure 3.1 as $T_g \propto \omega_g^2$. Note that the control law of Region 2 only requires measurement of the generator speed and does not need the estimation of the wind speed.

As can be seen in Figure 3.1, the transition from Region 1, where the generator torque is zero ($T_g = 0$) and no electrical power is extracted from the wind (but is used to accelerate the rotor for start-up) to Region 2 for optimizing power capture is made by linear interpolation. This transition region is denoted as Region 1.5. Similarly, a linear transition from Regions 2 to 3 occurs in the transition Region 2.5. To quantify these transitions, we firstly introduce the generator speed and torque limits of each region:

$$\omega_{g,x} = [\underline{\omega}_{g,x}, \overline{\omega}_{g,x}], \qquad T_{g,x} = [\underline{T}_{g,x}, \overline{T}_{g,x}], \tag{3.6}$$

where $\omega_{g,x}$ and $T_{g,x}$ denote the valid interval of region $x \in \{1, 1.5, 2, 2.5, 3\}$ with the lower $\underline{\omega}_{g,x}$ and upper bound $\overline{\omega}_{g,x}$ resp. $\underline{T}_{g,x}$ and $\overline{T}_{g,x}$. The specific numerical normalized values for the baseline controller of NREL 5-MW turbine from Jonkman et al. [2009] are given in Table 3.1. Thus, the control law for Region 1.5 to reach from the control region before cut-in wind speed (Region 1) to the power optimizing operation (Region 2) is given by the linear interpolation (see Figure 3.1)

$$\frac{T_g}{T_{g,rated}} = \underbrace{\frac{\overline{T}_{g,1.5} - \underline{T}_{g,1.5}}{\overline{\omega}_{g,1.5} - \underline{\omega}_{g,1.5}}}_{slope} \frac{\omega_{g,rated}}{T_{g,rated}} \frac{\omega_g - \underline{\omega}_{g,1.5}}{\omega_{g,rated}}. \tag{3.7}$$

Elimination of the rated values results in the simple form

$$T_g = f(\omega_g) = \frac{\overline{T}_{g,1.5} - \underline{T}_{g,1.5}}{\overline{\omega}_{g,1.5} - \underline{\omega}_{g,1.5}} \left(\omega_g - \underline{\omega}_{g,1.5}\right). \tag{3.8}$$

In addition to the control law of the start-up Region 1.5 (3.8), a transition needs to be implemented by interpolating between Regions 2 and 3. To avoid oscillations at one fixed switching point, e.g. $p_{2\rightarrow 2.5} = \{\underline{\omega}_{g,1.5}, \underline{T}_{g,1.5}\}$, a hysteresis is used instead.[2] The hysteresis distinguishes between at least two switching points for the transition from Regions 2 to 2.5 and back from 2.5 to 2. These switching points, which must be on the curve (3.5), are denoted by

$$p_{2\leftrightarrow 2.5} = \{\omega_{g,2\leftrightarrow 2.5}, T_{g,2\leftrightarrow 2.5}\}. \tag{3.9}$$

This results in the interpolation for torque control in Region 2.5 (see Figure 3.1)

$$T_g = f(\omega_g) = \frac{\overline{T}_{g,2.5} - T_{g,2\leftrightarrow 2.5}}{\overline{\omega}_{g,2.5} - \omega_{g,2\leftrightarrow 2.5}} \left(\omega_g - \omega_{g,2\leftrightarrow 2.5}\right) + T_{g,2\leftrightarrow 2.5}, \tag{3.10}$$

where

$$T_{g,2\leftrightarrow 2.5} = k_{opt}\, \omega^2_{g,2\leftrightarrow 2.5}, \tag{3.11}$$

which avoids sudden changes of the generator torque during the transition from Regions 2 to 2.5.

As illustrated in Figure 3.1 for Region 3, the generator torque (1) or generator power (2) is set constant at the rated value

$$\begin{aligned} (1): \quad & T_g = T_{g,rated} = \text{const.}, \\ (2): \quad & T_g = f(\omega_g) = P_{g,rated}/\omega_g, \end{aligned} \tag{3.12}$$

where $P_{g,rated}$ denotes the rated generator power. In contrast to regions 1.5 to 2.5, in Region 3 the collective pitch angle is adjusted with $\beta \geq 0°$ for power limitation. Let us now consider the second control variable, the collective pitch angle $\beta \geq 0°$. In the regions 1.5, 2, and 2.5 considered so far, the pitch angle was not adjusted and fixed at $\beta = 0$. To limit the power, the c_Q (Figure 2.13) resp. c_P (Figure 2.14) coefficient must be reduced as the wind speed increases by the blade-pitch angle in order to avoid further acceleration of the turbine. This results directly from the equation of motion (2.53).

In Jonkman et al. [2009] a PI gain-scheduling controller has been presented for the adjustment of the collective blade-pitch angle

$$\beta_{ref} = k_P(\beta)\,(\omega_{g,rated} - \omega_g) + k_I(\beta) \int_0^t (\omega_{g,rated} - \omega_g)\, d\tau, \tag{3.13}$$

2 In principle, hysteresis can be built in at any transition $x \leftrightarrow y$. However, it is essential for transitions in higher power ranges.

with

$$k_P(\beta) = \underbrace{2\,\frac{J\,\omega_{g,rated}}{n_g}\,\frac{\zeta\,\omega_n}{-\frac{\partial P}{\partial \beta}|_{\beta=0}}}_{k_P|_{\beta=0}=\text{const.}}\,\frac{1}{1+\frac{\beta}{\beta_k}}, \tag{3.14}$$

and

$$k_I(\beta) = \underbrace{\frac{J\,\omega_{g,rated}}{n_g}\,\frac{\omega_n^2}{-\frac{\partial P}{\partial \beta}|_{\beta=0}}}_{k_I|_{\beta=0}=\text{const.}}\,\frac{1}{1+\frac{\beta}{\beta_k}}, \tag{3.15}$$

where J is the lumped drive train inertia which is already introduced in (2.54), ω_n is the natural frequency, and ζ is the damping coefficient of the desired closed-loop system represented by a linear second-order system, and $\frac{\partial P}{\partial \beta}|_{\beta=0}$ is the pitch sensitivity of the generator power at pitch angle zero. This value is always negative because the power is reduced with a positive pitch angle change. Additional parameters are the support point value β_k of the linear regression of the pitch sensitivity function $\frac{\partial P}{\partial \beta}(\beta)$ proposed in Jonkman et al. [2009]. Note that β_{ref} as controller output of the control law (3.13) is the collective reference variable of all pitch drives (one drive each for one rotor blade) and β denotes the measured pitch angle. Table 3.2 lists all the parameters of the baseline controller necessary for its gain-scheduling coefficients $k_P(\beta)$ and $k_I(\beta)$ using a 1-DOF

Table 3.2 Baseline Controller parameters of NREL 5-MW wind turbine.

Symbol	Value	Unit	Description
k_{opt}	2.4186	$\text{Nms}^2/\text{rad}^2$	Control coefficient of Region 2 in (3.5)
J	$43.785 \cdot 10^6$	kg m^2	Drive train inertia (2.54), see Table D.2
$\omega_{g,rated}$	122.9087	rad/s	Rated generator speed given in Table D.2
ζ	0.7		Damping of desired closed-loop system
ω_n	0.6	rad/s	Natural freq. of desired closed-loop sys.
$\frac{\partial P}{\partial \beta}\|_{\beta=0}$	$-25.52 \cdot 10^6$	W/rad	Pitch sensitivity at pitch angle zero
β_k	8.0	deg	Value of the linear regression of pitch sensitivity function
$k_P\|_{\beta=0}$	0.018826		k_P controller gain in (3.14) for $\beta = 0$
$k_I\|_{\beta=0}$	0.008068		k_I controller gain in (3.15) for $\beta = 0$
$\omega_{c,f}$	0.25	rad/s	Corner frequency of the low-pass filter of the generator angular speed

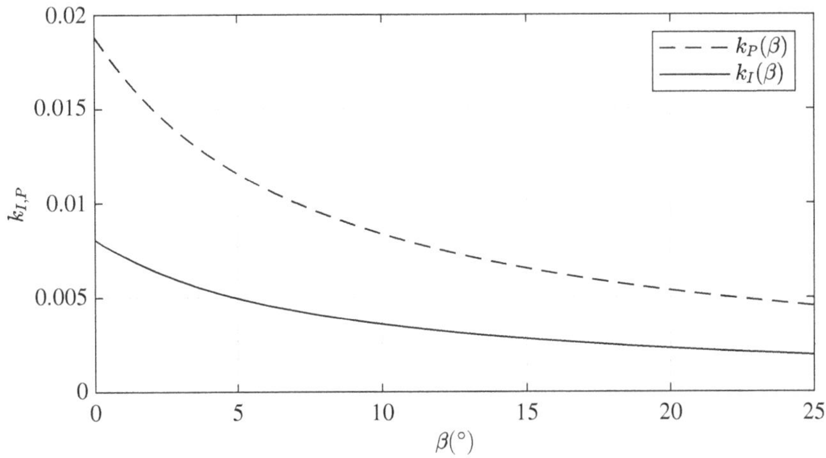

Figure 3.2 Controller coefficients k_p and k_I of the gain-scheduling control law (3.13) as a function of the measured pitch angle.

turbine model are presented in Jonkman et al. [2009]. For a better understanding of the gain-scheduling control law (3.13) in Region 3, the proportional (3.14) and integral coefficient (3.15) as a function of the blade-pitch angle are shown in Figure 3.2. One can see that both coefficients become smaller with increasing pitch angle, which compensates the increasing amount $|\frac{\partial P}{\partial \beta}|_{\beta \in [0, \beta_{max}]}|$ of the pitch sensitivity function as input gain of the plant.

For comparison with the model-based control approach as a focus of this book, simulations with the base-line controller determined by (3.5), (3.8), (3.10), (3.12), and (3.13) are now presented. It should be noted that the measurement signal of the generator speed ω_g in (3.5) and (3.13) for both the torque and pitch controllers is low-pass filtered to mitigate high-frequency excitation of the control systems. The low-pass filter as suggested in Jonkman et al. [2009] (time-discrete form) is given in the s-domain as

$$\omega_{gf}(s) = \frac{1}{s/\omega_{fc} + 1} \omega_g(s), \tag{3.16}$$

where ω_{gf} is the filtered generator speed and ω_{fc} denotes the corner frequency (the −3 dB point) of the low-pass filter. In general, filtering of signals in control systems is a common procedure in industrial practice, but it can significantly corrupt the loop dynamics of the controlled system assumed in the design. A serious problem is that the additional phase shift in open circuits caused by a poorly tuned extra filter can make the controlled system unstable. For this reason, in model-based design, requirements for mitigating the system's natural frequencies

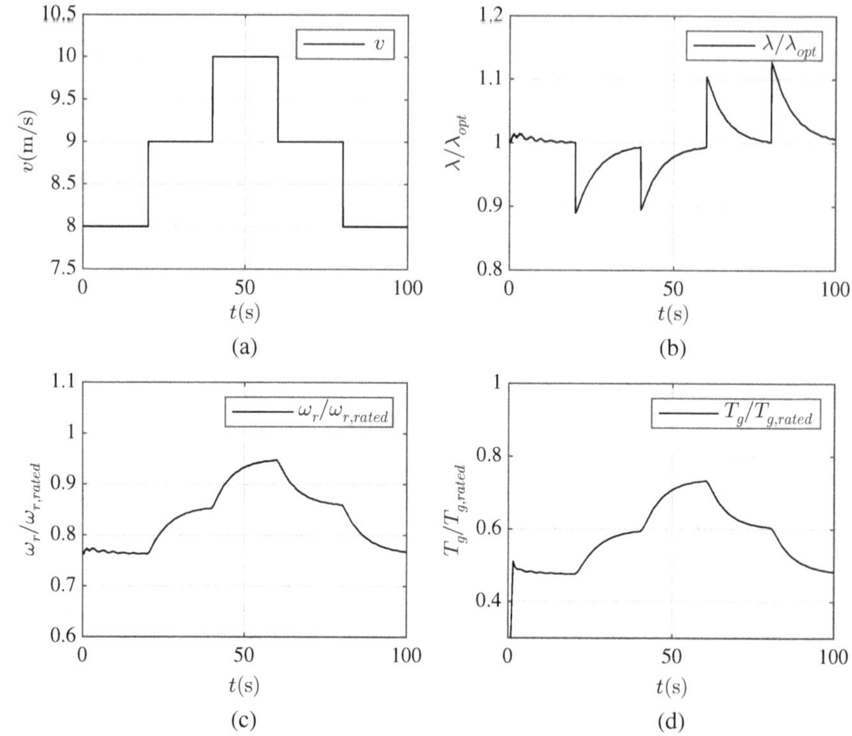

Figure 3.3 Simulation results of baseline controller validation with the FAST wind turbine model in the partial-load region with step-shaped wind excitation.

are represented by the design model from the beginning. That will be explained in Sections 3.3 and 3.4 with the TS models presented so far from Chapter 2.

Let us now turn to the simulation results using the FAST 5-MW wind turbine model subdivided into partial-load and full-load operation. Figure 3.3 presents simulations for the partial-load operating region with the baseline generator-torque controller. First, the turbine is excited with artificial wind speed jumps to analyze the convergence property of the power optimization control. The chosen wind speed values are always below the nominal wind speed so that the turbine operates in the partial-load region. The generator torque T_g set by the controller is shown in Figure 3.3d. For better comparability with other turbines and controllers, the torque is related to the rated value $T_{g,rated}$. In Figure 3.3c the rotor speed resulting from the difference between generator and rotor torque is shown as normalized value $\omega_r/\omega_{r,rated}$. The plot in Figure 3.3b illustrates the controller performance. Note that an ideal controller would result in a λ that is as close as possible to the optimal value λ_{opt}, i.e. $\lambda/\lambda_{opt} \approx 1$. However, due to the generator limitation and rotor inertia, the controller needs up to 20 seconds to reach the optimal point after a change of v. Section 3.3 will show how to improve the performance by utilizing the wind speed estimation. However, physical limitations due to the manipulated variable constraints are also given.

Second, to evaluate the baseline controller in full-load operating region, simulations were performed with turbulent wind excitation above the rated wind speed. The collective blade-pitch activity β caused by the controller and resulting rotor speed ω_r and generator power output T_g in relation to the rated values are shown in Figure 3.4. To evaluate the mechanical loads, the associated blade and tower motion of the 5-MW FAST turbine is shown in Figure 3.5.

We can clearly see a variation of the rotor speed and corresponding proportional generator torque (because $T_g = T_{g,rated}$) of maximum 7% in the full-load region ($\beta > 0, t \leq 245$ s) around the setpoints resp. nominal values. The question arises here if it is feasible to achieve a significant improvement – related to the mitigation of power fluctuations and mechanical loads – with a model-based controller design. We will address this point in Section 3.3.

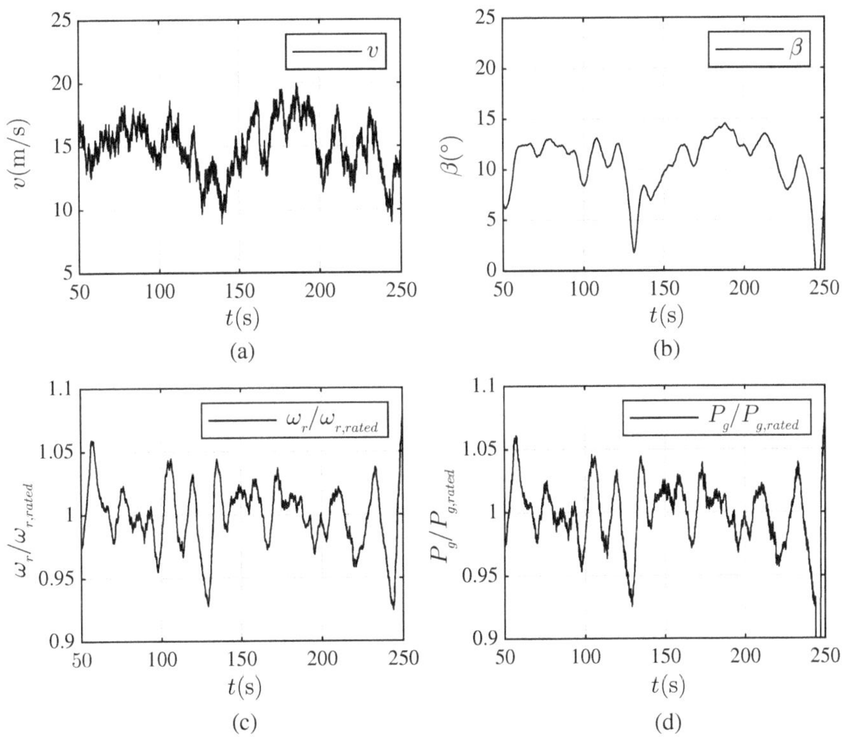

Figure 3.4 Simulation results of baseline controller validation with the FAST wind turbine model in the full-load region with turbulent wind excitation (normal turbulence model NTM016Aa).

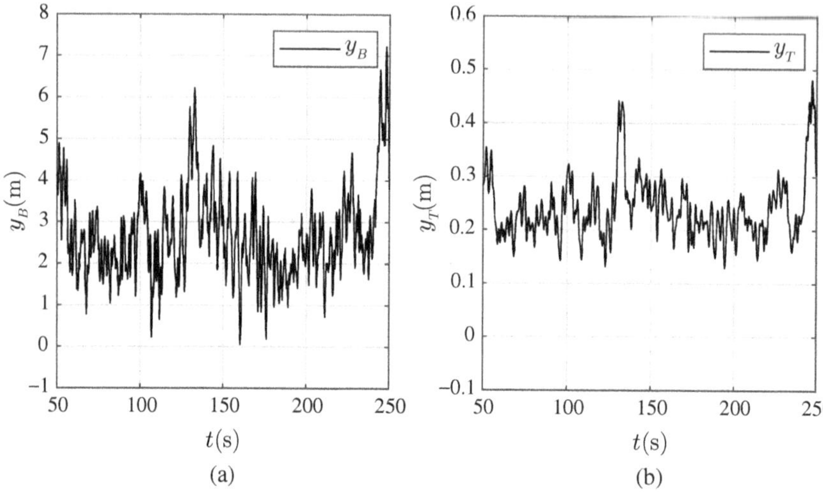

Figure 3.5 Simulation results of baseline controller validation with mechanical motion of the blade y_B and tower y_T oriented to the coordinates in Figure 2.16.

3.3 Model-Based Control of WTs

3.3.1 Control Scheme

An overview of the model-based controller structure is shown in Figure 3.6. Within the controller block all sub-controller are included the partial-load controller for Region 2, the full-load controller for Region 3, and the advanced active power control (APC) mode for grid support (see Figure 3.6). Optionally, a combined controller for both regions can be included. All necessary steps for the controller design such as formalization of control requirements, LMI-based synthesis, and system integration are described in this and Section 3.4.

3.3.2 Effective Wind Speed and State Observer

A central component of the proposed model-based controller scheme is the effective wind speed and system state estimator (see Figure 3.6). It is realized by an extended Luenberger observer in TS form

$$\dot{\hat{\tilde{\mathbf{x}}}} = \sum_{i=1}^{N_r=4} h_i(\hat{\mathbf{z}}) \left(\tilde{\mathbf{A}}_i \hat{\tilde{\mathbf{x}}} + \tilde{\mathbf{B}}\tilde{\mathbf{u}} + \mathbf{L}_i(\mathbf{y} - \hat{\mathbf{y}})\right),$$

$$\hat{\mathbf{y}} = \tilde{\mathbf{C}}\hat{\tilde{\mathbf{x}}}, \tag{3.17}$$

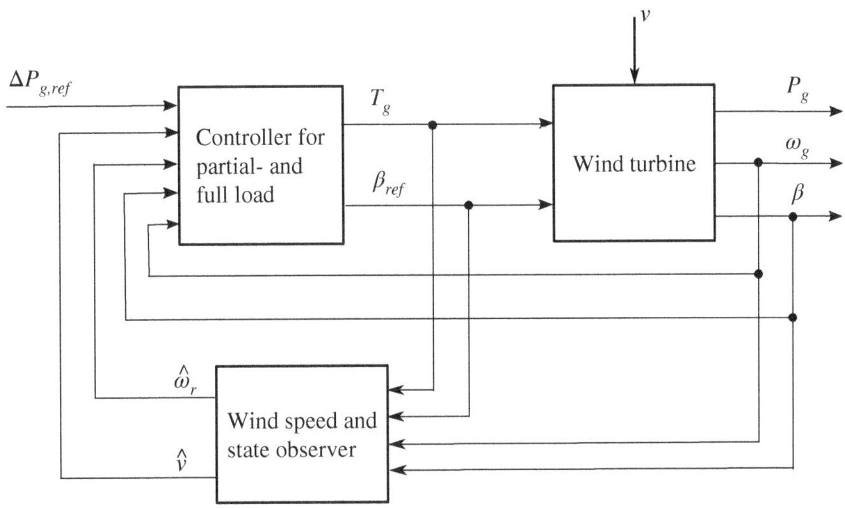

Figure 3.6 Overview of model-based wind turbine control scheme.

using 1-DOF wind turbine model (2.147), (2.148) augmented by first-order wind speed description (2.140) as disturbance model. The observer contains the system matrices of each sub-system $\tilde{\mathbf{A}}_i \in \mathbb{R}^{3\times3}$, the common input and output matrix $\tilde{\mathbf{B}} \in \mathbb{R}^{3\times2}$ and $\tilde{\mathbf{C}} \in \mathbb{R}^{2\times3}$. The reconstructed state vector with the effective wind speed to be estimated is given by $\hat{\tilde{\mathbf{x}}} = (\hat{\omega}_r, \hat{\beta}, \hat{v})^T$. The measurable and known values are included in the input vector and measurement vector $\tilde{\mathbf{u}} = (T_g, \beta_{ref}, v_0)^T$ and $\mathbf{y} = (\omega_g, \beta)^T$. The vector of premise variables $\hat{\mathbf{z}} = (\hat{\omega}_r, \beta, \hat{v})^T$ in (3.17) contains the estimated rotor speed, estimated effective wind speed, and the measurable pitch angle. Therefore the observer gains $\mathbf{L}_i$, $i = 1, 2, \dots, 4$ determined by specifying the eigenvalues of all error differential equations

$$\lambda_i = \text{eig}(\tilde{\mathbf{A}}_i - \tilde{\mathbf{L}}_i\tilde{\mathbf{C}}), \qquad i = 1, 2, \dots, 4, \tag{3.18}$$

in a desired pole region $S(\alpha, r, \theta)$ illustrated in Figure 3.9. As with the specification of the region of closed-loop eigenvalues, the observer gains $\mathbf{L}_i$ are to be calculated by the LMI-based optimization problem (B.58), see Appendix B.6.1. For the observer design, a pole region $S(\alpha, r, \theta)$ with the parameter set

$$\alpha = 0.3, \quad r = 50/3, \quad \theta = 66^\circ, \tag{3.19}$$

is chosen related to an a priori time constant $\tau_v = 4$ s of the disturbance model (2.140) proposed in Georg [2015]. In subsequent studies [Gauterin et al., 2015] it could be shown that the choice of τ_v also depends on the underlying mechanical model of the wind turbine. To increase the estimation sensitivity a design model integrating the blade dynamics is proposed in Gauterin et al. [2023] where the first-order parameter can be reduced up to $\tau_v = 0.2$ s.

Now let's start with the calculation of the observer matrices $\mathbf{L}_i$. That begins with defining the pole location region for the LMI supported design. The pole region is bounded, first to the right with $\alpha = 0.3$ caused by the pole location of the wind model with $s_{p_{\tau_v}} = 1/\tau_v = 0.25$, second by the maximum radius $r = 50/3$ which limits the natural frequency and indirectly the observer gains, and third by $\theta = 66°$ to take into account the request that weak damping is allowed to get fast reconstruction time. Note that the θ value can be calculated from the time domain specification based on the assumption of a dominant pole pair with an overshoot[3] of $M_p = 0.25$. The calculation is based on the step response analysis of the second-order system [Franklin et al., 1994] using the formula

$$\begin{aligned} M_p &= e^{-\pi\zeta/\sqrt{1-\zeta^2}}, & & 0 \le \zeta < 1, \\ \zeta &= \sin(90° - \theta), & & \{\theta \,|\, 0° < \theta \le 90°\} \rightarrow \{\zeta \,|\, 1 > \zeta \ge 0\}, \end{aligned} \tag{3.20}$$

which is plotted in Figure 3.7, where ζ denotes the damping ratio. A summary of both plots is shown in Figure 3.8. Here one can get a direct view of the functional relationship between the angle θ of the polar region and overshoot M_p.

With the specification of the desired pole region, the observer matrices can now be determined. The result of the convex optimization of (B.58) with the LMI Sedumi [Sturm, 1999] solver for the NREL 5-MW wind turbine (system parameter see Table D.2) and the defined pole region (3.19) is presented in Figure 3.10, where the eigenvalues (3.18) at the vertices of the TS observer defined by $\text{eig}(\tilde{\mathbf{A}}_i - \tilde{\mathbf{L}}_i\tilde{\mathbf{C}})$ for $i = 1, 2, \dots, 4$ are shown. As a result of the optimization, the following observer

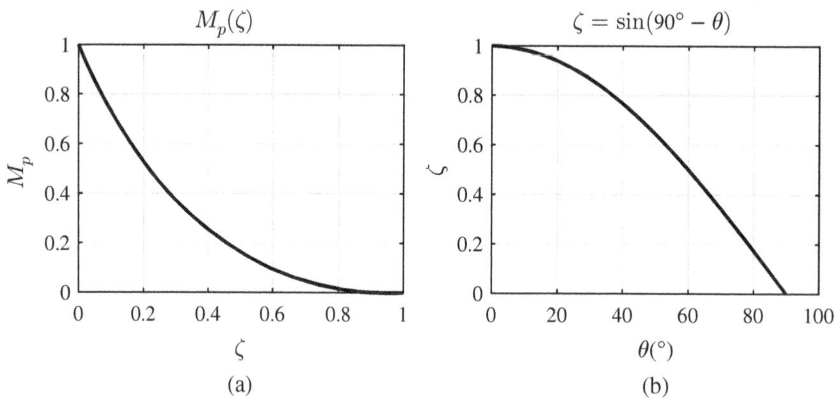

Figure 3.7 Overshoot M_p vs. damping ratio ζ for the second-order system (3.20) and damping ratio ζ as a function of region parameter θ, see Figure 3.9.

3 M_p as relative difference between the pick and final value.

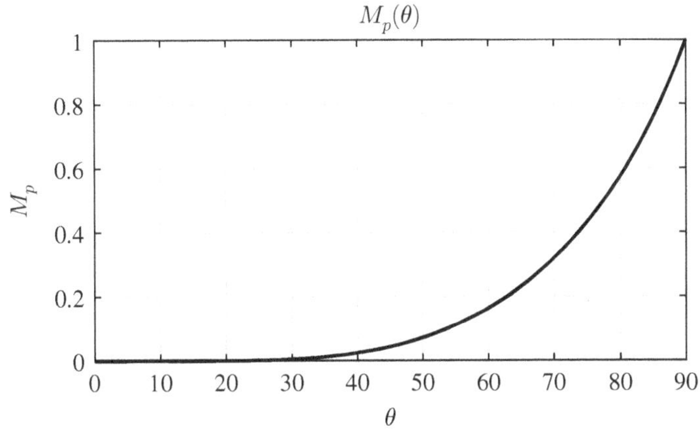

Figure 3.8 Overshoot M_p vs. region parameter θ (see Figure 3.9) related to the second-order system (3.20).

matrices were obtained:

$$\mathbf{L}_1 = \begin{pmatrix} 0.1 & -1.4 \\ 0.0 & 0.0 \\ 45.4 & -1753.5 \end{pmatrix}, \qquad \mathbf{L}_2 = \begin{pmatrix} 0.2 & 6.0 \\ 0.0 & 0.0 \\ 94.2 & 7523.2 \end{pmatrix},$$
$$\mathbf{L}_3 = \begin{pmatrix} 0.1 & -1.1 \\ 0.0 & 0.0 \\ 45.4 & -1406.6 \end{pmatrix}, \qquad \mathbf{L}_4 = \begin{pmatrix} 0.2 & -6.2 \\ 0.0 & 0.0 \\ 94.9 & -7806.6 \end{pmatrix}. \tag{3.21}$$

The validation of the wind speed observer (3.17) was carried for different wind speed profiles by the baseline controlled wind turbine (see Section 3.2) represented by the 1-DOF model (2.57). A more realistic and practical test is performed at the end of this section with the FAST simulation model of the 5-MW reference turbine [Jonkman, 2020]. The goal is to verify that the observer is robust to non-modeled dynamics and converges sufficiently fast.

But first, we consider the error dynamics without non-modeled dynamics by using the 1-DOF wind turbine model. The results are shown in Figures 3.11–3.13. In particular, the investigation by simulation is made for a wind gust, a turbulent wind, and an artificial staircase curve. Sufficient reconstruction performance is given for all three scenarios. It can be seen clearly in Figure 3.12 that the wind speed observer has a cold start capability. This is evidenced by the fact that the estimate converges to the true value within two seconds from time $t = 0$ s. The step responses of the step-shaped wind excitation in Figure 3.13 show the requirements for the dynamics of the error differential equation determined by the given pole region (3.19) are fulfilled. For example, the overshoot height of 0.2 is below the maximum allowable of $M_p = 0.76$. However, it can also be seen that the estimation

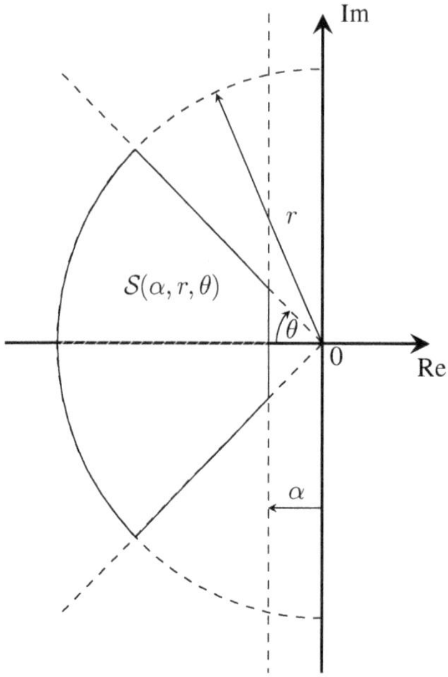

Figure 3.9 Pole region $\mathcal{S}(\alpha, r, \theta)$ defined by α, r, and θ.

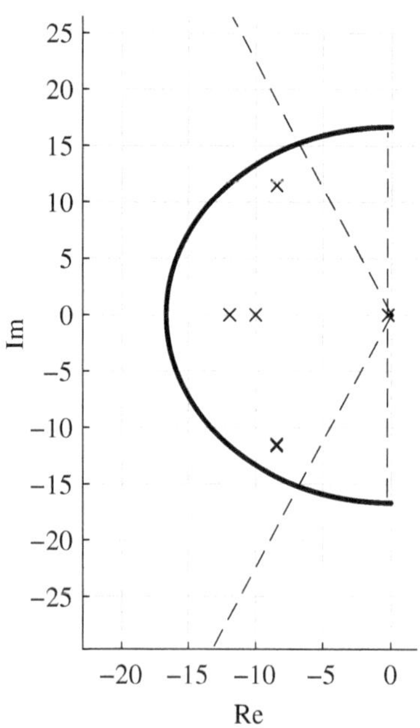

Figure 3.10 Eigenvalues of error Eq. (3.18).

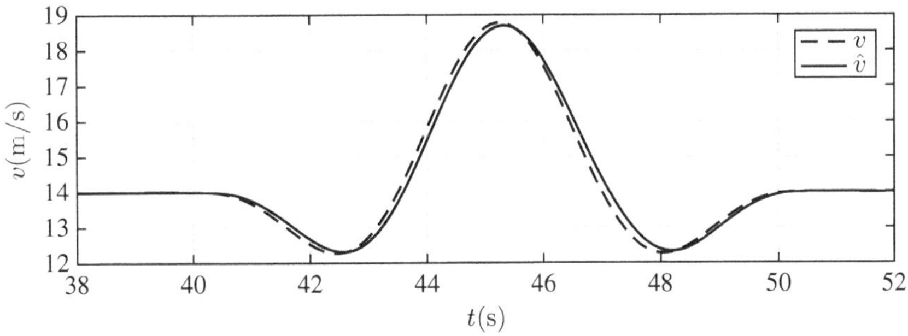

Figure 3.11 Comparison of the estimated wind speed $\hat{v} = \Delta\hat{v} + v_0$ for $v_0 = 14$ m/s by the true wind speed v far in front of the wind turbine rotor using wind gust profile.

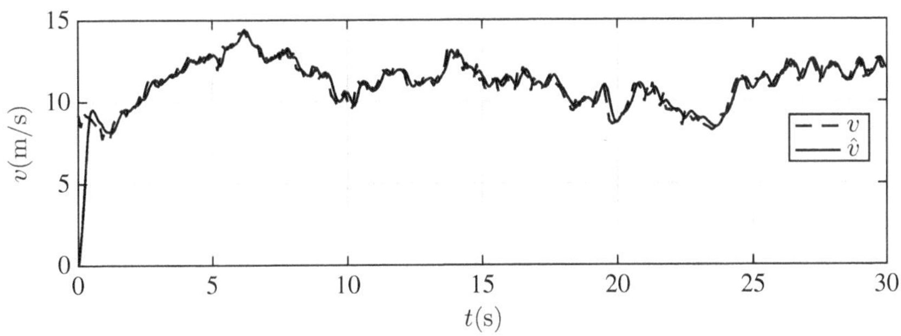

Figure 3.12 Comparison of the estimated wind speed $\hat{v} = \Delta\hat{v} + v_0$ for $v_0 = 12$ m/s by the true wind speed v far in front of the wind turbine rotor using turbulent wind speed profile
.

error increases with increasing wind speed, since the disturbance model refers to $v_0 = 3$ m/s.

To improve the performance of the wind speed observer an adaptive feed forward term $v_0(\beta, \omega_g)$ is considered as a third input of the observer law (3.17) with

$$\tilde{\mathbf{u}} = \left(T_g,\ \beta_{ref},\ v_0(\beta, \omega_g)\right)^T. \tag{3.22}$$

The forward term is a function of the interpolated stationary equilibrium points $\{\beta_i^e, v_i^e\}$ in the full-load region and results from the assumption of power optimized operation in the partial-load region. This gives the functional relationship valid for all regions:

$$v_0(\beta, \omega_g) = \begin{cases} 0 & \text{for} \quad \text{region 1} \\ 3 & \text{for} \quad \text{region 1.5} \\ \dfrac{R}{\lambda_{opt}\, n_g}\, \omega_g & \text{for} \quad \text{region 2 and 2.5} \\ f(\beta)|_{\{\beta_i^e, v_i^e\}} & \text{for} \quad \text{region 3,} \end{cases} \tag{3.23}$$

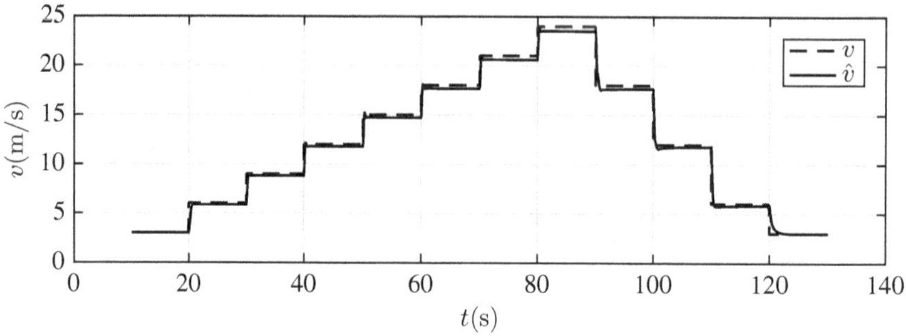

Figure 3.13 Comparison of the estimated wind speed $\hat{v} = \Delta\hat{v} + v_0$ for $v_0 = 3$ m/s by the true wind speed v far in front of the wind turbine rotor using a step-shaped test signal over the whole working range.

where $f(\beta)|_{\{\beta_i^e, v_i^e\}}$ denotes the lookup table determined by the equilibrium points $\{\beta_i^e, v_i^e\}$ (see Figure A.1). The estimation of the wind speed using the forward term (3.23) is shown in Figure 3.14. One can clearly see the improvement compared to the estimation without feed forward term only with a constant equilibrium point. The steady-state errors shown in Figure 3.13 have been significantly reduced for the most part, see the result in Figure 3.14.

Finally, it is worth mentioning that for the presented observer scheme the influence of non-measurable premise variables $\hat{\mathbf{z}}$ was not formally considered in the design process. For this purpose of investigation, we consider the observer law (3.17)

$$\dot{\hat{\tilde{\mathbf{x}}}} = \sum_{i=1}^{N_r=4} h_i(\hat{\mathbf{z}}) \underbrace{\left(\tilde{\mathbf{A}}_i \hat{\tilde{\mathbf{x}}} + \tilde{\mathbf{B}}\tilde{\mathbf{u}} + \mathbf{L}_i \tilde{\mathbf{C}}(\tilde{\mathbf{x}} - \hat{\tilde{\mathbf{x}}})\right)}_{\mathbf{f}_i(\hat{\tilde{\mathbf{x}}}, \tilde{\mathbf{x}}, \tilde{\mathbf{u}})}, \tag{3.24}$$

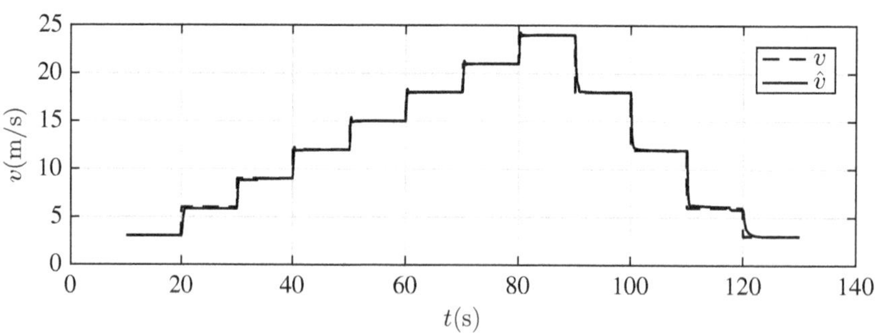

Figure 3.14 Comparison of the estimated wind speed $\hat{v} = \Delta\hat{v} + v_0(\beta, \omega_g)$ with (3.23) by the true wind speed v far in front of the wind turbine rotor using a step-shaped test signal over the whole working range.

where the outputs have been replaced by the measurement equations $\hat{\mathbf{y}} = \tilde{\mathbf{C}}\hat{\tilde{\mathbf{x}}}$, $\mathbf{y} = \tilde{\mathbf{C}}\,\tilde{\mathbf{x}}$ and add

$$\sum_{i=1}^{N_r=4} h_i(\mathbf{z})\mathbf{f}_i(\hat{\tilde{\mathbf{x}}}, \tilde{\mathbf{x}}, \tilde{\mathbf{u}}) - \sum_{i=1}^{N_r=4} h_i(\mathbf{z})\mathbf{f}_i(\hat{\tilde{\mathbf{x}}}, \tilde{\mathbf{x}}, \tilde{\mathbf{u}}) = 0.$$

This results in

$$\begin{aligned} \dot{\hat{\tilde{\mathbf{x}}}} &= \sum_{i=1}^{N_r=4} h_i(\mathbf{z})\mathbf{f}_i(\hat{\tilde{\mathbf{x}}}, \tilde{\mathbf{x}}, \tilde{\mathbf{u}}) - \Delta(\hat{\mathbf{z}}, \mathbf{z}, \hat{\tilde{\mathbf{x}}}, \tilde{\mathbf{x}}, \tilde{\mathbf{u}}), \\ \text{with} \quad & \Delta(\hat{\mathbf{z}}, \mathbf{z}, \hat{\tilde{\mathbf{x}}}, \tilde{\mathbf{x}}, \tilde{\mathbf{u}}) = \sum_{i=1}^{N_r=4} \left(h_i(\mathbf{z}) - h_i(\hat{\mathbf{z}})\right) \mathbf{f}_i(\hat{\tilde{\mathbf{x}}}, \tilde{\mathbf{x}}, \tilde{\mathbf{u}}). \end{aligned} \tag{3.25}$$

The related error differential equation $\dot{\tilde{\mathbf{e}}} = \dot{\tilde{\mathbf{x}}} - \dot{\hat{\tilde{\mathbf{x}}}}$ follows from the substitution of $\dot{\tilde{\mathbf{x}}}$ and $\dot{\hat{\tilde{\mathbf{x}}}}$ with (2.148) and (3.25). This leads to the differential equation of the estimation error with consideration of the non-measurable premise variables

$$\dot{\tilde{\mathbf{e}}} = \sum_{i=1}^{N_r=4} h_i(\mathbf{z}) \left(\tilde{\mathbf{A}}_i - \tilde{\mathbf{L}}_i\tilde{\mathbf{C}}\right) + \Delta(\hat{\mathbf{z}}, \mathbf{z}, \hat{\tilde{\mathbf{x}}}, \tilde{\mathbf{x}}, \tilde{\mathbf{u}}). \tag{3.26}$$

A proof of the exponential stability of that error differential Eq. (3.26) in contrast to (B.54) is not possible in general due to the state dependence of the $\Delta(\hat{\mathbf{z}}, \mathbf{z}, \hat{\tilde{\mathbf{x}}}, \tilde{\mathbf{x}}, \tilde{\mathbf{u}})$ term caused by $\hat{\mathbf{z}} \neq \mathbf{z}$. To solve this problem, however, some methods have been proposed in recent years. Either the error influence takes into account by the magnitude of the Lipschitz constants, which can lead to an extremely conservative design [Bergsten et al., 2001]. The conservatism can be partial reduced by certain known techniques in the TS framework [Ichalal et al., May 2011]. Or the error function may be transformed into a convex optimization problem by the sector nonlinearity approach for a special class of nonlinear functions proposed in Schulte [2016]. Further investigations can follow, but are not the focus of this study. At this point, it is sufficient to verify the stability and performance with the scenarios shown in this section and further investigations in combination with state controllers will be carried out in Section 3.3.3.

3.3.3 State Feedback Design for Partial- and Full-Load Operating Region

The proposed design decomposes the multivariable into two single-variable control problems. As already shown in Section 3.3, the control objectives (1) power optimization and (2) power limitation are decoupled and assigned to the partial-load and full-load region separately. For power optimization, the single-variable controller of the partial-load region adjusts only the generator torque. In order not to reduce the power factor c_P (see the c_P-λ curve in

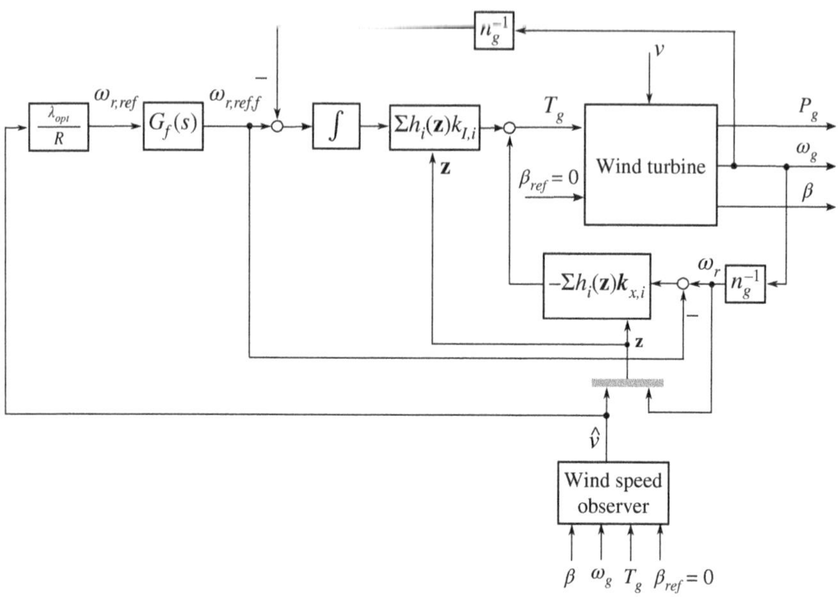

Figure 3.15 Control scheme for partial-load with state feedback, integral action, trajectory generation of the desired rotor speed and wind speed observer.

Figure 2.14), the reference value of the pitch angle β_{ref} remains at zero. The proposed control scheme of the partial-load region is illustrated in Figure 3.15.

The second single-variable controller is responsible for the full-load region of wind turbines. To limit the wind turbine power to the rated power at wind speeds above the rated wind speed, the reference pitch angle[4] is increased ($\beta_{ref} > 0$) and generator torque is held to the rated torque ($T_g = T_{g,rated}$) so that the generator power P_g is kept at the nominal value. The corresponding control scheme is shown in Figure 3.16. If negative change of control power is to be provided for grid frequency stabilization, the power is reduced, compared to the nominal value. For this purpose, a model-based design procedure in the TS framework for active power-tracking control is presented in Section 3.4.

The same structure of the control law and state feedback design methods is used for the partial- and full-load region. This simplifies the design and enables a further combination of both controllers into an overall scheme. Let us now consider the control law. It consists of a state and integral error feedback and is given as

$$u = -\sum_{i=1}^{N_r} h_i(\mathbf{z})\,\mathbf{k}_{x,i}^T\,\Delta\mathbf{x} + \sum_{i=1}^{N_r} h_i(\mathbf{z})\,k_{I,i}\,x_I, \tag{3.27}$$

4 As collective setpoint for the lower-level position control of the pitch drives.

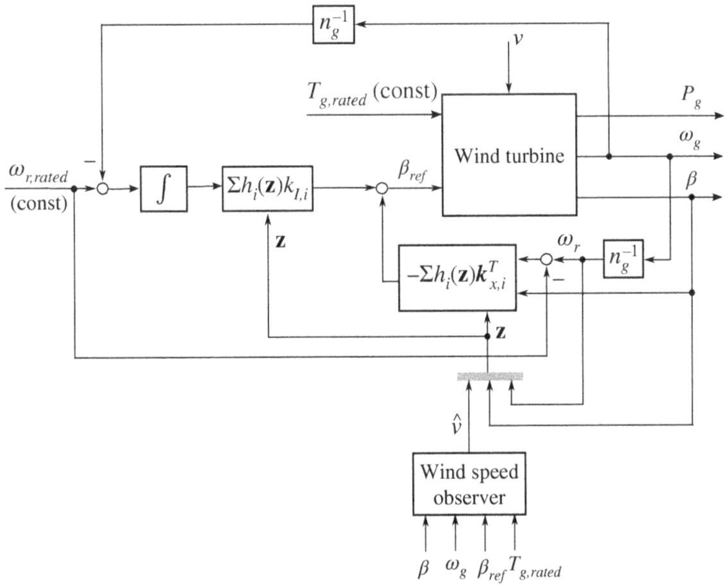

Figure 3.16 Control scheme for full-load operating region with state feedback, integral action, and wind speed observer.

where

$$x_I = \int_0^t \left(\omega_{r,ref} - \omega_r\right) d\tau = \int_0^t e\, d\tau, \tag{3.28}$$

with the control error $e = \omega_{r,ref} - \omega_r$ and the premise variables summarized in $\mathbf{z} = (\mathbf{x}^T, \hat{v})^T$. The difference state vector $\Delta\mathbf{x}$ of the full-load region controller is given as

$$\Delta\mathbf{x} = \begin{pmatrix} \omega_r - \omega_{r,ref} \\ \beta \end{pmatrix} = \begin{pmatrix} x_1 - x_{1,ref} \\ x_2 \end{pmatrix}, \tag{3.29}$$

see the scheme in Figure 3.16. This results from the corresponding design-oriented model of the full-load region (2.97). For the partial-load region, since the pitch angle is zero, it is not a state in the design model (2.108), only the rotor speed is feedback as shown in Figure 3.15.

$$\Delta x = \omega_r - \omega_{r,ref} = x_1 - x_{1,ref}. \tag{3.30}$$

It should be noted that the wind speed is an unmeasurable premise variable and is estimated by the disturbance observer proposed in Section 3.3.2, see the control scheme in Figures 3.15 and 3.16. The controller for power optimization and limitation just differs in the control variable $u = \{T_g, \beta_{ref}\}$ and also in generation of the reference value $\omega_{r,ref}$ to be tracked.

Thus, let us take a look at this in detail: (1) The partial-load controller (see Figure 3.15) adjusts the generator torque $u = T_g$ to track the rotor speed following $\omega_{r,ref} = \frac{\lambda_{rated}}{R} v$ with $\lambda_{rated} = \lambda_{opt}$, cf. (2.30), where the pitch angle is set constantly to zero $\beta_{ref} = 0$. According to (2.31), this results directly from the control objective of maximizing the rotor power P_r, where the reference rotor speed is tracked so that an optimum value $c_{p,max} = c_P(\lambda_{opt})$ is reached. Here, the dynamics is limited by the achievable actuating effort, generator torque. (2) On the other hand, the controller of the full-load region (see Figure 3.16) adjusts the pitch angle set point $\beta_{ref} \geq 0$ so that the rotor speed is kept to its rated value, i.e. $\omega_{r,ref} = \omega_{r,rated}$. To obtain rated power $P_{g,rated}$ if $v > v_{rated}$, the generator torque is simultaneously kept to its rated value.

The synthesis of the described control law (3.87) for both operating regions can be converted separately into a convex programming problem written in terms of LMI conditions proposed in Appendix B.5.1, see Theorem B.6. The major task now is to define the requirements and desired pole region of both closed-loop systems, which can be translated into LMI conditions of a constrained optimization problem.

3.3.3.1 Design of a Full-Load Region Control with State Feedback

The state feedback design in the full-load operating region based on the augmentation of state differential equation in (2.97) by the previously defined new state (3.28) results in[5]

$$\dot{\tilde{\mathbf{x}}} = \sum_{i=1}^{N_r=4} h_i(\mathbf{z})\tilde{\mathbf{A}}_i\,\tilde{\mathbf{x}} + \tilde{\mathbf{b}}\,u + (0,\,0,\,1)^T x_{1,ref}, \tag{3.31}$$

with $\tilde{\mathbf{x}} = (\mathbf{x}^T, x_I)^T$ and $x_{1,ref} = \omega_{r,ref}$, where

$$\tilde{\mathbf{A}}_i = \begin{pmatrix} \mathbf{A}_i & \mathbf{0}_{2\times 1} \\ -\mathbf{c}^T & 0 \end{pmatrix}, \quad \tilde{\mathbf{b}} = \begin{pmatrix} \mathbf{b} \\ 0 \end{pmatrix}, \tag{3.32}$$

with

$$\mathbf{A}_1 = \begin{pmatrix} \underline{f}_1 & \underline{f}_2 \\ 0 & -\frac{1}{\tau_\beta} \end{pmatrix}, \quad \mathbf{A}_2 = \begin{pmatrix} \underline{f}_1 & \overline{f}_2 \\ 0 & -\frac{1}{\tau_\beta} \end{pmatrix}, \quad \mathbf{A}_3 = \begin{pmatrix} \overline{f}_1 & \underline{f}_2 \\ 0 & -\frac{1}{\tau_\beta} \end{pmatrix},$$

$$\mathbf{A}_4 = \begin{pmatrix} \overline{f}_1 & \overline{f}_2 \\ 0 & -\frac{1}{\tau_\beta} \end{pmatrix}, \quad \mathbf{B} = \begin{pmatrix} 0 \\ \frac{1}{\tau_\beta} \end{pmatrix}, \qquad \mathbf{c}^T = (1\ 0).$$

The upper $\overline{f}_1$, $\overline{f}_2$ and lower bounds $\underline{f}_1$, $\underline{f}_2$ (2.88) of the sector functions $f_{1,2}(\mathbf{x}, v)$ (2.84) are calculated with the system parameters of the NREL wind turbine (see

5 The differentiation of the integral x_I (3.28) yields $\dot{x}_I = -x_1 + \omega_{r,ref}$, which gives $\mathbf{c}^T = (1, 0)$ in the augmented system matrix $\tilde{\mathbf{A}}_i$ and the reference input $x_{1,ref} = \omega_{r,ref}$ in (3.31).

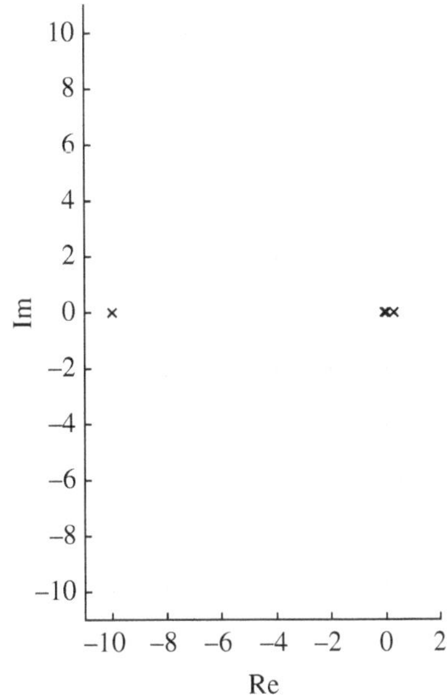

Figure 3.17 Eigenvalues of $\mathbf{A}_i$ related to the open-loop system (3.31).

Table D.2) related to the state-space and input range defined in (2.89):

$$\underline{f}_1 = -0.0942, \qquad \overline{f}_1 = 0.2745, \tag{3.33}$$

$$\underline{f}_2 = -0.6290, \qquad \overline{f}_2 = -0.1493. \tag{3.34}$$

The distribution of the open-loop poles of the augmented model (3.31) is shown for all $N_r = 4$ sub-models in Figure 3.17. Unstable poles result from the positive upper bound $\overline{f}_1 = 0.2745$ in $\mathbf{A}_3$ and $\mathbf{A}_4$. In relation to the open-loop poles and closed-loop requirements, we define now a desired pole region of the closed-loop system. For this purpose, the integral state feedback control law (3.27) is rewritten according to the augmented state vector

$$u = -\sum_{i=1}^{N_r} h_i(\mathbf{z})\,\tilde{\mathbf{k}}_i^T\,\Delta\tilde{\mathbf{x}}, \qquad \tilde{\mathbf{k}}_i^T = (\mathbf{k}_{x,i}^T, -k_{I,i}). \tag{3.35}$$

with

$$\Delta\tilde{\mathbf{x}} = \begin{pmatrix} \Delta\mathbf{x} \\ x_I \end{pmatrix} = \begin{pmatrix} x_1 - x_{1,ref} \\ x_2 \\ x_I \end{pmatrix} = \tilde{\mathbf{x}} - \begin{pmatrix} x_{1,ref} \\ 0 \\ 0 \end{pmatrix}. \tag{3.36}$$

Combined with the augmented model (3.31), this results in the closed-loop system

$$\dot{\tilde{\mathbf{x}}} = \sum_{i=1}^{N_r=4} h_i(\mathbf{z}) \left(\tilde{\mathbf{A}}_i\tilde{\mathbf{x}} - \tilde{\mathbf{b}}\tilde{\mathbf{k}}_i^T \left(\tilde{\mathbf{x}} - \mathbf{e}_1 x_{1,ref} \right) \right) + \mathbf{e}_3 x_{1,ref}. \tag{3.37}$$

with $\mathbf{e}_1 = (1, 0, 0)^T$ and $\mathbf{e}_3 = (0, 0, 1)^T$. With this we obtain

$$\dot{\tilde{\mathbf{x}}} = \sum_{i=1}^{N_r=4} h_i(\mathbf{z}) \left(\tilde{\mathbf{A}}_i - \tilde{\mathbf{b}}\tilde{\mathbf{k}}_i^T \right) \tilde{\mathbf{x}} + \sum_{i=1}^{N_r=4} h_i(\mathbf{z}) \left(\tilde{\mathbf{b}}\, \tilde{\mathbf{k}}_i^T\, \mathbf{e}_1 + \mathbf{e}_3 \right) x_{1,ref}. \tag{3.38}$$

The closed-loop dynamics is determined by the weighted combination of the system matrices $\tilde{\mathbf{A}}_i - \tilde{\mathbf{b}}\tilde{\mathbf{k}}_i^T$, $i = 1, \ldots, N_r$, where the final value of the rotor speed is governed by the reference value $x_{1,ref} = \omega_{r,ref}$.

Now we will perform the controller design. The computation of $\tilde{\mathbf{k}}_i^T$ in (3.35) is done using the LMI region criteria (B.37) of Theorem B.6 with (B.41) for control effort reduction. Justifiable parameters of the pole region $S(\alpha, r, \theta)$ are defined by

$$\alpha = 0.1, \quad r = 2.5, \quad \theta = 18°. \tag{3.39}$$

The selected pole region follows from the requirements of the closed-loop system with the maximum overshoot $M_p = 0.043$ where $\zeta = \sin(90° - \theta) = 1/\sqrt{2}$, see Figure 3.7. The approximate relations result from the aggregation of the closed-loop dynamics by a second-order system.

After this preliminary work, the appropriate LMI criteria of the previous pole region specification $S(\alpha, r, \theta)$ are applied to calculate the controller coefficients $\tilde{\mathbf{k}}_i^T = \tilde{\mathbf{m}}_j^T\, \mathbf{X}^{-1}$ for the NREL 5-MW wind turbine benchmark (for system parameter see Table D.2). The result of the convex optimization with the LMI Sedumi [Sturm, 1999] solver is shown in Figure 3.18 on the basis the eigenvalues (3.18) at the vertices of the TS closed-loop system defined by $\text{eig}(\tilde{\mathbf{A}}_i - \tilde{\mathbf{b}}\tilde{\mathbf{k}}_i^T)$ for $i = 1, \ldots, 4$.

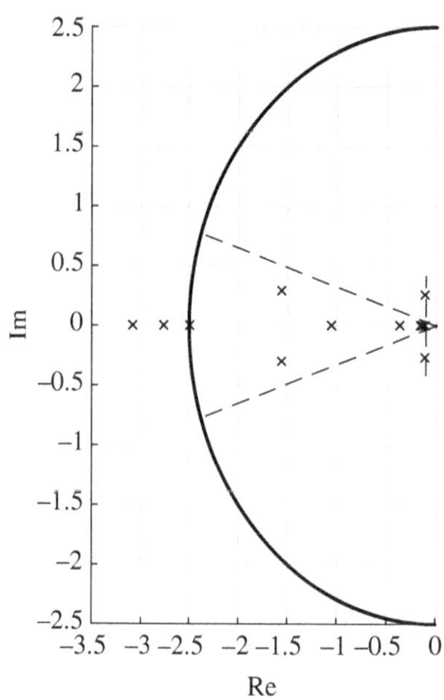

Figure 3.18 Eigenvalues of $(\tilde{\mathbf{A}}_i - \tilde{\mathbf{b}}\tilde{\mathbf{k}}_i^T)$ related to the closed-loop system (3.38).

The related controller matrices are as follows

$$\tilde{\mathbf{k}}_1^T = \Big(\underbrace{-0.4094 \; -0.6864}_{\mathbf{k}_{x,1}^T} \; \underbrace{0.0451}_{-k_{I,1}}\Big), \quad \tilde{\mathbf{k}}_2^T = \Big(\underbrace{-0.7763 \; -0.6828}_{\mathbf{k}_{x,2}^T} \; \underbrace{0.0960}_{-k_{I,2}}\Big),$$
$$\tilde{\mathbf{k}}_3^T = \Big(\underbrace{-0.6799 \; -0.6020}_{\mathbf{k}_{x,3}^T} \; \underbrace{0.0656}_{-k_{I,3}}\Big), \quad \tilde{\mathbf{k}}_4^T = \Big(\underbrace{-1.1629 \; -0.6422}_{\mathbf{k}_{x,4}^T} \; \underbrace{0.1649}_{-k_{I,4}}\Big). \tag{3.40}$$

For a realistic controller validation, as already introduced in Section 2.2.5, the aeroelastic 5-MW wind turbine benchmark modeled in NREL's FAST v8.16 is used. The up to 24 degrees of freedom of the FAST wind turbine model implemented in MATLAB®-Simulink [Singh et al., 2014] allows a sufficiently accurate simulation of the mechanical structural dynamics of large wind turbines. A limitation of the pitch rate (angular velocity of β) was added to the simulation model. In modern wind turbines, this value is typically 8 degrees per second [OAT, 2021]

$$\frac{d\beta}{dt} \leq 8^\circ/\text{s}. \tag{3.41}$$

A simulation of the closed-loop according to Figure 3.16 with a turbulent wind speed curve far above the rated wind speed (i.e. the turbine is only operated in the full-load region) is shown in Figure 3.19. It clearly shows that the pitch control limits the rotor speed and thus the power with constant generator torque $T_g = T_{g,rated}$ regardless of fluctuating wind speeds. Therefore, the characteristic of the rotor speed at constant generator torque is directly proportional to the generator power. For a better comparison, the rotor speed is related to the rated rotor speed (see Table D.2) resp. is given in per unit (pu). The deviation of the rotor speed is still far below the permitted deviation from the rated value of $\pm 10\%$. Further simulation results which takes into account the rotor blade and tower motion is shown in Figure 3.20. The motion of the turbine structure is represented by the fore–aft tower top deflection y_T, and the flap-wise bending of the blades at the tip is determined by the position of the rotor blade tip y_B (see the schematic illustration of the DOFs in Figure 2.16). The maximum blade deflections of the proposed controller are below the permissible distances to the tower, and the tower deflection is far below the permissible deflection [Burton et al., 2021].

3.3.3.2 Control-Design of Partial-Load Operating Region

Let us now dedicate to the controller design for the partial-load operating region. The control scheme was shown previously in this section, see Figure 3.15. As with the full-load region, the control law is based on an integral state feedback

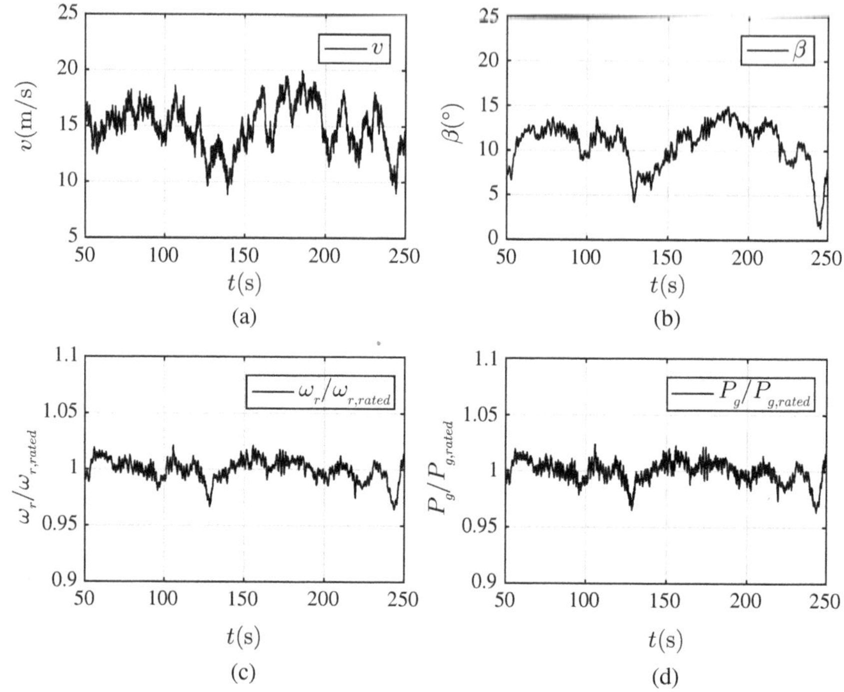

Figure 3.19 Simulation results of full-load region controller validation (3.35) for the calculated controller coefficients (3.40) using NREL's FAST wind turbine model excited by turbulent wind speed (normal turbulence model NTM016Aa) .

controller. In order to achieve the control objective of aerodynamic power optimization, the reference rotor speed is calculated with the use of the estimated wind speed $\hat{v}$

$$\omega_{r,ref} = \frac{\lambda_{opt}}{R}\hat{v}. \tag{3.42}$$

As with the previous controller design, the related control-oriented model (2.108) is augmented by the state x_I (3.28)

$$\dot{\tilde{\mathbf{x}}} = \begin{pmatrix} \dot{x} \\ \dot{x}_I \end{pmatrix} = \begin{pmatrix} \sum_{i=1}^{N_r=2} h_i(\mathbf{z})a_i x + bu \\ x_{ref} - x \end{pmatrix}, \tag{3.43}$$

with $a_1 = \underline{f}$, $a_2 = \overline{f}$, and $b = -n_g/J$, where the sector function $f \in [\underline{f}, \overline{f}]$ is specified in (2.101). It follows the final TS form

$$\dot{\tilde{\mathbf{x}}} = \sum_{i=1}^{N_r=2} h_i(\mathbf{z}) \underbrace{\begin{pmatrix} a_i & 0 \\ -1 & 0 \end{pmatrix}}_{\tilde{\mathbf{A}}_i} \tilde{\mathbf{x}} + \underbrace{\begin{pmatrix} b \\ 0 \end{pmatrix}}_{\tilde{\mathbf{b}}} u + \underbrace{\begin{pmatrix} 0 \\ 1 \end{pmatrix}}_{\mathbf{e}_2} x_{ref}. \tag{3.44}$$

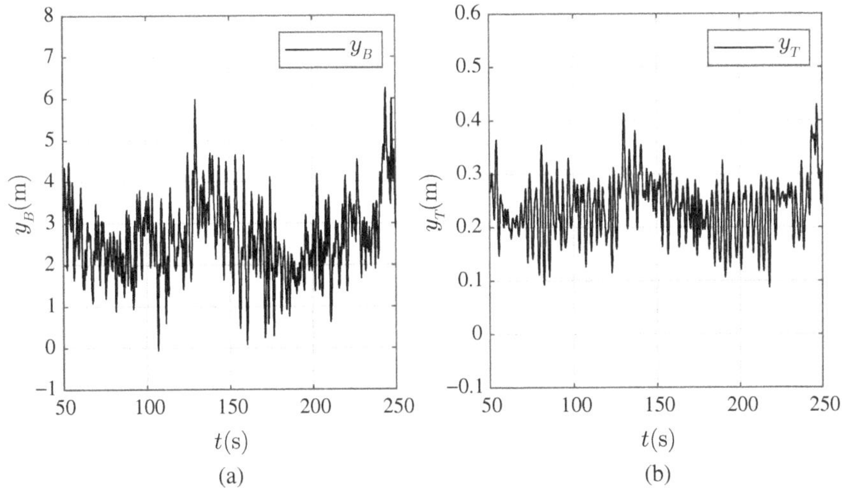

Figure 3.20 Simulation results related to Figure 3.19: Mechanical motion of the blade and tower represented by y_B and y_T, see the specification of coordinates in Figure 2.16.

with $\mathbf{z} = (\omega_r, v)^T$. The general integral state feedback control law (3.87) applied to the partial-load region is

$$u = -\sum_{i=1}^{N_r=2} h_i(\mathbf{z})\tilde{\mathbf{k}}_i^T \Delta\tilde{\mathbf{x}}, \qquad \mathbf{k}_i^T = \left(k_x, -k_I\right), \tag{3.45}$$

with

$$\Delta\tilde{\mathbf{x}} = \tilde{\mathbf{x}} - \begin{pmatrix} x_{ref} \\ 0 \end{pmatrix} = \begin{pmatrix} x \\ x_I \end{pmatrix} - \begin{pmatrix} x_{ref} \\ 0 \end{pmatrix}. \tag{3.46}$$

Substituting the control law (3.45) into (3.44) leads to the closed-loop system

$$\dot{\tilde{\mathbf{x}}} = \sum_{i=1}^{N_r=2} h_i(\mathbf{z}) \left(\tilde{\mathbf{A}}_i \tilde{\mathbf{x}} + \tilde{\mathbf{b}} \left(-\tilde{\mathbf{k}}_i^T \Delta\tilde{\mathbf{x}}\right)\right) + \mathbf{e}_2 x_{ref}. \tag{3.47}$$

A reformulation results in the final closed-loop system of operating region 2

$$\dot{\tilde{\mathbf{x}}} = \sum_{i=1}^{N_r=2} h_i(\mathbf{z}) \left(\tilde{\mathbf{A}}_i - \tilde{\mathbf{b}}\tilde{\mathbf{k}}_i^T\right)\tilde{\mathbf{x}} + \sum_{i=1}^{N_r=2} h_i(\mathbf{z}) \left(\tilde{\mathbf{b}}\tilde{\mathbf{k}}_i^T \mathbf{e}_1 + \mathbf{e}_2\right) x_{ref}, \tag{3.48}$$

where $\mathbf{e}_1 = (1, 0)^T$. It consists of the dynamics of the closed system and an input part to reach the desired reference value.

Now, as for the full-load region, the controller gain $\tilde{\mathbf{k}}_i$ in (3.48) is determined by the LMI region criteria (B.37) in combination with (B.41). Let us first consider the poles of the augmented open-loop system (3.44) to define the pole region of the closed-loop system (3.48). These are used to define the pole region of the closed-loop system. Indeed, the larger the distance between the closed-loop poles and the open loop, the greater the necessary control effort. The distribution of the open-loop eigenvalues for the parameter set of the NREL 5-MW wind turbine (see Appendix D.2) is given

$$\begin{aligned} i = 1: &\quad \lambda_1 = 0, \quad \lambda_2 = 1.05 \cdot 10^{-4} \\ i = 2: &\quad \lambda_1 = 0, \quad \lambda_2 = 0.17, \end{aligned} \tag{3.49}$$

and illustrated in Figure 3.21. Note that the positive eigenvalues result from the positive-valued sector function (2.100) in the control-oriented model and the zero eigenvalues from the augmentation by the x_I state (3.28).

Let us now study the influence of the desired conical-shaped pole region of the closed-loop dynamics. Four settings defined in Table 3.3 are studied for the following controller LMI-based synthesis. The controller gains as a result of the calculation with the LMI criteria (B.37), (B.41) using the LMI Lab Solver from Mathwork [Mathworks, 2021] are listed in Table 3.3. The closed-loop eigenvalues are illustrated in Figure 3.22. Note that all eigenvalues are located in the desired pole regions where the related controller gain decreases with the shift of the pole region toward the imaginary axis (see Table 3.3).

Let us now consider the simulation results, which are first performed using the simple 1-DOF model of the 5-MW wind turbine as plant model (2.57) and a stepped change of the wind velocity excitation as shown in Figure 3.23. Plotted are the wind speed estimation with different controller and the same observer gains proposed in Section 3.3.2. The differences are not significant which are caused by different controller outputs as plant inputs T_g. The estimation of the effective wind speed is

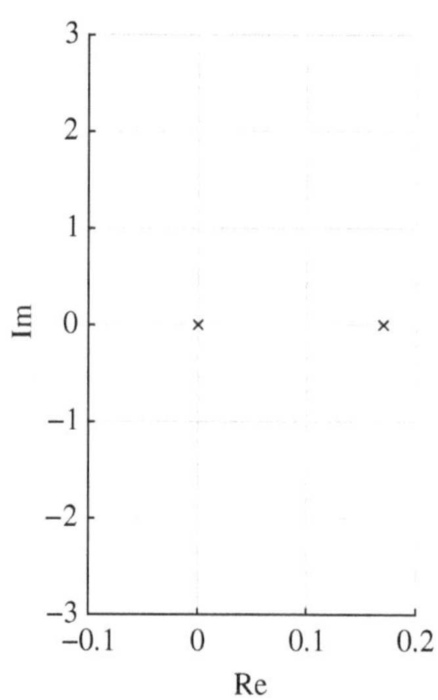

Figure 3.21 Eigenvalues of $\mathbf{A}_i$ related to the open-loop system (3.44) with the predefined operating range (2.106) .

Table 3.3 Comparison of LMI-solver synthesis results for different pole region settings $S(\alpha, r, \theta)$, see Figure 3.22.

No.	α	r	θ		$k_{x,i}$	$k_{I,i}$
#1	1.0	3	36°	$i = 1:$	$-1.4104 \cdot 10^6$	$-1.2354 \cdot 10^6$
				$i = 2:$	$-1.4523 \cdot 10^6$	$-1.2257 \cdot 10^6$
#2	0.5	2	36°	$i = 1:$	$-8.6261 \cdot 10^5$	$-4.1462 \cdot 10^5$
				$i = 2:$	$-8.9268 \cdot 10^5$	$-4.1114 \cdot 10^5$
#3	0.4	1	36°	$i = 1:$	$-5.4907 \cdot 10^5$	$-1.9276 \cdot 10^5$
				$i = 2:$	$-6.1486 \cdot 10^5$	$-2.0519 \cdot 10^5$
#4	0.2	0.5	36°	$i = 1:$	$-2.8146 \cdot 10^5$	$-0.5131 \cdot 10^5$
				$i = 2:$	$-3.5895 \cdot 10^5$	$-0.5761 \cdot 10^5$

sufficient in all cases to determine the reference rotor speed with (3.42) supporting optimum power operation. The reference signal filter in the tracking control scheme is not used[6] in this case, i.e. $G_f(s) = 1$, see Figure 3.15. Due to the tracking control, the rotor speed follows $\omega_{r,ref}$, so that the optimal tip-speed ratio λ_{opt} and thus an optimal c_P denoted as $c_{P,max}$ is reached (see Figure 2.14). Note that the differences of the λ curves related to the different controller coefficients are

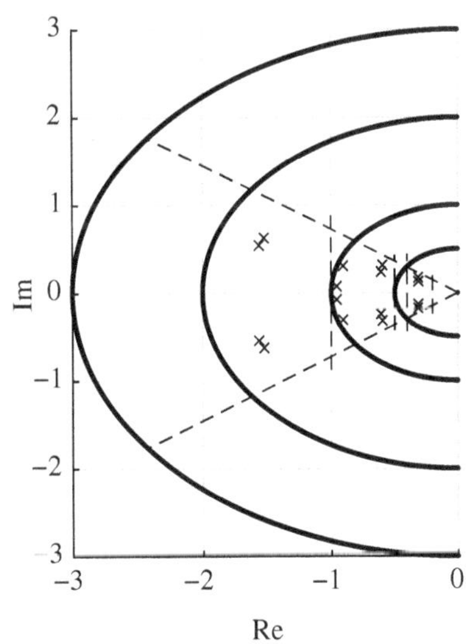

Figure 3.22 Calculated eigenvalues of $(\tilde{\mathbf{A}}_i - \tilde{\mathbf{b}}\tilde{\mathbf{k}}_i^T)$ related to the closed-loop system (3.48) with $\tilde{\mathbf{k}}_i^T = (k_{x,i}, k_{I,i})$ for #1 , … #4 regions, see Table 3.3.

6 In the case of controller validation using the FAST turbine model, the filter is used as a low pass, which will be shown later.

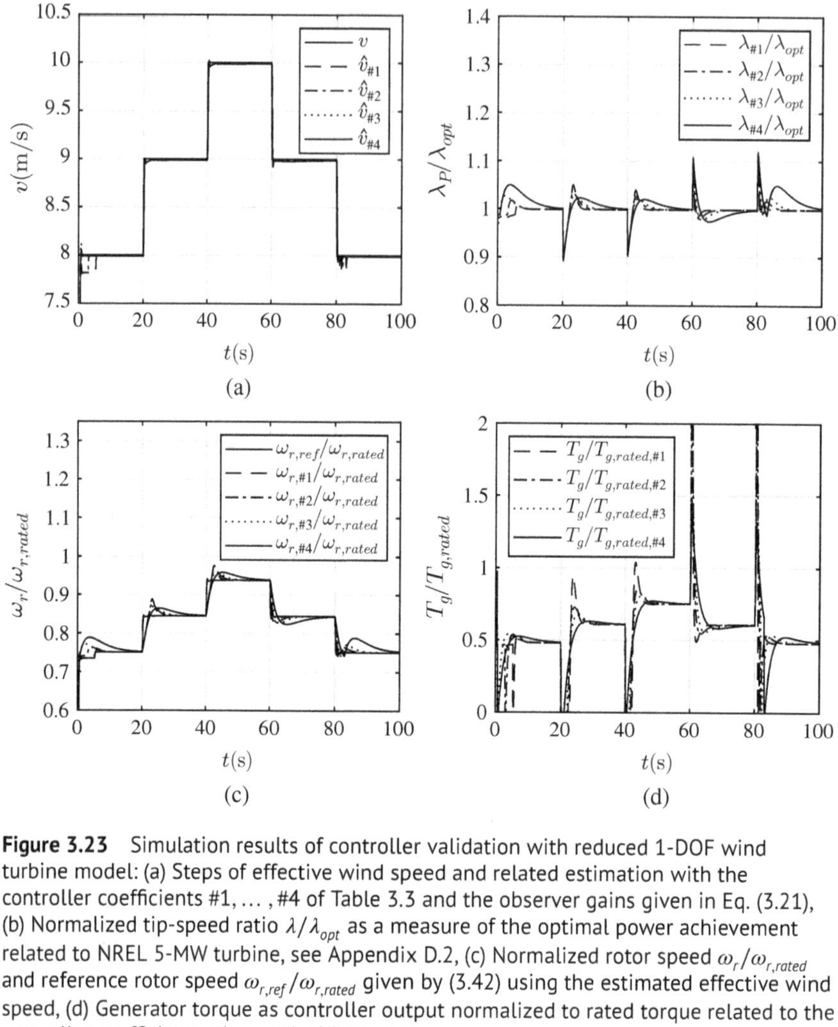

Figure 3.23 Simulation results of controller validation with reduced 1-DOF wind turbine model: (a) Steps of effective wind speed and related estimation with the controller coefficients #1, … , #4 of Table 3.3 and the observer gains given in Eq. (3.21), (b) Normalized tip-speed ratio λ/λ_{opt} as a measure of the optimal power achievement related to NREL 5-MW turbine, see Appendix D.2, (c) Normalized rotor speed $\omega_r/\omega_{r,rated}$ and reference rotor speed $\omega_{r,ref}/\omega_{r,rated}$ given by (3.42) using the estimated effective wind speed, (d) Generator torque as controller output normalized to rated torque related to the controller coefficients #1, … , #4 of Table 3.3 .

small. Figure 3.23c shows the governed rotor speed $\omega_{r,j}$ using the four different controller settings $j = \#1, \dots, \#4$. Due to the relation (3.42) the reference signal $\omega_{r,ref}$ to be tracked is proportional to the estimated wind speed. One can see a significant influence of the controller setting on the tracking performance. A good balance between a not too high overshoot amplitude and settling time is the controller setting #2. However, if one considers the necessary controller effort, one sees that only controller #1 remains within a valid limit. The generator torque increases only for a short time to 1.5 times the rated torque.

Based on the investigations with the reduced model, we now use the FAST wind turbine model for a much more realistic controller validation that takes the aero-elastic, structure-dynamics, and inflow dynamics into account. First, it must be noted that the observer parameterization must be newly tuned due to the non-modeled inflow and structural dynamics in the design model (2.100). For this purpose, the parameters of the desired polar region were shifted so that the overall error dynamics are slower. A compromise was found using a new set of pole region parameters

$$\alpha = 0.3, \quad r = 10, \quad \theta = 20^\circ, \tag{3.50}$$

where the observer sensitivity is reduced without making the estimation too slow. The renewed LMI design with the parameter set from (3.50) leads to the following observer matrices

$$\begin{aligned} \mathbf{L}_1 &= \begin{pmatrix} 0.0791 & -0.0519 \\ 0.0 & 0.0 \\ 22.0217 & 16.2907 \end{pmatrix}, \quad \mathbf{L}_2 = \begin{pmatrix} 0.1239 & 0.3871 \\ 0.0 & 0.0 \\ 12.1135 & 124.7059 \end{pmatrix}, \\ \mathbf{L}_3 &= \begin{pmatrix} 0.0786 & -0.2066 \\ 0.0 & 0.0 \\ 21.8613 & -67.5790 \end{pmatrix}, \quad \mathbf{L}_4 = \begin{pmatrix} 0.1239 & 1.3425 \\ 0.0 & 0.0 \\ 12.2085 & 482.6077 \end{pmatrix}, \end{aligned} \tag{3.51}$$

where the feedback gains $\mathbf{L}_i$ are between one and two orders of magnitude smaller than the original design in (3.21). The simulation result with the new observer gains, but with the previously calculated controller gains #4 presented in Table 3.3, is shown in Figure 3.24. Here, the same staircase-shaped undisturbed wind speed curve is used as in the simulation with the reduced wind turbine model. It can be seen well that after each wind jump the controller is able to operate the turbine again in the power optimum after approx. 10 seconds. Thus, the tip-speed ratio will return to the optimal value of λ_{opt} (see Figure 3.24b). The result is achieved with a moderate change of the generator torque. To avoid the excitation of tower oscillations around the tower eigenfrequency of

$$\omega_{0,T} = \sqrt{\frac{k_T}{m_T}}, \tag{3.52}$$

the reference signal $\omega_{r,reff}$ was filtered by the second-order low-pass filter denoted as $G_f(s)$

$$\omega_{r,reff}(s) = G_f(s)\,\omega_{r,ref}(s), \tag{3.53}$$

with

$$G_f(s) = \frac{1}{(\frac{s}{\omega_{c,f}} + 1)(\frac{s}{\omega_{c,f}} + 1)}, \tag{3.54}$$

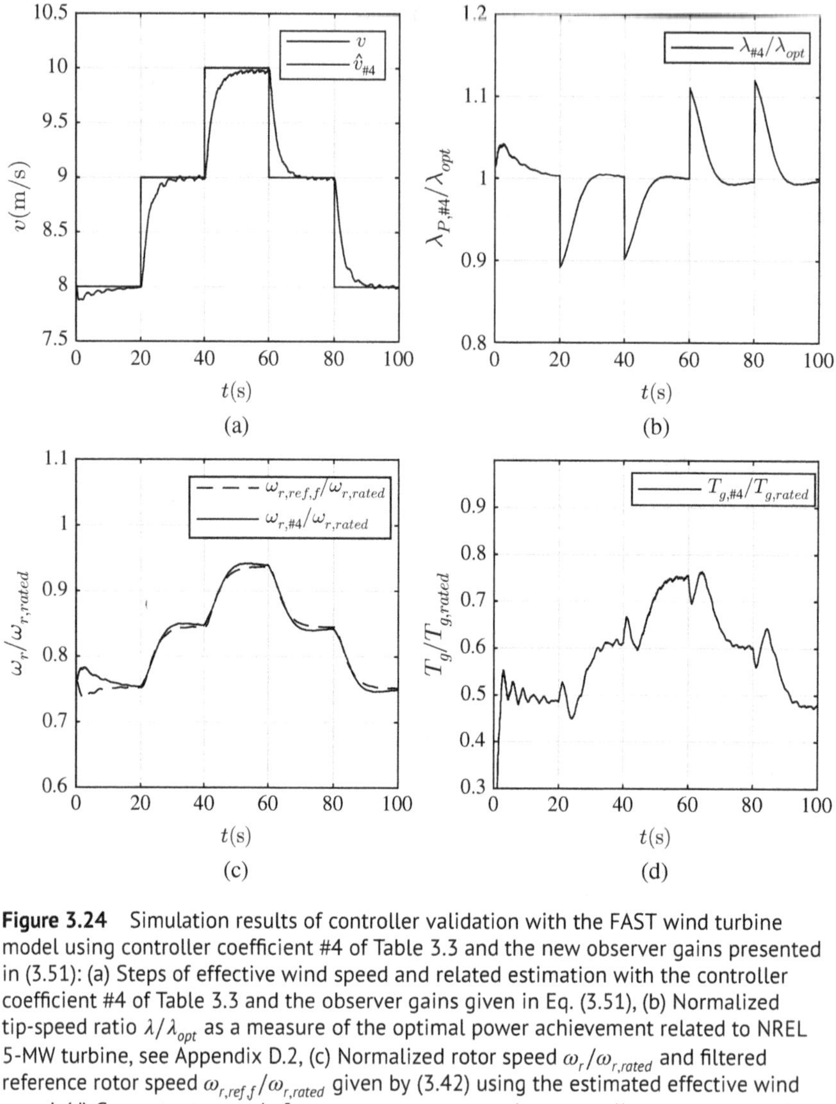

Figure 3.24 Simulation results of controller validation with the FAST wind turbine model using controller coefficient #4 of Table 3.3 and the new observer gains presented in (3.51): (a) Steps of effective wind speed and related estimation with the controller coefficient #4 of Table 3.3 and the observer gains given in Eq. (3.51), (b) Normalized tip-speed ratio λ/λ_{opt} as a measure of the optimal power achievement related to NREL 5-MW turbine, see Appendix D.2, (c) Normalized rotor speed $\omega_r/\omega_{r,rated}$ and filtered reference rotor speed $\omega_{r,ref,f}/\omega_{r,rated}$ given by (3.42) using the estimated effective wind speed, (d) Generator torque (reference generator torque) as controller output normalized to rated torque.

where $\omega_{c,f}$ denotes the cut-off frequency. For the NREL 5-MW wind turbine (based on the parameters from Table D.2 in Appendix D.2), a frequency value of

$$\omega_{c,f} = 0.5\,\text{rad/s}, \tag{3.55}$$

was selected, which provides the necessary distance from the tower eigenfrequency calculated with (3.52)

$$\omega_{0,T} = \sqrt{\frac{1981900\,\frac{\mathrm{N}}{\mathrm{m}}}{436865\,\mathrm{kg}}} = 2.1299\,\frac{\mathrm{rad}}{\mathrm{s}}, \quad f_{0,T} = 0.339\,\mathrm{Hz}. \tag{3.56}$$

The reference signal filter has already been introduced in the control scheme from Figure 3.15 but was set to one in the previous simulation with the reduced turbine model without tower dynamics. As determined by simulation studies, the reference signal filtering is necessary in the design model without feedback of the tower dynamics. Conversely, this means that if the tower dynamics are included in the controller design model, an additional reference value filter, which is not initially included in the design process, can be omitted via the active damping of the tower oscillation. Controllers which compensate tower and blade fatigue are briefly addressed for the TS-LMI design framework in Section 3.6.

3.3.3.3 Control Concept for Bumpless Transition Between Operating Regions

For a bumpless control between partial-load and full-load, the control laws of these regions are merged using the TS approach. To this purpose, a coupling filter for bumpless transition is introduced

$$\begin{pmatrix} \tilde{\beta}_{ref} \\ \tilde{T}_g \end{pmatrix} = \begin{pmatrix} f_1(\omega_r) & 0 \\ 0 & f_2(T_g) \end{pmatrix} \begin{pmatrix} \beta_{ref} \\ T_g \end{pmatrix} = \mathbf{F}(\omega_r, T_g) \begin{pmatrix} \beta_{ref} \\ T_g \end{pmatrix}, \tag{3.57}$$

where

$$\begin{aligned} f_1(\omega_r) &= w_{11}(\omega_r)\underline{f}_1 + w_{21}(\omega_r)\bar{f}_1, \\ f_2(T_g) &= T_{g,max}\frac{1}{T_g} = w_{12}(T_g)\underline{f}_2 + w_{22}(T_g)\bar{f}_2 \end{aligned}$$

according to the index notation of w_{ij} in (A.7) with the lower and upper sector bounds

$$\underline{f}_1 = 0, \quad \bar{f}_1 = 1, \quad \underline{f}_2 = \frac{T_{g,max}}{T_{g,max}} = 1, \quad \bar{f}_2 = \frac{T_{g,max}}{T_{g,min}}.$$

Due to the convex sum condition $w_{1j}(\cdot) + w_{2j}(\cdot) = 1$ determined by (A.7), illustrated in Figure 3.25 and Figure 3.26, the filter matrix (3.57) can be written as

$$\mathbf{F}(\omega_r, T_g) =$$
$$\begin{pmatrix} \left(w_{12}(T_g) + w_{22}(T_g)\right)\left(w_{11}(\omega_r)\underline{f}_1 + w_{21}(\omega_r)\bar{f}_1\right) & 0 \\ 0 & \left(w_{11}(\omega_r) + w_{21}(\omega_r)\right)\left(w_{12}(T_g)\underline{f}_2 + w_{22}(T_g)\bar{f}_2\right) \end{pmatrix}$$

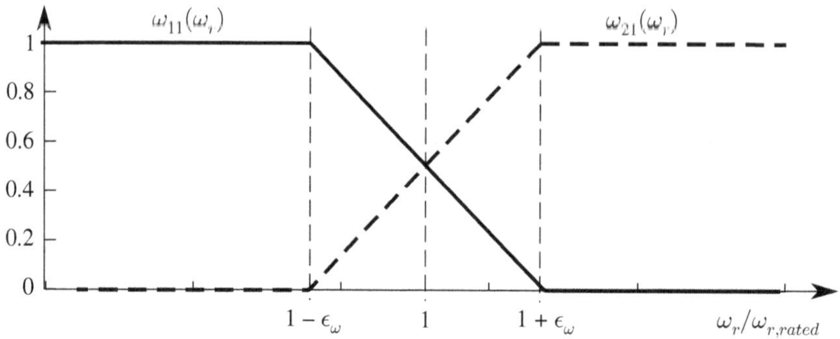

Figure 3.25 Weighting functions $w_{i1}(\omega_r)$ for $i = 1, 2$ with the overlapping factor ϵ_ω.

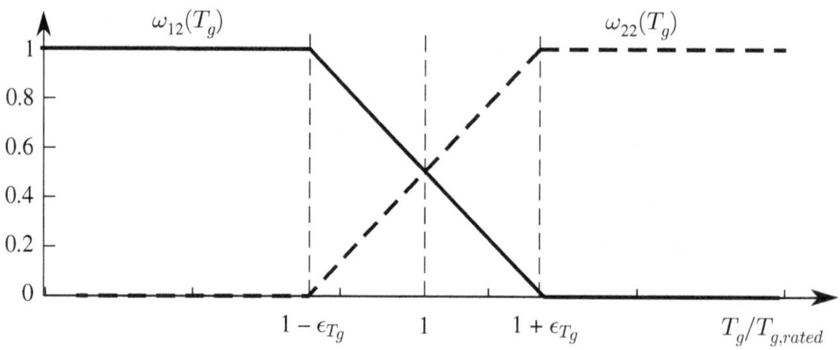

Figure 3.26 Weighting functions $w_{i2}(T_g)$ for $i = 1, 2$ with the overlapping factor ϵ_{T_g}.

and rearranged in

$$\begin{aligned}\mathbf{F}(\omega_r, T_g) = &\underbrace{w_{11}(\omega_r)w_{12}(T_g)}_{h_1^a(\omega_r,T_g)}\underbrace{\begin{pmatrix}\underline{f}_1 & 0\\ 0 & \underline{f}_2\end{pmatrix}}_{\mathbf{F}_1} + \underbrace{w_{11}(\omega_r)w_{22}(T_g)}_{h_2^a(\omega_r,T_g)}\underbrace{\begin{pmatrix}\underline{f}_1 & 0\\ 0 & \bar{f}_2\end{pmatrix}}_{\mathbf{F}_2}\\ &+\underbrace{w_{21}(\omega_r)w_{12}(T_g)}_{h_3^a(\omega_r,T_g)}\underbrace{\begin{pmatrix}\bar{f}_1 & 0\\ 0 & \underline{f}_2\end{pmatrix}}_{\mathbf{F}_3} + \underbrace{w_{21}(\omega_r)w_{22}(T_g)}_{h_4^a(\omega_r,T_g)}\underbrace{\begin{pmatrix}\bar{f}_1 & 0\\ 0 & \bar{f}_2\end{pmatrix}}_{\mathbf{F}_4},\end{aligned}$$

which gives the compact sum form

$$\mathbf{F}(\omega_r, T_g) = \sum_{i=1}^{N_r^a=4} h_i^a(\mathbf{z}^a)\,\mathbf{F}_i \tag{3.58}$$

with $\mathbf{z}^a = (\omega_r, T_g)^T$. Note that the index a is introduced for the functions h_i^x in order to be able to differentiate between function type $x \in \{a, b, c\}$ in the following.

Control law of the partial-load region from (3.45) with that nomenclature is

$$u_1 = -\sum_{i=1}^{N_r^b} h_i^b(\mathbf{z}^b)\,(k_{x,i}^b,\ 0)\,\Delta\mathbf{x} + \sum_{i=1}^{N_r^b} h_i^b(\mathbf{z}^b)\,k_{I,i}^b\,x_I, \tag{3.59}$$

where $u_1 = T_g$ with $\mathbf{z}^b = (\omega_r, \hat{v})^T$. And the control law of the full-load region from (3.35) with $u_2 = \beta$ as controller output is

$$u_2 = -\sum_{i=1}^{N_r^c} h_i^c(\mathbf{z}^c)\,(\mathbf{k}_{x,i}^c)^T\,\Delta\mathbf{x} + \sum_{i=1}^{N_r^c} h_i^c(\mathbf{z}^c)\,k_{I,i}^c\,x_I. \tag{3.60}$$

with $\mathbf{z}^c = (\omega_r, \beta, \hat{v})^T$. The membership functions $h_i^b(\mathbf{z}^b)$ and $h_i^c(\mathbf{z}^c)$ correspond to the previously introduced functions contained in the control laws (3.45) and (3.35). Both are related to the common vector $\Delta\mathbf{x}$ (3.29) and integral state (3.28)

$$\Delta\mathbf{x} = \begin{pmatrix} \omega_r - \omega_{r,ref} \\ \beta \end{pmatrix}, \qquad x_I = \int_0^t \left(\omega_{r,ref} - \omega_r\right) d\tau\,.$$

These are merged into a common control law

$$\begin{pmatrix} u_1 \\ u_2 \end{pmatrix} = -\sum_{i=1}^{N_r^b}\sum_{j=1}^{N_r^c} h_i^b(\mathbf{z}^b)h_j^c(\mathbf{z}^c) \left[\begin{pmatrix} (k_{x,i}^b,\ 0) \\ (\mathbf{k}_{x,j}^c)^T \end{pmatrix}\Delta\mathbf{x} + \begin{pmatrix} k_{I,i}^b \\ k_{I,j}^c \end{pmatrix} x_I\right] \tag{3.61}$$

and finally combined with the bumpless coupling filter (3.58)

$$\begin{pmatrix} \tilde{u}_1 \\ \tilde{u}_2 \end{pmatrix} = -\sum_{m=1}^{N_r^a}\sum_{i=1}^{N_r^b}\sum_{j=1}^{N_r^c} h_m^a(\mathbf{z}^a)h_i^b(\mathbf{z}^b)h_j^c(\mathbf{z}^c)\mathbf{F}_m \left[\begin{pmatrix} (k_{x,i}^b,\ 0) \\ (\mathbf{k}_{x,j}^c)^T \end{pmatrix}\Delta\mathbf{x} + \begin{pmatrix} k_{I,i}^b \\ k_{I,j}^c \end{pmatrix} x_I\right]. \tag{3.62}$$

3.4 Model-Based Power Tracking

3.4.1 Introduction

In the previously presented schemes, the controlled wind turbine was either power-optimized in the partial-load or power-limited in the full-load operating region. In the steady-state, the electrical power fed into the grid is solely a function of the effective wind speed. Below[7] the rated wind speed v_{rated}, the power increases cubically with the wind speed ($P_g \propto v^3$, $v < v_{rated}$), and above, the wind turbine generator (power) is controlled at the rated power ($P_g \approx P_{g,rated}$, $v > v_{rated}$). To support the frequency control by wind turbines, the power fed into the grid

7 Neglecting the transition from regions 2 to 3.

must be adjustable. In general, the ability of generators to follow power set point commands is called APC or demanded power tracking (DPT) that is only partially supported by conventional wind turbines. In order to ensure stable and reliable operation even with an increasing share of renewable energy[8] in power systems, it is necessary to enhance regulated wind turbines with power-tracking functionality (capability) [van Kuik et al., 2016]. For this reason, promising research results have been published in recent years. First, based on the commonly applied gain-scheduled PI controller (e.g. used in the base-line controller in Section 3.2) an extension to provide power-tracking functionality has been proposed in Aho et al. [2016], Galinos et al. [2019], and Shan et al. [2020]. Second, the extension of the operating envelope through dynamic governance of the generated power has been considered in model-based approaches in Ibáñez et al. [2020], Inthamoussou et al. [2016], and Pöschke et al. [2017, 2020]. Experimental validation of power-tracking control concepts by wind tunnel tests with a scaled wind turbine was provided in Kim et al. [2018] (extended gain-scheduled PI controller) and Pöschke et al. [2022] (multi-variable model-based controller). Following these ideas, a model-based control concept is presented that extends the existing TS design models to be able to design controllers with power-tracking functionality. This means that the usual division of the control regions is no longer used. In the actual full-load operation, a reduction of the generator torque is allowed and in the partial-load operation, the maximum power does not have to be reached.

Before the power-tracking operating strategies by generator torque and speed control of wind turbines is addressed, the conventional speed governor concept of synchronous machines driven by steam turbines and the automatic generation control (AGC) is briefly discussed. The study of dynamic behavior of conventional power plants for grid support provides us the basis for designing a higher-level wind farm controller, which usually dispatches power setpoints to each wind turbine in the farm. Thereby, each WT must be able to maintain the setpoints within the available power envelope $P_g(v)$, see Figure 3.27.

3.4.2 Governor Concept of Synchronous Machines

A simplified steam turbine (so-called Type A model) and governor model taken from Sauer et al. [2018] are given as

$$\begin{aligned} \tau_{ch}\frac{dT_M}{dt} &= -T_M + P_{SV}, \\ \tau_{SV}\frac{dP_{SV}}{dt} &= -P_{SV} + P_C - \frac{1}{R_D}\left(\frac{\omega}{\omega_s} - 1\right), \quad 0 \le P_{SV} \le P_{SV,max}, \end{aligned} \tag{3.63}$$

8 Instead of disconnecting renewable energy systems from the electrical grid in case of overfrequencies events and thereby possibly aggravating the existing problem [Das et al., 2018], wind turbines as well as photovoltaic (PV) power plants must be equipped with decentralized controls schemes, e.g. demanded power point tracking that continuously adjust the current generated power as a reaction to the grid frequency.

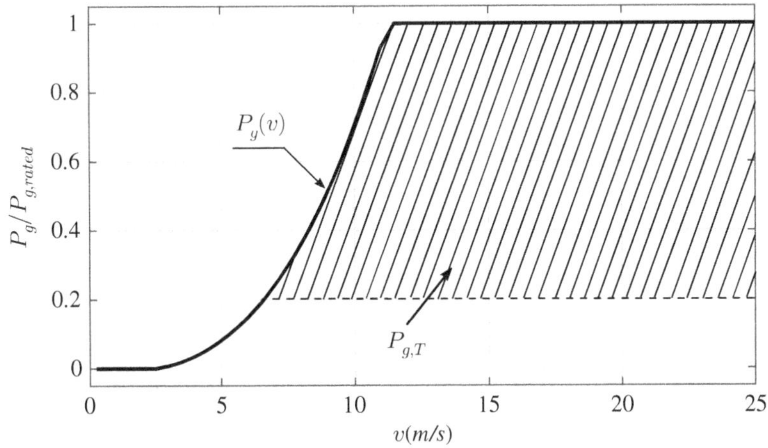

Figure 3.27 Power curve $P_g(v)$ of a wind turbine (controlled FAST NREL 5-MW turbine without tracking) and the power-tracking area (dashed area) defined by $P_{g,T} \in \{P_g \,|\, 0.2\, P_{g,rated} \leq P_g \leq P_g(v)\}$.

where all quantities are in p.u. with T_M as mechanical torque of the steam turbine, P_{SV} as the servo valve position (bounded and rate limited), the electrical radians per second defined by

$$\omega = \frac{p}{2}\,\omega_{shaft}, \tag{3.64}$$

with the number of poles p and the shaft speed ω_{shaft} of the synchronous machine. The model parameters are distinguished in the plant parameters with the rated synchronous speed ω_s, the time constant of the steam turbine model τ_{ch}, time constant of the servo valve actuator τ_{sv}, and the parameter of the governor and AGC. These are the power offset P_C at $\omega = \omega_s$ and the speed regulation quantity R_D as

$$R_D = \frac{2\pi \cdot \text{droop}}{\omega_s}, \tag{3.65}$$

where the parameter "droop" is a speed droop expressed in Hz/p.u. megawatt. The model (3.63) in the Laplace domain is

$$T_M = \frac{1}{\tau_{ch}s + 1}\,\frac{1}{\tau_{sv}s + 1}\left(P_C - \frac{1}{R_D}\left(\frac{\omega}{\omega_s} - 1\right)\right). \tag{3.66}$$

Thus, the manipulated variable T_M is a function of the control error (proportional behavior) with ω_s as the setpoint, ω as the actual variable, and the time-variable P_C. The offset P_C is the result of the integration of the area control error (ACE)

$$P_C = P_C^{ED} + \gamma \int_0^t ACE\, d\tau, \tag{3.67}$$

with

$$ACE = P_{export,sch} - P_{export,act}, \tag{3.68}$$

where P_C^{ED} denotes the economic dispatch set point approximately update every five minutes, γ is a constant dispatch factor of the specific generator, $P_{export,sch}$ denotes the summation of scheduled real power export and $P_{export,act}$ as the summation of actual real power export. Updating of the PC offset value via (3.67) and (3.68) is called AGC. In response to load changes, the AGC is a control system for adjusting the reference power of multiple generators at different power plants simultaneously.

In order to achieve similar behavior with the integration of wind turbines or farms[9], the power of the generator must be adjusted according to

$$P_{g,ref} = \left(P_C - \frac{1}{R_D}\left(\frac{\omega}{\omega_s} - 1\right)\right). \tag{3.69}$$

The reference power calculation in (3.69) is obtained from the negligence of two first-order dynamics of the steam turbine and governor model in (3.66), which are quantified in Sauer et al. [2018] (p. 133) with $\tau_{ch} = \tau_{sv} = 0.1$ s and the multiplication by ω_s (p.u. = 1). The simplification is justifiable since the speed of conventional turbines is nearly $\omega \approx \omega_s$. Note that with synchronous machines, the speed difference $\omega - \omega_s$ leads to a change in the phase angle δ, which results in a new power flow, e.g. for a single-area two-bus system, where δ_2 denotes the angle at the second bus it is proportional to $\sin(\delta - \delta_2)$. The details are presented in Chapter 5. What is important here is that we can use the wind turbine (primary power conversion) to provide power according to the external demand $P_{g,ref}$, which is then fed by the inverter (secondary power conversion) into the AC grid. The distinction between primary and secondary conversion follows the structure that we have illustrated at the beginning in Figure 1.3.

3.4.3 Control-Oriented Models and Schemes

The power-tracking controller design in this section uses the 1-DOF model with the nonlinear aero-dynamic map proposed in Section 2.3.2. To meet the requirements for load and vibration mitigation in the mechanical structure (in the tower, blades and/or drive train), 2- and 3-DOF models presented in Section 2.3.3 can be used by analogy in additional studies. The basic structure of the presented control scheme is invariant to this extension. In the following the control areas (1) below ($v < v_{rated}$) and (2) above the rated wind speed ($v \geq v_{rated}$) are considered.

---

9 In practice, a higher-order wind farm controller distributed the power to $i = 1, \ldots, N$ turbines simplest with $P_{g,ref,i} = P_{g,ref}/N$. More effective methods consider aerodynamic wakes to optimize the overall performance of a wind farm.

3.4.3.1 Control-Oriented Model and Control Scheme for $v \geq v_{rated}$

We begin with the model utilized for controller synthesis above the rated wind speed. Unlike the previously proposed control scheme in Section 3.3.3, controllers with power-tracking functionality adjust the generator torque in all operating regions. As a result, the valid workspace must be extended by T_g values below $T_{g,rated}$. Hence the function f_1 in (2.83) see (2.84) contains additionally the variable T_g instead of the constant $T_{g,rated}$ (used for the full-load region controller):

$$f_1(\mathbf{x}, T_g, v) = \frac{1}{2}\left(\frac{\rho\,\pi\,R^3\,v^2\,c_{Q,1}(\mathbf{x}, v)}{J} - \frac{n_g\,T_g}{J}\right)\frac{1}{x_1}, \qquad T_{g,min} \leq T_g \leq T_{g,max},$$
$$f_2(\mathbf{x}, v) = \frac{\rho\,\pi\,R^3\,v^2\,c_{Q,2}(\mathbf{x}, v)}{2J}\,\frac{1}{x_2}, \tag{3.70}$$

with $T_{g,min} = T_{g,rated}$, where the $c_{Q,i}$-maps are still given in (2.81). The corresponding state-space model is then

$$\dot{\mathbf{x}} = \begin{pmatrix} f_1(\mathbf{x}, T_g, v) & f_2(\mathbf{x}, v) \\ 0 & -\frac{1}{\tau_\beta} \end{pmatrix}\mathbf{x} + \begin{pmatrix} 0 \\ \frac{1}{\tau_\beta} \end{pmatrix} u, \tag{3.71}$$

where

$$u = \beta_{ref}, \qquad \mathbf{x} = \left(x_1, x_2\right)^T = \left(\omega_r, \beta\right)^T.$$

Using the sector-nonlinearity (SL) approach (see Appendix A.2.1), as before in Section 2.3.2, we get an equivalent plant description in TS form that is valid in a predefined operating region of the controller. To be able to fulfil this task, we define

$$\begin{aligned} &\omega_r \in [\underline{\omega}_r, \overline{\omega}_{r,rated}] = [0.8\,\omega_{r,rated}, 1.4\,\omega_{r,rated}] \quad \text{rad/s} \\ &\beta \in [\underline{\beta}, \overline{\beta}] = [0, 0.45] \quad \text{rad} \\ &T_g \in [\underline{T}_g, \overline{T}_g] = [0.2\,T_{g,rated}, T_{g,rated}] \quad \text{Nm} \\ &v \in [10, 25] \quad \text{m/s}, \end{aligned} \tag{3.72}$$

related to the permissible intervals of the rotor speed, blade pitch angle, and wind speed. The lower bound of the generator torque correlates with the definition of the $P_{g,T}$ range in Figure 3.27. It should be noted that in the case of turbines with gearboxes, the generator torque should not be below an acceptable range; otherwise, additional loads will occur due to insufficient tension in the gearbox. The control-oriented model valid for $v \geq v_{rated}$ results in

$$\dot{\mathbf{x}} = \sum_{i=1}^{N_r=4} h_i(\mathbf{z})\,\mathbf{A}_i\,\mathbf{x} + \mathbf{b}\,u, \qquad \mathbf{y} = \mathbf{c}^T\,\mathbf{x}, \tag{3.73}$$

where $\mathbf{z} = (\omega_r, \beta, v, T_g)^T$ with

$$\begin{aligned}
\mathbf{A}_1 &= \begin{pmatrix} \underline{f}_1 & \underline{f}_2 \\ 0 & -\frac{1}{\tau_\beta} \end{pmatrix}, \quad \mathbf{A}_2 = \begin{pmatrix} \underline{f}_1 & \bar{f}_2 \\ 0 & -\frac{1}{\tau_\beta} \end{pmatrix}, \quad \mathbf{A}_3 = \begin{pmatrix} \bar{f}_1 & \underline{f}_2 \\ 0 & -\frac{1}{\tau_\beta} \end{pmatrix}, \\
\mathbf{A}_4 &= \begin{pmatrix} \bar{f}_1 & \bar{f}_2 \\ 0 & -\frac{1}{\tau_\beta} \end{pmatrix}, \quad \mathbf{b} = \begin{pmatrix} 0 \\ \frac{1}{\tau_\beta} \end{pmatrix}, \quad \mathbf{c}^T = \begin{pmatrix} n_g & 0 \\ 0 & 1 \end{pmatrix},
\end{aligned} \tag{3.74}$$

and the same structure of h_i (2.93), but with the new weighting functions

$$\begin{aligned}
w_{1,1} &= \frac{\bar{f}_1 - f_1(\mathbf{x}, T_g, v)}{\bar{f}_1 - \underline{f}_1}, \qquad w_{2,1} = \frac{f_1(\mathbf{x}, T_g, v) - \underline{f}_1}{\bar{f}_1 - \underline{f}_1}, \\
w_{1,2} &= \frac{\bar{f}_2 - f_1(\mathbf{x}, v)}{\bar{f}_2 - \underline{f}_2}, \qquad w_{2,2} = \frac{f_2(\mathbf{x}, v) - \underline{f}_2}{\bar{f}_2 - \underline{f}_2}.
\end{aligned} \tag{3.75}$$

The min/max values of f_j in (3.74) and (3.75) are denoted in the same way as in (2.86)

$$\bar{f}_j = \max_{\substack{\mathbf{x} \in [\underline{\mathbf{x}}, \bar{\mathbf{x}}] \\ T_g \in [\underline{T}_g, \bar{T}_g] \\ v \in [\underline{v}, \bar{v}]}} f_j, \quad \underline{f}_j = \min_{\substack{\mathbf{x} \in [\underline{\mathbf{x}}, \bar{\mathbf{x}}] \\ T_g \in [\underline{T}_g, \bar{T}_g] \\ v \in [\underline{v}, \bar{v}]}} f_j \quad \text{for} \quad j = 1, 2. \tag{3.76}$$

Following numerical values are obtained for the 5-MW FAST turbine model given in Appendix D.2:

$$\underline{f}_1 = -0.0942, \qquad \bar{f}_1 = 0.3297, \tag{3.77}$$

$$\underline{f}_2 = -0.6290, \qquad \bar{f}_2 = -0.1493, \tag{3.78}$$

Based on this model, a controller structure is proposed, which is shown in the control scheme in Figure 3.28. With the assumption that the default power as $P_{g,ref}$ is less than or equal to the available power of the turbine. If $\omega_{r,ref}$ is set to the fixed nominal speed $\omega_{r,ref} = \omega_{r,rated} = const.$, the power is tracked by varying the generator torque alone. Another degree of freedom in power tracking is provided by the variation of $\omega_{r,ref}$. The turbine can be operated in a derated mode to provide positive control power. The different power-tracking modes are analyzed in detail in Section 3.4.4. Before continuing with this, let's look at the control scheme and the related control-design model for wind speeds below the rated wind speed.

3.4.3.2 Control-Oriented Model and Scheme for $v < v_{rated}$

The appropriate control scheme shown in Figure 3.29 is an extension of the partial-load operating control structure (without tracking) proposed in

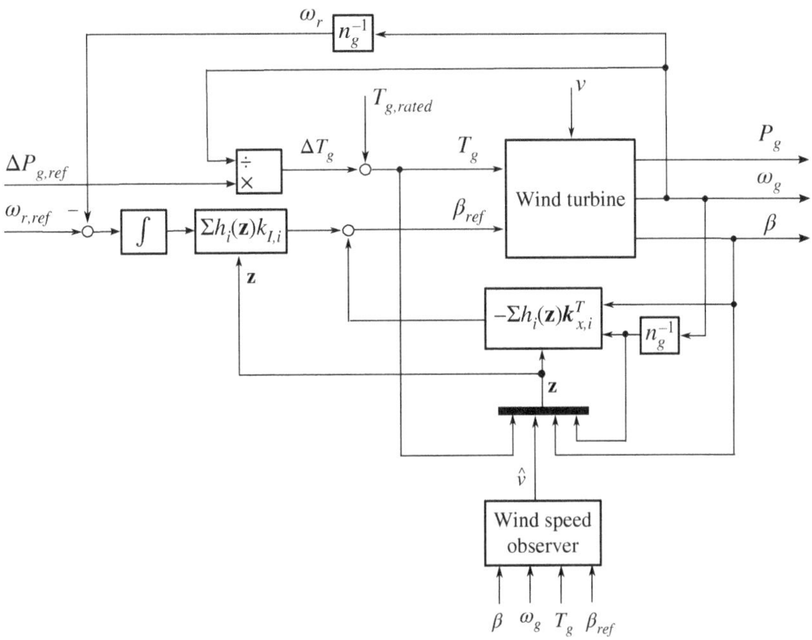

Figure 3.28 Control scheme for power tracking above the rated wind speed.

Figure 3.15. In addition to the generator torque as a manipulated variable, the reference pitch angle is adapted for active power reduction[10] via a map-based feedforward control. The requested power change $\Delta P_{g,ref}$ is first converted into a change of the power factor Δc_P using the estimated effective wind speed in Eq. (2.4)

$$\Delta c_P = \frac{\Delta P_{g,ref}}{P_{wind}(\hat{v})}, \qquad P_{wind} = \frac{1}{2}\rho\,\pi\,R^2\,\hat{v}^3. \tag{3.79}$$

Then, an inverse characteristic curve c_P^{-1} is used to calculate the reference pitch angle for the underlying pitch drive system of the wind turbine. The inverse based on the c_P curve (Figure 2.14), where all Δc_P values at the constant $\lambda = \lambda_{opt}$ value are mapped into a characteristic curve as a function of the pitch angle. Illustration of that curve shown in Figure 3.30 is integrated as a lookup table of the controller structure in Figure 3.29. With this feedforward control, the turbine can quickly reduce power without changing the speed control loop. The proposed approach

10 Active in sense of automatic controlled power reduction not caused by wind power reduction also called derating operation, where the turbine operates at a lower power than it could deliver based on the wind conditions and its power curve also used for wake and fatigue reduction in wind farms [Juangarcia et al., 2018].

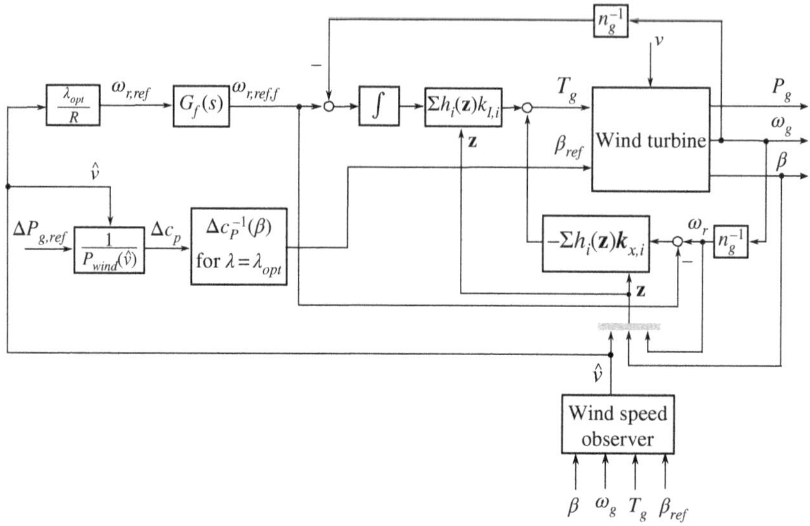

Figure 3.29 Control scheme for power tracking below the rated wind speed .

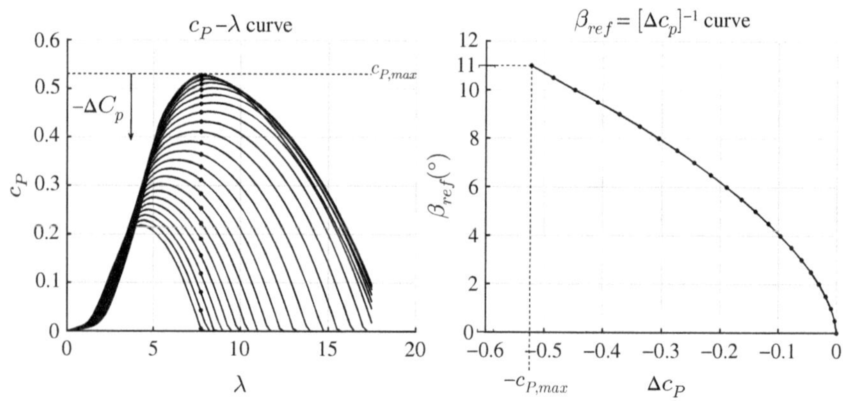

Figure 3.30 Visualization of $\Delta c_P = c_P(\beta, \lambda = \lambda_{opt}) - c_{P,max}$ for the set point calculation $\beta_{ref} = f(\Delta c_P)$.

has the advantage that when Δc_P is regulated back to zero, the turbine is again operated at the optimum set point without controlling additional transitions.

To calculate the coefficients $k_{I,i}$ and $\mathbf{k}_{x,i}^T$ of the speed control loop, we refer to the model already proposed in Section 2.3.2 with (2.99)–(2.108) extended by the pitch angle β. The extension means that β is not fixed to $\beta = 0$ (as in (2.100)), but contains the relevant control range for power tracking resp. reduction. However, β

is not adjusted by the closed control loop but is only feedforwarded by β_{ref}. This is accordingly shown in the control diagram in Figure 3.29 and has the consequence that the state vector of the design model does not but only the premise variable $\mathbf{z}$ have to be extended

$$\dot{x} = \sum_{i=1}^{N_r=2} h_i(\mathbf{z})\, a_i\, x + b\, u, \qquad y = x, \tag{3.80}$$

with $\mathbf{z} = (z_1, z_2, z_3)^T = (\omega_r, \beta, v)^T$, where $a_1 = \bar{f}_1$ and $a_2 = \underline{f}_1$ denoted as the lower and upper bounds of

$$f_1(\omega_r, v) = \frac{1}{2}\frac{\rho\,\pi\,R^3}{J}\, v^2\, c_Q(\lambda(\omega_r, v), \beta)\,\frac{1}{\omega_r}, \tag{3.81}$$

which are recalculated due to the new premise variable β with

$$\bar{f}_j = \max_{\substack{\omega_r \in [\underline{\omega}_r, \bar{\omega}_r] \\ \beta \in [\underline{\beta}, \bar{\beta}] \\ v \in [\underline{v}, \bar{v}]}} f_j, \qquad \underline{f}_j = \min_{\substack{\omega_r \in [\underline{\omega}_r, \bar{\omega}_r] \\ \beta \in [\underline{\beta}, \bar{\beta}] \\ v \in [\underline{v}, \bar{v}]}} f_j. \tag{3.82}$$

For the NREL wind turbine (Appendix D.2), the lower and upper bounds of ω_r and v are related to (2.106) and the choice of pitch angle interval based on the characteristic map (see Figure 3.30)

$$\beta \in [\,0,\, 0.1920\,]\ \text{rad} = [\,0,\, 11\,]\ \text{deg}. \tag{3.83}$$

From this follows

$$\underline{f}_1 = 0.0, \qquad \bar{f}_1 = 0.170, \tag{3.84}$$

where the lower bound is zero and different from (2.109). This is no surprise because the increase in pitch angle causes that the c_Q value in (3.81) drops to zero. Let's add a note about the design model that was used. If load mitigation control objectives should be addressed in the design, the model (3.80) can be extended by different degrees of freedom such as tower motion and elastic drive train. This is still considered as a task in Section 3.6.

Before starting with the model-based controller design, the system-related limitation of power generation must be taken into account. It is obvious that wind turbine generators can only provide power equal to or below the power c_P-weighted power from the wind, see Eq. (2.31). The available instantaneous power for feeding into the grid is shown in Figure 3.27. Corresponding to the area under the power curve denoted as $P_{g,T}$, a straightforward algorithm for converting an higher-level instance power request $P_{g,req}$ into the reference value $\Delta P_{g,ref}$ for the controller scheme shown in Figures 3.28 and 3.29 is specified by:

$$\Delta P_{g,ref} = \begin{cases} 0 & , \quad P_{g,req} > P_{g,rated} \wedge \hat{v} > v_{rated} \\ P_{g,req} - P_{g,rated} & , \quad P_{g,req} \leq P_{g,rated} \wedge \hat{v} > v_{rated} \\ 0 & , \quad P_{g,req} > P_{g,av} \wedge \hat{v} \leq v_{rated} \\ P_{g,req} - P_{g,av} & , \quad P_{g,req} \leq P_{g,av} \wedge \hat{v} \leq v_{rated} \end{cases}, \tag{3.85}$$

Thus $\Delta P_{g,ref} \leq 0$ is always valid, even if it is controlled from full-load operation. In derated operation mode ($P_{g,av} < P_{g,rated}$), positive control power can also be provided $\Delta P_{g,ref} > 0$. If the request cannot be followed, the function (3.85) returns zero. To make this unambiguous for higher-level network control systems, a return flag can be used as $b_{pa} = \{0, 1\}$, where "1" denotes that power is available. The function (3.85) with this additional feedback channel is illustrated as a system block in Figure 3.31 that complements the power-tracking control scheme in Figures 3.28 and 3.29. If the maximum available power is below the rated power, i.e. $v < v_{rated}$, the adjustable difference can no longer be calculated with $P_{g,rated}$, instead it must be determined from the moving average $P_{g,av}$ of the injected instantaneous power

$$P_{g,av}(t) = \frac{1}{N} \sum_{i=0}^{N-1} P_g(t - i\,T_s), \tag{3.86}$$

where T_s denotes a constant sampling time.

3.4.4 State Feedback and Feedforward Design for Power Tracking

The mixed state feedback and feedforward design for power tracking is divided into the operating region, first, valid for $v \geq v_{rated}$ related to the model (3.71) and control scheme illustrated in Figure 3.28 and, second, related to (3.80) with the scheme in Figure 3.29 valid for $v < v_{rated}$.

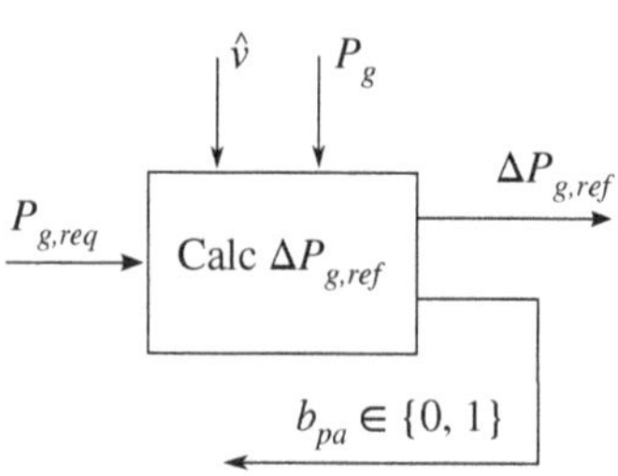

Figure 3.31 Calculation of difference reference power $\Delta P_{g,ref}$ as a function of requested power $P_{g,req}$, available instantaneous power P_g, and estimated effective wind speed $\hat{v}$ with acknowledge bit b_{pa} for transmission system operator (TSO)/distribution system operator (DSO) .

3.4.4.1 Controller Design for $v \geq v_{rated}$

Let's start with control synthesis for wind turbine operation above the rated wind speed. Without power tracking, this would correspond to the full-load operating region. Due to the possible power reduction, it is no longer suitable to speak of a full-load region. In order to determine a suitable range of closed-loop poles, we first look at the distribution of the open-loop poles (eigenvalues of $\mathbf{A}_i$ in (3.74)) of the FAST 5-MW turbine given as $\lambda_1 = (0, -0.0942, -10.0000)$, $\lambda_2 = (0, -0.0942, -10.0000)$, $\lambda_3 = (0, 0.3297, -10.0000)$, and $\lambda_4 = (0, 0.3297, -10.0000)$ illustrated in Figure 3.32. Unstable poles result from the positive value $\overline{f}_1$ in the system matrices (3.74). The controller gain $\tilde{\mathbf{k}}_i^T = (\mathbf{k}_{x,i}^T, -k_{I,i})$ in the control law for power tracking (see the whole scheme in Figure 3.28)

$$u = -\sum_{i=1}^{N_r} h_i(\mathbf{z}) \left(\mathbf{k}_{x,i}^T, -k_{I,i}\right) \begin{pmatrix} x_1 - x_{1,ref} \\ x_2 \\ x_I \end{pmatrix}, \tag{3.87}$$

with $\mathbf{z} = (\omega_r, \beta, v, T_g)^T$ and

$$x_I = \int_0^t \left(\omega_{r,ref} - \omega_r\right) d\tau, \tag{3.88}$$

is obtained using suitable LMI region criteria. It based on the same augmented plant model (3.31) as in Section 3.3, but with the new upper and lower bounds in $\mathbf{A}_i$ (3.74) related to the power-tracking task. To calculate $\tilde{\mathbf{k}}_i$, we first tried to apply the

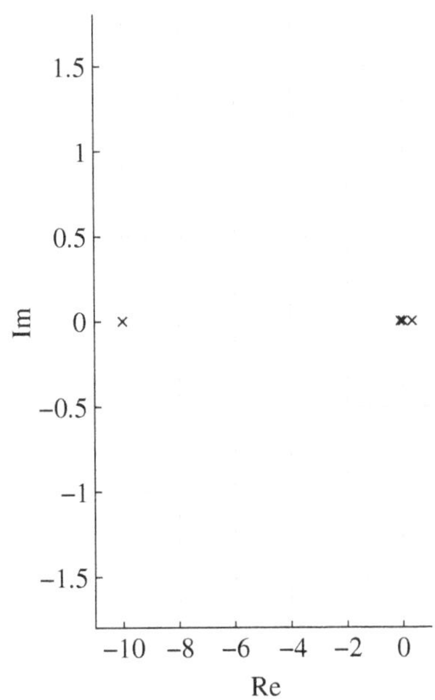

Figure 3.32 Eigenvalues of $\mathbf{A}_i$ related to the open-loop system (3.73).

LMI criterion (B.37) with the same range limits from the one used for the full-load range. However, it has been shown using the cone shaped region constraint (see Figure B.3) and the previously proposed limits in (3.39) that no solution is feasible. But, a LMI solution could be found for the strip-shaped specification of the pole region $S(\alpha_{\min}, \alpha_{\max})$

$$\alpha_{\min} = 0.05, \qquad \alpha_{\max} = 2.0, \tag{3.89}$$

using the constraints (B.35) of Theorem B.5 and (B.36)

$$\tilde{\mathbf{k}}_i^T = \tilde{\mathbf{m}}_i^T \mathbf{X}^{-1}, \quad i = 1, \dots, 4.$$

The feasible solution for the control scheme in Figure 3.28 with feedback gains

$$\tilde{\mathbf{k}}_1^T = \Big(\underbrace{-0.6680 \quad -0.6962}_{\mathbf{k}_{x,1}^T} \; \underbrace{0.0482}_{-k_{I,1}} \Big), \quad \tilde{\mathbf{k}}_2^T = \Big(\underbrace{-0.3730 \quad -0.8389}_{\mathbf{k}_{x,2}^T} \; \underbrace{0.0283}_{-k_{I,2}} \Big),$$
$$\tilde{\mathbf{k}}_3^T = \Big(\underbrace{-0.7674 \quad -0.6957}_{\mathbf{k}_{x,3}^T} \; \underbrace{0.0460}_{-k_{I,3}} \Big), \quad \tilde{\mathbf{k}}_4^T = \Big(\underbrace{-0.5832 \quad -0.8303}_{\mathbf{k}_{x,4}^T} \; \underbrace{0.0320}_{-k_{I,4}} \Big). \tag{3.90}$$

is illustrated in Figure 3.33. Here, due to the lack of damping barriers, it can be clearly seen that the calculated control loops are less damped than in Figure 3.18. This has, on the one hand, made the optimization problem feasible. Still, on the other hand, it also leads to the fact that the requirements of closed-loop damping,

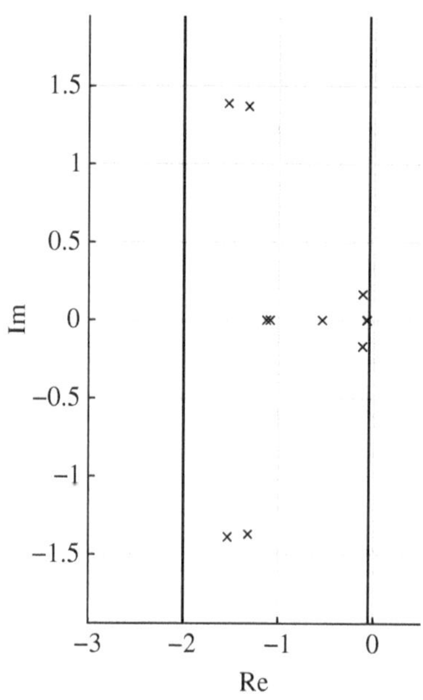

Figure 3.33 Eigenvalues $(\tilde{\mathbf{A}}_i - \tilde{\mathbf{b}}\tilde{\mathbf{k}}_i^T)$ related to (3.38).

i.e. overshoot M_p in the time domain (see Figure 3.8), are not explicitly considered in the design. Therefore, it is necessary to verify the performance based on simulation. This is done in the following.

Simulation results for a single load case, e.g. for a turbulent wind speed profile, are shown in Figure 3.34. For the sake of comparison, the same wind field has been used as in the simulation of the full-load region without power tracking (cf. Figures 3.19 and 3.20). The requested power output $P_{g,ref} = P_{g,rated} + \Delta P_{g,ref}$ in Figure 3.34 is specified with

$$\Delta P_{g,ref} = -P_{g,rated} \cdot \begin{cases} 0\ , & t < 80\,\mathrm{s} \\ 0.8, & 80\,\mathrm{s} \leq t < 120\,\mathrm{s} \\ 0.4, & 120\,\mathrm{s} \leq t < 160\,\mathrm{s} \\ 0.1, & 160\,\mathrm{s} \leq t < 200\,\mathrm{s} \\ 0.8, & t \geq 200\,\mathrm{s}. \end{cases} \tag{3.91}$$

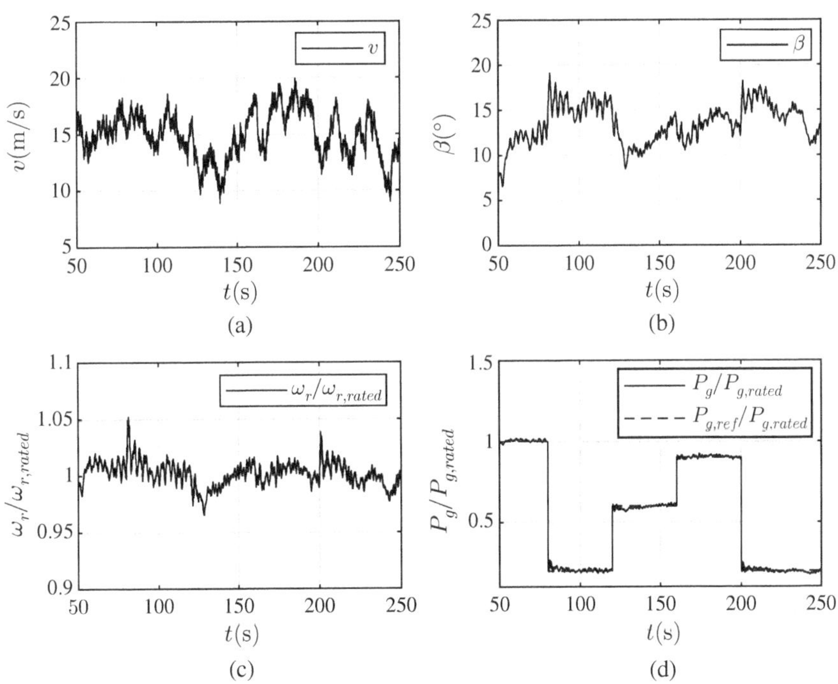

Figure 3.34 Simulation results of power-tracking controller validation (control scheme in Figure 3.28) for the calculated controller coefficients (3.84) using NREL's FAST wind turbine model excited by turbulent wind speed (normal turbulence model NTM016Aa).

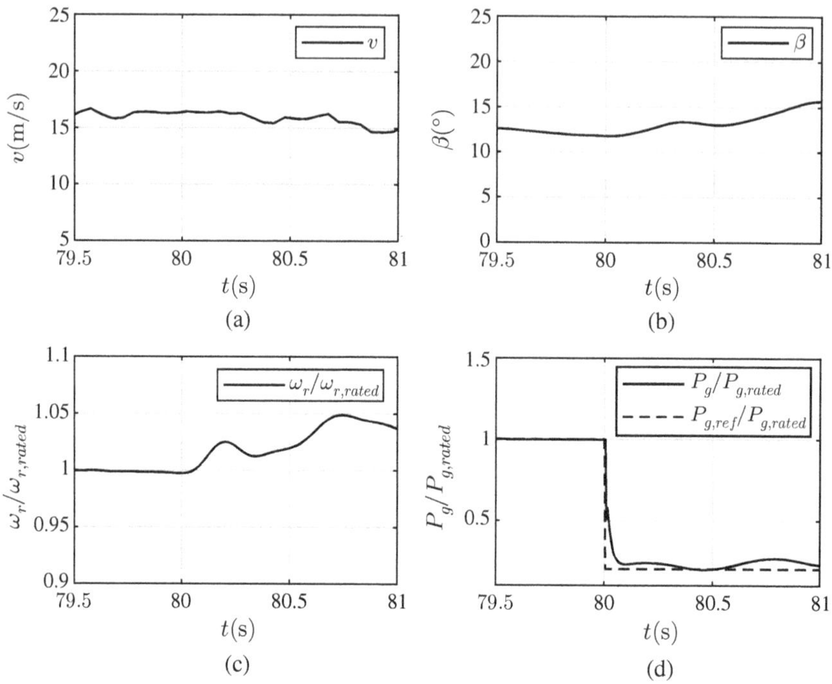

Figure 3.35 Zoom into simulation of Figure 3.34 in time interval $t = [79.5, 81]$.

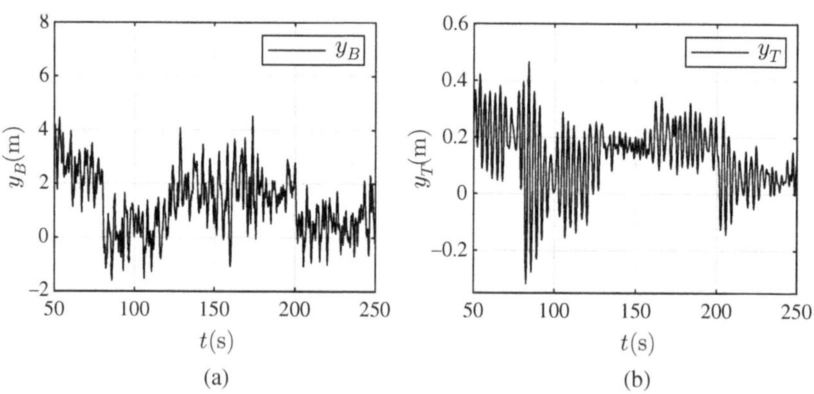

Figure 3.36 Simulation results related to Figure 3.34: Mechanical motion of the blade and tower represented by y_B and y_T, see the specification of coordinates in Figure 2.16.

Here, P_g follows the reference value $P_{g,ref}$ without overshoots (see Figure 3.34) and small delays shown in the zoom plot in Figure 3.35. As a result, the generated power can be adjusted rapidly by the change of generator torque. One can clearly see the effect of automatic control. In order to prevent the increase of the rotor speed due to the reduction of the generator torque at $t = 80$ s and $t = 200$ s, the pitch angle β is automatically increased. However, due to the rotor inertia and control variable constraint, the speed increases briefly to 5% above the nominal speed. But this is generally acceptable. On the other hand, it should be noted that due to the rapid changes of the pitch angle, not only the $c_Q(\lambda, \beta)$ coefficient of the torque relation of the rotor (2.32) changes, but also the $c_T(\lambda, \beta)$ coefficient in the thrust force relation (2.33). This excites a tower oscillation as clearly shown in Figure 3.36b. Due to the abrupt reduction of the thrust force at $t = 80$ s, the tower is deflected by up to $y_T = -0.25$ m. This oscillation can be mitigated by a model-based controller design in which the tower degree of freedom is included. For this purpose, the 2-DOF model (2.114) from Section (2.3.3) with the state vector $\mathbf{x} = \left(y_T, \dot{y}_T, \omega_r, \beta\right)^T$ is converted into a TS form with $\mathbf{z} = (\omega_r, \beta, v, T_g)^T$ analogous to the TS model with 1-DOF (3.73). Alternatively, in addition to the TS framework presented here with sector-nonlinearities, a 2-DOF TS model based on Taylor linearization (see Appendix A.2.2) can be used. Simulation results and the complete design task from TS modeling to LMI-based design are presented in Pöschke et al. [2020]. Further experimental validation with a model wind turbine in a wind tunnel is published in Pöschke et al. [2022].

3.4.4.2 Controller Design for $v < v_{rated}$

For the actual controller design, the same closed-loop structure and integral state feedback control law (3.45) can be used as for the partial load without power-tracking capability (Figure 3.15).

$$u = -\sum_{i=1}^{N_r=2} h_i(\mathbf{z})\,\tilde{\mathbf{k}}_i^T\,\Delta\tilde{\mathbf{x}}, \qquad \mathbf{k}_i^T = \left(k_x, -k_I\right),$$

in accordance with proposed model (3.80) with $\mathbf{z} = (\omega_r, \beta, v)^T$. Due to the additional pitch angle variation, only the premise variable has been extended by this quantity. The capability of power tracking results only from the feedforward of the pitch angle calculated from the inverted c_P-λ curve shown in Figure 3.30. Experienced wind turbine control engineers might argue that the accuracy of that steady-state map is not sufficient to set the required power level without additional feedback. To follow Occam's razor[11] we treat first a control scheme without an extra external power control loop. This is justified because in power plants a fast power change for frequency stabilization, i.e. the $d\Delta P_g/dt$, is more

11 Also known as the principle of parsimony.

important than the exact setting of a requested power value. After the design, FAST simulation is used to investigate how much the control error deviates without power feedback.

With the model (3.80) using the FAST turbine parameters, the pole location region of the desired closed-loop behavior

$$\alpha = 0.2., \quad r = 0.5, \quad \theta = 36^\circ,$$

and with the same LMI criteria (B.37), (B.41) as for the partial-load controller design, the controller coefficients are obtained:

$$\begin{aligned} k_{x,1} &= -2.8141 \cdot 10^5, \quad k_{I,1} = -0.5134 \cdot 10^5, \\ k_{x,2} &= -3.5918 \cdot 10^5, \quad k_{I,2} = -0.5763 \cdot 10^5. \end{aligned} \tag{3.92}$$

Comparing these values with the coefficients from Table 3.3, there is just a difference in the third and fourth decimal digits. That's not surprising, since, as already mentioned, the $\mathbf{A}_i$ parameters of the plant model differ only marginally, see (2.109) vs. (3.84).

Finally, we come to the controller validation in the low wind speed region. In order to evaluate the effect of reference power variation, the FAST 5-MW NREL

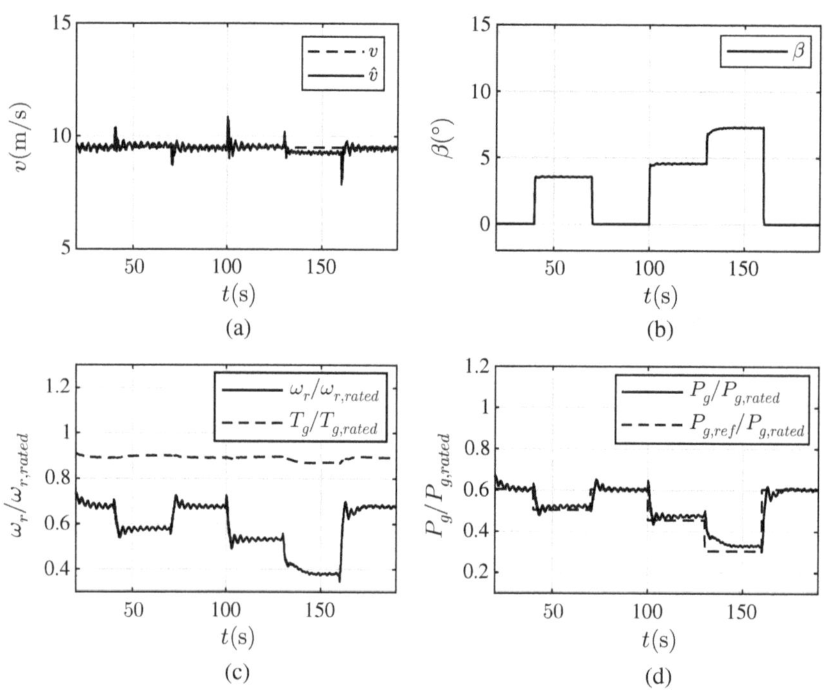

Figure 3.37 Power-tracking controller validation ($v < v_{rated}$ control scheme in Figure 3.29) with (3.86) using NREL's FAST wind turbine model.

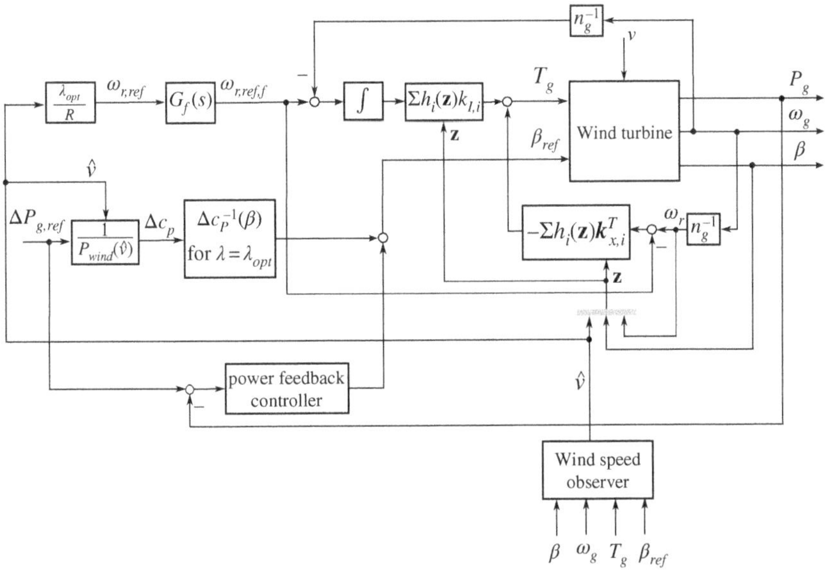

Figure 3.38 Control scheme for power tracking with power feedback controller.

wind turbine was excited with a constant wind speed of 9.5 m/s, which is below the rated wind speed v_{rated}. It can be seen in Figure 3.37 how, due to the requested power reduction, the pitch angle is adjusted in such a way that the instantaneous power follows the reference value as the product of the generator speed and generator torque. Due to the constant wind speed, the rotor speed resp. generator speed is kept at a constant value by $\omega_{r,ref} = (\lambda_{opt}/R)\hat{v}$. Thus, power variation is decoupled from power optimization. If $\Delta P_{g,ref} = 0$ is chosen, the turbine follows the optimization strategy. For $\Delta P_{g,ref} < 0$, the pitch angle is adjusted by feedforward control in such a way that the power is reduced as required. However, it can also be seen that due to inaccuracies of the static map, the control scheme in Figure 3.29 does not lead to an exact steady-state behavior. An additional feedback controller proposed in Figure 3.38 may provide a solution to this problem. Further potential for improvement lies in the fine-tuning of the wind speed observer. Here, a reduced estimation dynamic could help to enhance the performance.

3.5 Summary

This chapter presents the controller structures for wind turbines, focused on primary power conversion, including the model-based design process. With the control-oriented models derived in Chapter 2 in combination with formal requirements, the controller coefficients were calculated, and the fulfillment of the control performance was validated by the wind turbine simulation program FAST. The first section of the chapter has addressed conventional control objectives for variable speed wind turbines, subdividing the operating regions into the partial-load (power optimization) and full-load region (power limitation). It was shown how the formal design could be enhanced to achieve further control objectives, such as mitigating of loads in the mechanical structure.

The second part is dedicated to the power-tracking capability of wind turbines. This becomes more and more important for future power systems because the inertia reduction caused by high renewable penetration has to be compensated by virtual inertia. To achieve this, wind turbines must be able to follow a rapidly changing power request. For the high wind speed region ($v \geq v_{rated}$) a control scheme was proposed, where the generator torque is additionally regulated in dependence on the measured rotor speed. The rotor speed is kept constant at the nominal speed, and the power change for fast guiding behavior is achieved by the generator torque variation only. During the low wind speed operating ($v < v_{rated}$), the pitch angle is additionally adjusted depending on the requested power. For this purpose, the estimated effective wind speed is used to determine the reference pitch angle by inverting the power curve. As in the first part, the applicability and achievable performance (maximum gradient of the power change considering the limits of the mechanical loads) have been realistically validated by simulation runs with FAST.

3.6 Problems

3.1 Consider the influence of parameter variation on the wind speed observer (3.17). How sensitive is the reconstruction quality related to τ_v of the disturbance model?

3.2 In the full-load operating region, the control objective of tower vibration-damping has to be taken into account. For this purpose, the control scheme in Figure 3.16 is to be modified to include the feedback of the tower deflection y_T. For the formal design in the TS framework, use the 2-DOF model and derive the controller in the same way as using 1-DOF in Section 3.4.

3.3 A feedback loop should be added to the controller structure in Figure 3.29 for power tracking below the rated wind speed to improve the set point accuracy. Design the power controller as marked in Figure 3.38 and verify the stability of the entire cascaded controller structure.

3.4 According to the "no-free-lunch" principle, the ability of power tracking was obtained by additional feedback loops. For the high wind speed region ($v \geq v_{rated}$), to be able to follow a requested power change, the generator torque is additionally regulated in dependence on the measured rotor speed, see Figure 3.28. By means of simulations, the performance of the enhanced controller structure could be validated, illustrated in Figures 3.34–3.36. However, a formal proof of stability is missing. Carry out a formal verification of the speed feedback in the reference value calculation based on LMI criteria.

3.7 Bibliography

A detailed overview of the control methods used in wind turbines would go beyond the scope of this book. In summary, however, both model-free (still mainly used in industry), robust linear model-based [Geyler and Caselitz, 2008] and nonlinear model-based approaches [Inthamoussou et al., 2016; Ostergaard et al., 2009; Pöschke et al., 2020] are used to synthesize controllers containing feedforward and feedback loops. The nonlinear model-based approach uses the description of complex wind turbine dynamics to design closed-loop controllers for individual regions [Díaz de Corcuera et al., 2015] or for all regions [Inthamoussou et al., 2016], [Pöschke et al., 2020] with the required closed-loop reference dynamics. The nonlinear approach presented here differs from Pöschke et al. [2020] by the modeling process and treatment of the control regions. Herein, the TS fuzzy models for control synthesis are created using the sector nonlinearity approach. Note the difference between this consideration to Pöschke et al. [2020] is the TS modeling and the treatment of the control regions. The TS fuzzy models for control synthesis are created in this book using the sector nonlinearity approach. It allows for a generic modeling based on given analytical descriptions. In Pöschke et al. [2020], on the other hand, local linear models of the consequence part are obtained by numerical linearization using the aero-elastic simulator FAST [Jonkman et al., 2005]. The membership functions are formed over products of local triangular weighting functions. Further, a unified controller synthesis for all regions is proposed in Ostergaard et al. [2009] and Pöschke et al. [2020]. In contrast, due to the structural implications of using the sector-nonlinearity approach, this work divides the control into partial-load and full-load regions according to the control objectives.

4

Modeling and Control of Photovoltaic Power Plants

4.1 Introduction

Grid-connected photovoltaic (PV) systems shown in Figure 4.1 can generally be divided into primary and secondary parts of power conversion. The primary part contains the PV generator to convert solar radiation into electrical DC power and the DC–DC converter for power conditioning and maximum power point tracking (MPPT). System inputs of the PV generator are the irradiation S and the controllable voltage v_{pv}. The output of primary converter is the power injected into the secondary power conversion over DC link. To convert the DC voltage into a three-phase AC voltage, an inverter with LC filter is used. Note that the secondary power conversion is not studied in this chapter but in the following since it is a part of PV and wind power plants with full-converter systems.[1] It should be also mentioned that the generic system under consideration in Figure 4.1 shows the topology of a PV system with a central MPPT tracker at the array level. Since we do not study partial shading in the system analysis and controller design, the distributed DC–DC converters (e.g. at string level) can be combined into the model of a central converter without any limitation of generality.

Classification of PV systems is conventionally done via the power capacity [Ramakumar and Bigger, 1993]. A distinction is made between small-scale systems with less than 50 kW, intermediate-size systems with 50 kW up to 1000 kW, and large-scale PV systems with power production of more than 1 MW. An example of a small-scale system with $P \approx 40$ kWp (142 modules with 280 kWp each) is given in Figure 4.2. A picture of an intermediate-size field installation is illustrated in Figure 4.3. On the cover of this book, a large-scale PV power plant is depicted. Further details on the structure, topology, and technical characteristics of photovoltaic power systems are presented in Xiao [2017].

1 Since wind turbines with doubly fed induction generator (DFIG) differ from the full converter topology due to the direct connection of the generator stator winding and the partial conversion between generator rotor and the grid.

Advanced Control of Grid-Integrated Renewable Energy Power Plants: LMI-based Design in the Takagi-Sugeno Framework, First Edition. Horst Schulte.

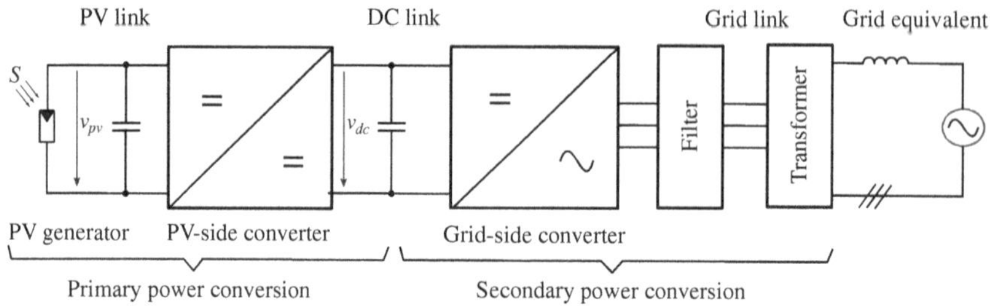

Figure 4.1 Two-stage power conversion with centralized maximum power point tracker.

Figure 4.2 Rooftop system at HTW Berlin (H building).

4.2 Modeling of PV Generators

Photovoltaic generators contain thousands and thousands of photovoltaic cells. Once integrated into an electrical circuit, each cell converts the incoming solar irradiation (direct and indirect) into an electric current. Neglecting wiring, converter losses, and power mismatch, the generated power of PV systems is the number of all included cells multiplied by the generated power of a single cell. This assumption is useful to summarize all the interconnected PV cells to derive

Figure 4.3 PV power plant at the railroad line Buchloe–Munich.

the mathematical model of the overall PV generator for systems analysis and controller design.

The basic units of a PV generator are the PV modules, also called solar panels. A standard PV module contains 48–73 cells in series connection mounted in a framework for protection and installation. PV generators are commonly assembled by configurations of modules in series and parallel. Solar modules are connected in series to increase the output voltage, frequently referred to as a "string." PV strings in a parallel connection form an array in which the power capacity can be built up to thousands or even millions of a watt. In large-scale PV systems that take over more power plant tasks, an array is divided into multiple subarrays.

The mathematical model of the PV generator results from the aggregation of all PV modules and aggregation of all single cells of a module where the current and voltage are suitably multiplied. For this, the single cell is modeled by an equivalent circuit consisting of an irradiation-dependent current source, a model of the diode D, and shunt resistors R_h, as shown in Figure 4.4. The irradiation S with the physical unit $\mathrm{W/m^2}$ is related to the direct (normal to the PV cell array) and the diffuse irradiation. The typical profile of S over a sunny day in Central Europe is shown in Figure 4.5. A distinction is made between the orientation of the measuring unit in south, east, and west. Let us now focus on the

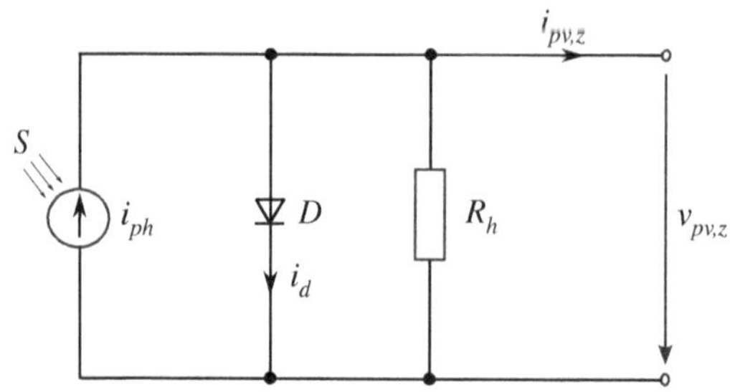

Figure 4.4 Simplified single-diode model (SSDM) of a PV cell.

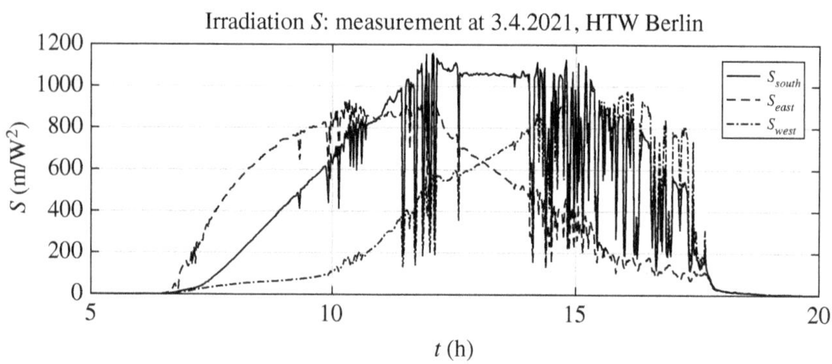

Figure 4.5 Irradiation profile from April 3, 2021, in Berlin (sunny day).

derivation of the mathematical model. The application of Kirchhoff's circuits laws results in

$$i_{pv,z} = i_{ph} - i_d - \frac{v_{pv,z}}{R_h}, \tag{4.1}$$

with $i_{ph,z}$ and i_d as the light and diode current, where $v_{pv,z}$ denotes the output voltage of a single PV cell. The I–V characteristic of a photovoltaic cell is determined by the diode current as nonlinear function of the PV terminal voltage $v_{pv,z}$ given by

$$i_d = i_s \left[e^{\frac{v_{pv,z}}{A_n\, v_{T,STC}}} - 1 \right], \quad v_{T,STC} = \frac{k\, T_{c,STC}}{q}, \tag{4.2}$$

where i_s represents the saturation current of the diode's diffusion effect

$$i_s = \frac{i_{ph} - \frac{v_{oc}}{R_h}}{e^{\frac{v_{oc}}{A_n\, v_{T,STC}}} - 1} \tag{4.3}$$

with v_{oc} as open-circuit voltage of the diode model and $v_{T,STC} = 25.7$ mV as thermal voltage of p-n junction at standard test conditions (STC). The others parameters are the shunt resistor R_h to model the cell losses, k as the Boltzmann constant with $k = 1.381 \times 10^{-23}$ J/K, A_n as the constant ideality factor, and elementary charge $q = 1.602 \times 10^{-19}$ As. The irradiation and temperature dependence of light current i_{ph} and open-circuit voltage v_{oc} are given by the relations

$$i_{ph} = \frac{S}{S_{STC}} i_{ph,sc,STC} \left(1 + \alpha_T (T_c - T_{c,STC})\right), \tag{4.4}$$

and

$$v_{oc} = v_{oc,STC} \left(1 + \beta_T (T_c - T_{c,STC})\right), \tag{4.5}$$

where S denotes the instance irradiation value and T_c the actual operation cell temperature. The constant S_{STC} denotes the irradiation at STC with $S_{STC} = 1000$ W/m^2, and $i_{ph,sc,STC}$ denotes the short-circuit current at STC. The temperature-dependent models (4.4) and (4.5) contain $\alpha_T > 0$ and $\beta_T < 0$ as temperature coefficients and the value $T_{c,STC} = 298\,°$K as the temperature at STC. Thus, the PV cell model results from Eqs. (4.1)–(4.5) and will be summarized as

$$i_{pv,z} = i_{ph}(S, T_c) - i_s(S, T_c) \left[e^{\frac{v_{pv,z}}{A_n\, v_{T,STC}}} - 1\right] - \frac{v_{pv,z}}{R_h}. \tag{4.6}$$

Here, the quantities $i_{ph}(S, T_c)$ and $i_s(S, T_c)$ are specified with the dependent variables to better distinguish them from the constant model parameters. Example of physical parameter, required to calculate the proposed PV cell model, is given in Table D.3 with the STC values listed in Table D.1.

To calculate the PV generator from individual cells, we assume a single PV module with all cells connected in series denoted as $N_{cell,s} = N_{cell}$. The individual modules are in turn connected in series (PV strings) and in parallel to a PV generator, where $N_{mod,p}$ denotes the number of modules in parallel and $N_{mod,s}$ the number of modules in series. Thus, the PV generator is composed in total of $N_s = N_{cell,s} N_{mod,s}$ PV cells in series, and these series connections are connected N_p-fold in parallel where $N_p = N_{mod,p}$. Assuming the solar irradiation and cell temperature are identical for each solar cell the aggregation of all cells in one PV generator is given as shown in Figure 4.6. The corresponding mathematical model of the PV generator with the current

$$i_{pv} = N_p\, i_{pv,z} \tag{4.7}$$

and voltage

$$v_{pv} = N_s\, v_{pv,z} \tag{4.8}$$

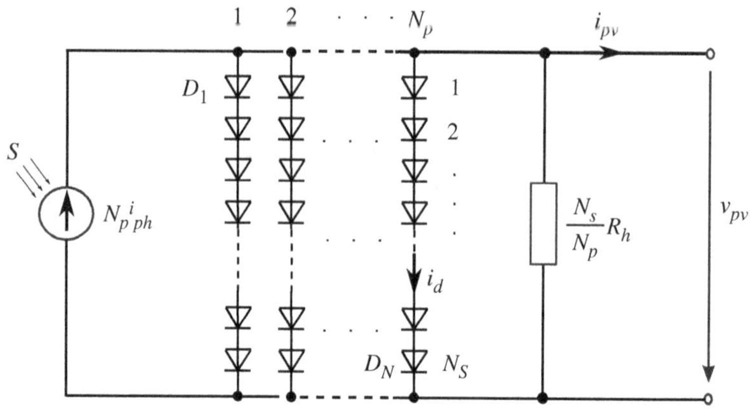

Figure 4.6 Equivalent circuit of a PV generator aggregated from $N = N_s N_p$ single-diode models.

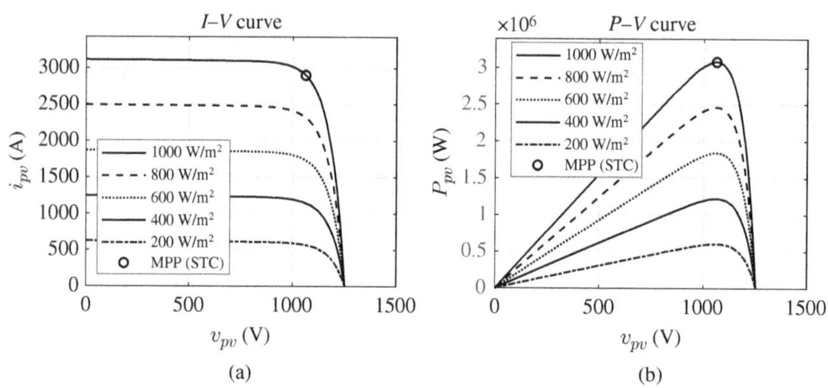

Figure 4.7 Aggregated *I*–*V* (a) and *P*–*V* (b) curves of a 3083 kW (related to STC) power plant with constant cell temperature $T_{c,STC} = 25\,°\text{C}$ under variation of irradiation $S = \{200, 400, 600, 800, 1000\ \text{W/m}^2\}$.

at the terminal is given as

$$\begin{aligned} i_{pv} &= f_{pv}(v_{pv}, S, T_c) \\ &= N_p\, i_{ph}(S, T_c) - N_p\, i_s(S, T_c) \left[e^{\frac{v_{pv}}{N_s\, v_{T,STC}\, A_n}} - 1 \right] - \frac{N_p}{N_s} \frac{v_{pv}}{R_h}, \end{aligned} \tag{4.9}$$

where i_s is determined by (4.3) and i_{ph} by (4.4).

The function curves of (4.9) shown in Figures 4.7 and 4.8 based on the parameters of a 3-MW PV generator with monocrystalline silicon solar cells are given in Table D.4. Both *I*–*V* curves clearly show the characteristic feature of a PV generator as a current source. The current is almost equal for constant irradiation and increases linearly with increasing irradiation of

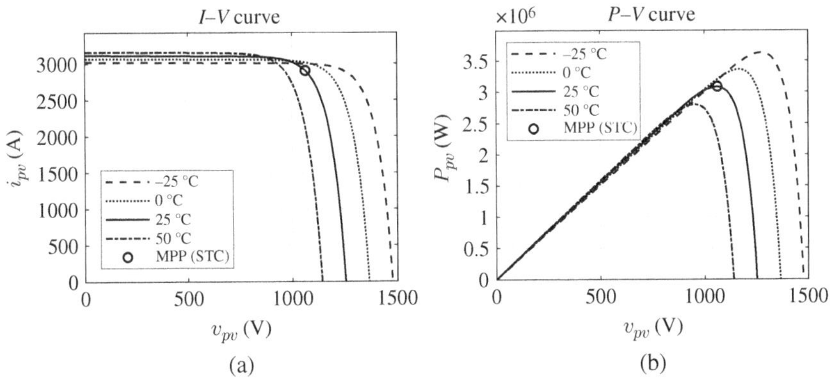

Figure 4.8 Aggregated I–V (a) and P–V (b) curves of a 3083 kW (related to STC) power plant with constant irradiation S_{STC} = 1000 W/m^2 under variation of T_c = {−25, 0, 25, 50 °C}.

S_{STC} = {200 W/m^2, … , 1000 W/m^2} as long as the PV voltage v_{pv} is below the sum of all diode forward voltage. If v_{pv} exceeds the threshold of forward voltages, the current drops significantly until the current is zero at the open-circuit voltage. As expected, this behavior is represented by a linear increase in power for the P–V curves until a maximum power point (MPP) is reached. By varying the radiation, a slight variation of the MPP related to the PV voltage v_{pv} can be seen in Figure 4.7 in the P–V diagram. The variation of the MPP is significantly greater with changing the PV cell temperature. In Figure 4.8 a significant shift of the MPP in the lower voltage region can be seen with rising temperature from 0° to 50°. Also it is seen that the MPP value decreases with increasing temperatures.

To hold the point of maximum power regardless of the current temperature and irradiation, the so-called MPP tracker is used. As a higher-level controller, the MPP tracker calculates the voltage reference value of the lower-level v_{pv} voltage controller in order to be able to follow the point of optimum power output even with variable irradiation and cell temperature. Thereby the PV voltage is adjusted through a DC–DC converter.

4.3 DC–DC Converter Modeling

4.3.1 Overview of Switched-Mode Converter

In general, the circuit topology of switched-mode DC–DC converters has a significant effect on the adjustable input and output voltage ratio to the output to input current ratio, respectively. For DC–DC converters in PV systems shown in Figure 4.1, the input voltage is the PV generator terminal voltage v_{pv} and the output is the DC link voltage v_{dc}. Three second-order switched-mode DC–DC converters, the buck (step-down), boost (step-up), and buck–boost (step-up/down) converter, are examined in detail.

All have in common that there are two energy storage elements, an inductance L and capacitance C combined with a diode and an electronic ideal switch Q with no switching and conduction losses. The high-frequency pulse width modulated (PWM) switching $u_{pwm} \in \{0, 1\}$ in the kHz range is not modeled, but an equivalent duty cycle $d \in [0, 1]$ is used instead as the system input. Thus, all three types can be described with a second-order state-space model where the states are defined by the current through the switched inductance denoted as i_L and the voltage at the capacitance v_c. Here the capacitance can be parallel to the input or output terminal. Figures 4.9–4.11 illustrate the equivalent circuit diagrams of these three converter types. Note the dynamics and restrictions of the grid-side inverter are not modeled here. It is assumed that the DC power P_{dc} is equal to the three-phase AC power $P_{3\phi}$ fed by the inverter ($P_{dc} = P_{3\phi}$) or we assume that the capacity is very large:

$$\dot{v}_{dc} = \frac{1}{C_{link}\, v_{dc}}(P_{dc} - P_{inv}) \approx 0, \qquad v_{dc,0} = v_{dc}(t = 0).$$

Thus the voltage in the DC bus v_{dc} at the DC link capacitor C_{link} (not shown in Figure 4.10) is constant $v_{dc} = v_{dc,0} = 900$ V. However, the fluctuating DC link voltage will be considered later in the overall system simulation in Chapter 6 and can be considered in the controller design by a H_∞ condition.

The buck converter of Figure 4.9 steps down voltage from its input v_{pv} to its output v_{dc} proportional to the duty cycle[2] denoted as d. The ideal static transmission

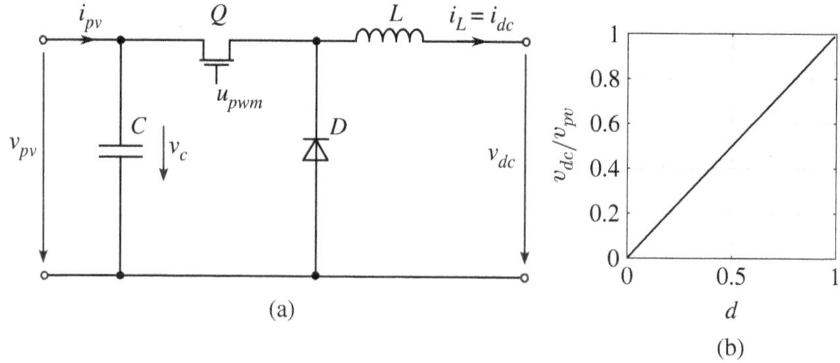

Figure 4.9 DC–DC buck converter with static transfer function (4.10).

2 The d is also called duty ratio.

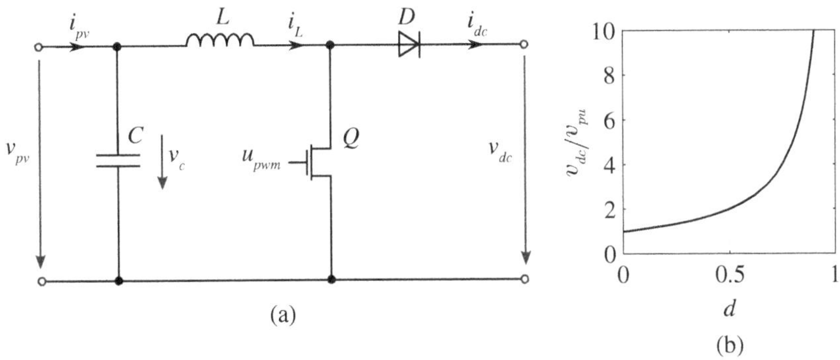

Figure 4.10 DC–DC boost converter with static transfer function (4.14).

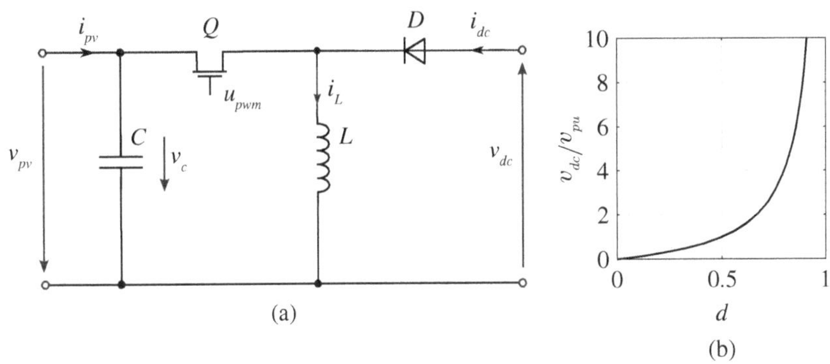

Figure 4.11 DC–DC buck–boost converter with static transfer function (4.15).

characteristic, also called conversion ratio $M(d)$, see Erickson [2007], is given by

$$d = \frac{v_{dc}}{v_{pv}}, \quad d \in [0, 1]. \tag{4.10}$$

The duty cycle d defines the ratio between the pulse duration denoted as T_{on} or pulse width (PW) and the total period denoted as T_{sw} of a rectangular waveform with amplitude one

$$d := \frac{T_{on}}{T_{sw}}, \quad d \in [0, 1]. \tag{4.11}$$

Instead of the total period duration, the switching frequency

$$f_{sw} = \frac{1}{T_{sw}} \tag{4.12}$$

is often specified. This rectangle waveform is called pulse width modulated (PWM) signal denoted as u_{pwm} and can be described as

$$u_{pwm}(t) = \begin{cases} 1, & \frac{t}{T_{sw}} - \text{floor}\left(\frac{t}{T_{sw}}\right) \leq \frac{T_{on}}{T_{sw}}, \\ 0, & \text{else}, \end{cases} \tag{4.13}$$

where floor is a function $f(x)$ with $x \in \mathbb{R}$ that gives as output the greatest integer less than or equal to x, e.g. floor(1.2) = 1.

As opposed to the book converter with (4.10), for the other two switched DC–DC converters, the ideal static transfer functions $v_{dc}/v_{pv} = M(d)$ are nonlinear.[3] The static input–output characteristic of the boost convert is represented by

$$\frac{v_{dc}}{v_{pv}} = \frac{1}{1-d}, \quad d \in [0, 1), \tag{4.14}$$

see Figure 4.10, and for buck–boost converter

$$\frac{v_{dc}}{v_{pv}} = \frac{d}{1-d}, \quad d \in [0, 1) \tag{4.15}$$

which is illustrated in Figure 4.11. From relation (4.14) follows that the boost converter (step-up converter) steps up voltage while stepping down current from its input to its output. The buck–boost converter can do both as indicated by its name: The output voltage magnitude is either greater than or less than the input voltage magnitude. The principle circuit diagram of this converter type is shown in Figure 4.10.

4.3.2 Generic State-Space Average Models

Kirchhoff's equations obtain the state-space descriptions of the three switching DC–DC converters. Here, Q is modeled as an ideal switch, and the diode is modeled as an ideal current-flow-dependent ideal switch as well. The closed switch refers to the ON state, and the open switch refers to the OFF state. The switching operation is assumed to be lossless and timeless. The diode has no voltage drop in the forward direction and blocks ideally like an open switch in the reverse direction.

Let's start with the buck converter modeling shown in Figure 4.9. For the ON-state, Kirchhoff's equations are as follows

$$i_{pv} - i_c - i_L = 0, \quad i_c = C\,\frac{dv_c}{dt} \tag{4.16}$$

---

3 The ideal static transfer functions for the DC–DC types as a function of d are valid if the induction current $i_L > 0$. In the literature this is called continuous mode [Cuk and Middlebrook, 1977]. It is still shown that this is always valid in PV power plant normal operation.

and

$$-v_c + v_L + v_{dc} = 0, \quad v_L = L\,\frac{di_L}{dt}, \quad v_{pv} = v_c \tag{4.17}$$

With the introduction of the state variables

$$x_1 := v_c \qquad x_2 := i_L \tag{4.18}$$

as variables of the energy storage $E_k = \frac{1}{2}\,L\,i_L^2$ and $E_p = \frac{1}{2}\,C\,v_c^2$, we obtain the buck converter model valid for the ON-state

$$\begin{aligned}\dot{x}_1 &= -\frac{1}{C}\,x_2 + \frac{1}{C}\,i_{pv},\\ \dot{x}_2 &= \ \frac{1}{L}\,x_1 - \frac{1}{L}\,v_{dc}.\end{aligned} \tag{4.19}$$

If the switch Q is open, Kirchhoff's equations result in

$$i_{pv} - i_c = 0, \quad i_c = C\,\frac{dv_c}{dt} \tag{4.20}$$

and

$$v_L + v_{dc} = 0, \quad v_L = L\,\frac{di_L}{dt}, \tag{4.21}$$

which gives with (4.18) the state-space representation for the OFF-state

$$\begin{aligned}\dot{x}_1 &= \frac{1}{C}\,i_{pv},\\ \dot{x}_2 &= -\frac{1}{L}\,v_{dc}.\end{aligned} \tag{4.22}$$

On the basis of the abstraction level represented by the state equations (4.19)–(4.21) and the proposed function of the PWM signal (4.13) we obtain a hybrid state-space model description of the switched buck-type DC–DC converter

$$\begin{aligned}\dot{x}_1 &= -\frac{1}{C}\,x_2\,u_{pwm} + \frac{1}{C}\,i_{pv},\\ \dot{x}_2 &= \ \frac{1}{L}\,x_1\,u_{pwm} - \frac{1}{L}\,v_{dc},\end{aligned} \tag{4.23}$$

where the zero value $u_{pwm} = 0$ corresponds to the open electronic switch, i.e. OFF-state and $u_{pwm} = 1$ corresponds to the closed electronic switch, i.e. ON-state: $u_{pwm} = \{0, 1\} \triangleq \{\text{OFF-state, ON-state}\}$.

Using the same procedure based on Kirchhoff's equations, the state models of the boost converter for the ON-state

$$\begin{aligned}\dot{x}_1 &= -\frac{1}{C}\,x_2 + \frac{1}{C}\,i_{pv},\\ \dot{x}_2 &= \frac{1}{L}\,x_1\end{aligned} \tag{4.24}$$

and OFF-state

$$\dot{x}_1 = -\frac{1}{C}x_2 + \frac{1}{C}i_{pv},$$

$$\dot{x}_2 = \frac{1}{L}x_1 - \frac{1}{L}v_{dc} \tag{4.25}$$

are determined. Both Eqs. (4.24) and (4.24) are combined in a hybrid model to describe the boost-type DC–DC converter as switching system

$$\dot{x}_1 = -\frac{1}{C}x_2 + \frac{1}{C}i_{pv},$$

$$\dot{x}_2 = \frac{1}{L}x_1 - \frac{1}{L}v_{dc}\left(1 - u_{pwm}\right) \tag{4.26}$$

with $u_{pwm} = \{0, 1\} \triangleq \{\text{OFF-state, ON-state}\}$. The last of the three systems to be modeled is the buck–boost converter shown in Figure 4.11. As before, a hybrid system is derived from the ON and OFF mode state models, describing the switching buck–boost-type converter

$$\dot{x}_1 = -\frac{1}{C}x_2\, u_{pwm} + \frac{1}{C}i_{pv},$$

$$\dot{x}_2 = \frac{1}{L}x_1\, u_{pwm} - \frac{1}{L}v_{dc}\left(1 - u_{pwm}\right) \tag{4.27}$$

with $u_{pwm} = \{0, 1\} \triangleq \{\text{OFF-state, ON-state}\}$. By comparing the previously derived hybrid models it can be seen well how the buck–boost converter (4.27) is composed of the combination of buck (4.23) and boost converter system (4.26). In the final step, we will derive the state-space average models from the determined switched systems. It allows us to have a more straightforward control design where the average states are used instead of the switched states. The derivation is based on the assumption that within a switching period the external quantities i_{pv} and v_{vd} can be assumed as constant values. Instead of the switching signal u_{pwm} with the discrete values $u_{pwm} \in \{0, 1\}$, see (4.13), the average value over a period is used

$$d = \frac{1}{T_{sw}} \int_0^{T_{sw}} u_{pwm}(t)dt = \frac{T_{on}}{T_{sw}} \tag{4.28}$$

with (4.11). As a consequence, d does not change within a switching period, which means that the switching frequency must be large compared to the LC circuit dynamics. The exact analytical derivation is given in Middlebrook and Cuk [1976]. It is based on the superposition of solutions of the ON- and OFF-state-space system.

Finally, based on the previous considerations (state-space averaging technique), the generic state-space average models of lossless[4] switched-mode DC–DC converter are given:

- State-space average model of second-order buck converter

$$\dot{\mathbf{x}} = \begin{pmatrix} 0 & -\frac{1}{C}d \\ \frac{1}{L}d & 0 \end{pmatrix} \mathbf{x} + \begin{pmatrix} \frac{1}{C} & 0 \\ 0 & -\frac{1}{L} \end{pmatrix} \begin{pmatrix} i_{pv} \\ v_{dv} \end{pmatrix}. \tag{4.29}$$

- State-space average model of second-order boost converter

$$\dot{\mathbf{x}} = \begin{pmatrix} 0 & -\frac{1}{C} \\ \frac{1}{L} & 0 \end{pmatrix} \mathbf{x} + \begin{pmatrix} \frac{1}{C} & 0 \\ 0 & -\frac{1}{L}(1-d) \end{pmatrix} \begin{pmatrix} i_{pv} \\ v_{dv} \end{pmatrix}. \tag{4.30}$$

- State-space average model of second-order buck–boost converter

$$\dot{\mathbf{x}} = \begin{pmatrix} 0 & -\frac{1}{C}d \\ \frac{1}{L}d & 0 \end{pmatrix} \mathbf{x} + \begin{pmatrix} \frac{1}{C} & 0 \\ 0 & -\frac{1}{L}(1-d) \end{pmatrix} \begin{pmatrix} i_{pv} \\ v_{dv} \end{pmatrix}, \tag{4.31}$$

with $\mathbf{x} = \left(v_c, i_L\right)^T$, see (4.18), where the duty cycle D in (4.29)–(4.31) is the controllable input variable and i_{pv} and v_{dv} are the non-controllable inputs. The input i_{pv} current will be replaced in Section 4.4 by the PV generator model (4.9) as a nonlinear voltage-, radiation-, and cell temperature-dependent current source $i_{pv} = f_{pv}(x_1, S, T_c)$ where $x_1 = v_c = v_{pv}$ is valid for all converter types.

4.3.3 Circuit Design and Simulation

To prepare the scenarios for simulation-based comparison between the switching and average system, the appropriate sizing of L and C components is determined. It based on the required reduction of the current i_L and voltage v_c ripple quantified by a peak-to-peak value denoted by Δi_L and Δv_c. In the buck converter circuit, during the ON-state increases the current by

$$\Delta i_L = \frac{v_{pv} - v_{dc}}{L} T_{on}.$$

From this with the steady-state relations (4.10)–(4.12) follows the maximum change of the induction current

$$\Delta i_L = \frac{\frac{v_{dc}}{d} - v_{dc}}{L} d\, T_{sw} = \frac{v_{dc}(1-d)}{L f_{sw}}.$$

Thus, the value of the inductance L can be calculated from

$$L = \frac{v_{dc}^e (1 - d^e)}{\Delta i_L f_{sw}} \tag{4.32}$$

4 In steady state: $P_{pv} = P_{dc}$, where $P_{pv} = v_{pv}\, i_{pv}$ and $P_{dc} = v_{dc}\, i_{dc}$.

related to the selected operating point $(d^e, v^e_{dc}, i^e_{pv})$. The capacitance is designed according to the maximum acceptable voltage change in a switching interval

$$\Delta v_c = -\frac{1}{C} i_c T_{on} = -\frac{1}{C}(i_{pv} - i_L) d T_{sw} = -\frac{1}{C}\left(i_{pv} - \frac{i_{pv}}{d}\right) d T_{sw}.$$

using the steady-state solution of the buck converter (4.29) with

$$\dot{x}_1 = -\frac{1}{C} D x_2 + \frac{1}{C} i_{pv} \overset{!}{=} 0 \quad \Rightarrow \quad D = \frac{i_{pv}}{i_L}.$$

From this, we obtain

$$C = \frac{i^e_{pv}(1 - d^e)}{\Delta v_c f_{sw}}, \tag{4.33}$$

which is related to the same operating point as in (4.32). Note that in case of a wider operating interval, the worst-case scenario in (4.32) and (4.33) is used with $v^e_{dc} = \max[v_{dc}]$, $i^e_{pv} = \max[i_{pv}]$, and $d^e = \min[d]$. With the same in mind, we come to the calculations for L and C of the boost

$$L = \frac{v^e_{pv} d^e}{\Delta i_L f_{sw}}, \qquad C = \frac{1}{8}\frac{\Delta i^e_L}{\Delta v_c f_{sw}}, \tag{4.34}$$

and buck–boost converter

$$L = \frac{v^e_{pv} d^e}{\Delta i_L f_{sw}}, \qquad C = \frac{i^e_{pv}(1 - d^e)}{\Delta v_c f_{sw}}. \tag{4.35}$$

Like before, the components must be designed for a chosen operating point $(d^e, v^e_{pv}, i^e_{pv})$. For a larger range, as it occurs in demanded power point-tracking operation (see Section 4.5), a worst-case estimation within the specification has to be chosen.

Let us now turn to a comparison by simulation between a switching and an averaging DC–DC converter model. The buck converter to be investigated for a PV system with $P_{MPP} = 3083$ kW is shown in Figure 4.12. We assume that the

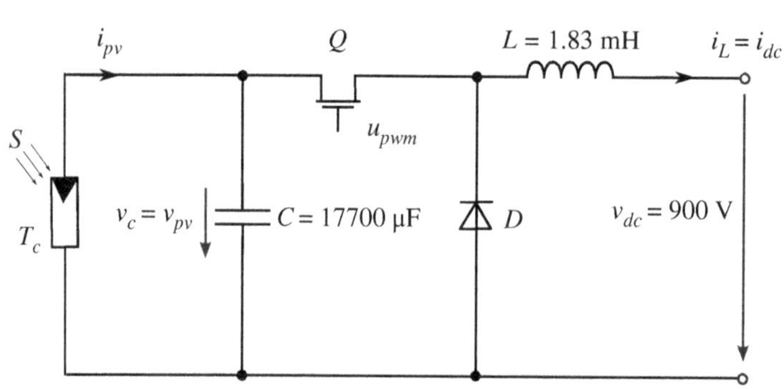

Figure 4.12 Buck converter combined with a PV generator.

switching frequency of the circuit is $f_{sw} = 5$ kHz, where a current and voltage ripple is required which should be below $\Delta v_c = 5$ V and $\Delta i_L = 15$ A. On the output side, the converter is connected to the constant DC-link voltage $v^e_{dc} = 900$ V. On the input side, the current i_{pv} is fed by a current source which corresponds to the PV generator model (4.9). Here we consider only the voltage dependence, where the irradiation and cell temperature are kept at the STC conform values of $S = S_{STC}$ and $T_c = T_{c,STC}$. Further, the operating point is set to the maximum operating point related to the STC condition, see Figure 4.7 with the marked MPP:

$$v^e_{pv} = 1062.2\,\text{V}, \quad i^e_{pv} = 2902.1\,\text{A}, \quad d^e = 0.8473.$$

For the previously defined circuit parameters, the components can be designed with (4.32) and (4.33). Thus we obtain

$$L = 1.8329\,\text{mH}, \qquad C = 17700\,\mu\text{F}.$$

Based on the components calculated above, the primary converter system of a PV power plant shown in Figure 4.12 will now be specified for further investigations. First, we compare simulations based on the switching and averaging model of the buck converter combined with a PV generator as a current source. Figure 4.13

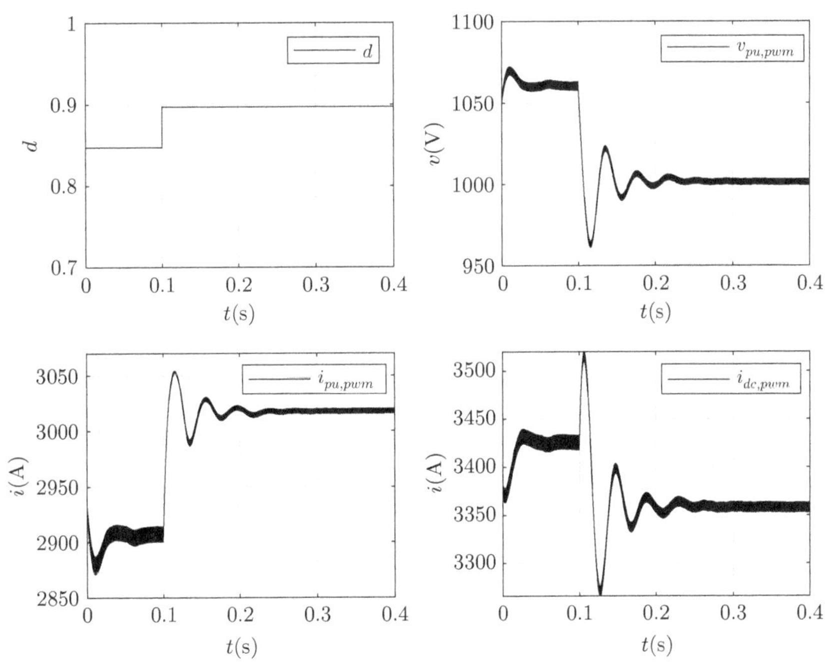

Figure 4.13 DC–DC buck converter system states with $x_1 = v_c = v_{pv}$, $x_2 = i_L = i_{dc}$, and PV generator current $i_{pv} = f(v_{pv})$ using the switched model (4.23).

shows the simulation results, the dynamic behavior of the system states $x_1 = v_c = v_{pv}$, $x_2 = i_L = i_{dc}$ and PV current $i_{pv} = f(v_{pv})$ caused by the change of duty cycle at $t = 0.1$ second from $d = d^e$ to $d = d^e + 0.05 = 0.8973$. It can be clearly seen that the increase of duty cycle leads to a drop of the input voltage v_{pv} for a constant output voltage v_{dc}. In the steady state of the simulation, the mean values are given by

$$t < 0.1 \; : \; v_{pv} = \frac{v_{dc}}{d^e} = 1062\,\text{V}, \qquad t > 0.1 \; : \; v_{pv} = \frac{v_{dc}}{d^e + 0.05} = 1003\,\text{V}.$$

Note that due to the voltage-dependent current source represented by the model of the PV generator, the PV current i_{pv} increases when v_{pv} is reduced to the left of MPP, see the static curve in Figure 4.7 and the resulting dynamic behavior in Figure 4.13. Comparison of the currents in steady-state regions also follows the static calculations

$$t < 0.1 \; : \; i_{dc} = \frac{i_{pv}}{d^e} = 3426\,\text{A}, \qquad t > 0.1 \; : \; i_{dc} = \frac{v_{pv}}{d^e + 0.05} = 3363\,\text{A}.$$

Finally the comparison between system states $x_1 = v_c = v_{pv}$ and $x_2 = i_L = i_{dc}$ from the switched (4.23) and average value-based model (4.29) is shown in Figure 4.14.

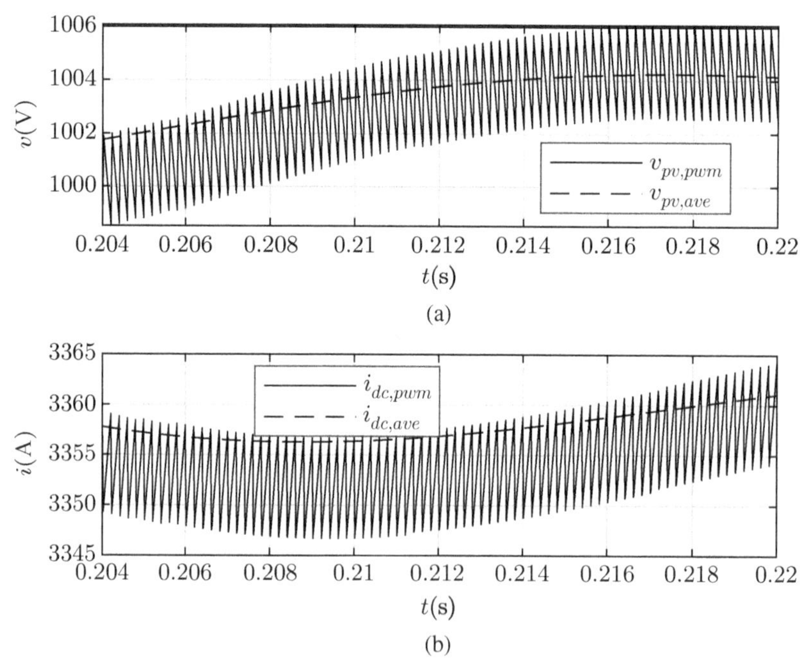

Figure 4.14 Comparison of the DC–DC buck converter system states $x_1 = v_c = v_{pv}$ and $x_2 = i_L = i_{dc}$ between the switched model (4.23) (zoom of Figure 4.13) and average value-based model (4.29).

One can see well that the mean values $v_{pv,ave}$ and $i_{pv,ave}$ always stay within the envelope of the rippled values $v_{pv,pwm}$ and $i_{pv,pwm}$ related to the switched model. This is not only valid for the zoom range but also for the whole simulation time, but this could not be shown because the average value in Figure 4.14 would be masked by the ripple gradients.

4.4 Voltage Control of PV Generators

In this section, the model-based controller design using linear matrix inequality (LMI) synthesis is presented. First, a generic Takagi–Sugeno (TS) model formulation of the aggregated PV generator DC–DC system by Taylor linearization and membership interpolation (see Appendix A.2.2) as control-oriented model is derived. With that generic structure, models with different depths, i.e. different sub-models, are examined regarding accuracy and controller design capability. Then, two approaches to controller synthesis are proposed: A one-step approach with simultaneous controller coefficient calculation with respect to the desired s-domain constraints and a two-step approach. The former method has already been used in the controller design for wind turbines in Chapter 3. The second controller design technique, which has not been presented yet, exploits the local validity of the sub-models. In the first step, the corresponding controller coefficients are determined locally for each single ith sub-model. After, in the second step, a formal proof of stability is performed by finding a common positive definite matrix of an associated quadratic Lyapunov function (LF) of all interconnected sub-models and controller gains. Finally, the designed controller is validated in an overall simulation for a width operating range with a variation of the reference voltage, irradiation, and cell temperature.

4.4.1 Control-Oriented State-Space Model

The dynamic model derived below is used for the model-based design of the PV voltage regulator, intended to be valid in a wide working range. Using the state-space average model of the buck converter (4.29) as an example, the substitution of the PV current by the aggregated PV generator model (4.9) as a current source $i_{pv} = f_{pv}(v_{pv}, S, T_c)$ results in

$$\begin{aligned} \dot{x}_1 &= -\frac{1}{C} x_2 u_1 + \frac{1}{C} f_{pv}(x_1, S, T_c), \\ \dot{x}_2 &= \frac{1}{L} x_1 u_1 - \frac{1}{L} u_2, \end{aligned} \tag{4.36}$$

with $\mathbf{x} = (x_1, x_2)^T = (v_{pv}, i_L)^T$, where $v_{pv} = v_c$ (see Figure 4.9) and $\mathbf{u} = (u_1, u_2)^T = (d, v_{dc})^T$. The irradiation S and PV cell temperature T_c are represented as time-varying parameters in the model due to slower dynamics compared to the $\mathbf{u}$ and $\mathbf{x}$. As described in the introduction, the model must contain the duty cycle $u_1 := d$ as a controllable input. The DC link voltage $u_2 := v_{dc}$ is applied as a non-controllable input and initially set to a constant value for the following decentralized design. If necessary, the requirement for minimizing the effect due to fluctuating DC link voltage can be integrated into the design by a H_∞ criterion, see Appendix B.5.3.1. The fluctuation is caused by the imbalance between the generated and injected power. Since this will be discussed in Chapter 6, we will deal with it later.

As mentioned, the control-oriented model shall be based on a TS form consisting of a weighted combination of locally valid linear state-space models. These are calculated by Taylor linearization around selected equilibrium points $i = 1, \dots, N_r$. The well-known procedure is briefly described in Appendix A.2.2 and applied to (4.36). From this follows the ith linear model

$$\Delta\dot{\mathbf{x}}_i = \underbrace{\begin{pmatrix} \frac{1}{C}\dfrac{\partial f_{pv}(x_1^e, S^e, T_c^e)}{\partial x_1} & -\frac{u_1^e}{C} \\ \frac{u_1^e}{L} & 0 \end{pmatrix}}_{\mathbf{A}_i} \Delta\mathbf{x}_i + \underbrace{\begin{pmatrix} -\frac{x_2^e}{C} & 0 \\ \frac{x_1^e}{L} & -\frac{1}{L} \end{pmatrix}}_{\mathbf{B}_i} \Delta\mathbf{u}_i \tag{4.37}$$

with the states $\mathbf{x}_i^e = (x_{1,i}^e, x_{2,i}^e)^T$ and inputs $\mathbf{u}_i = (u_{1,i}^e, u_{2,i}^e)^T$ at the ith equilibrium point, related environmental conditions S_i^e and $T_{c,i}^e$ and the distance $\Delta\mathbf{x}_i = \mathbf{x}_i - \mathbf{x}_i^e$, $\Delta\mathbf{u}_i = \mathbf{u}_i - \mathbf{u}_i^e$ to the ith equilibrium point. Calculation of the equilibrium points is done in a straightforward way by:

1. Selection of an arbitrary PV voltage values within $v_{pv} \in [V_{dc}, v_{oc}]$, where
$$x_{1,i}^e = v_{pv,i} \tag{4.38}$$
assuming that the DC voltage at the output of the buck converter is constant $V_{dc} \equiv \text{const}$.
2. Calculation of the related operating point of the input
$$u_{1,i}^e = d_i^e = \frac{V_{dc}}{x_{1,i}^e} \quad \text{and} \quad u_{2,i}^e = V_{dc} \quad \forall i, \tag{4.39}$$
which results in $\Delta u_{2,i} = V_{dc} - u_{2,i}^e = 0$
3. Calculation of current i_L at the equilibrium point
$$x_{2,i}^e = i_L^e = \underbrace{f_{pv}(x_{1,i}^e, S_i^e, T_{c,i}^e)}_{i_{pv}} \frac{1}{d_i^e} \tag{4.40}$$
for specified irradiation S_i^e and cell temperature $T_{c,i}^e$.

Hence, the steady-state operating points $\{\mathbf{u}_i^e, \mathbf{x}_i^e\}$ of (4.41) are defined by the terminal voltage at the PV generator, aggregated irradiation, and cell temperature. Due to the constant voltage at the right-hand side of the buck converter, see Figure 4.12, the control-oriented local models have a single input $\Delta u_i := \Delta u_{1,i} = u_1 - u_{1,i}^e$ with $u_i := u_{1,i}$ and $u := u_1$.

$$\Delta\dot{\mathbf{x}}_i = \underbrace{\begin{pmatrix} \frac{1}{C}\frac{\partial f_{pv}(x_1^e, S^e, T_c^e)}{\partial x_1} & -\frac{u_1^e}{C} \\ \frac{u_1^e}{L} & 0 \end{pmatrix}}_{\mathbf{A}_i}\Delta\mathbf{x}_i + \underbrace{\begin{pmatrix} -\frac{x_2^e}{C} \\ \frac{x_1^e}{L} \end{pmatrix}}_{\mathbf{b}_i}\Delta u_i. \tag{4.41}$$

To merge the set of linear models into a TS model with a common state vector, we rewrite (4.41) as

$$\dot{\mathbf{x}} = \mathbf{A}_i\mathbf{x} + \mathbf{b}_i u \underbrace{- \mathbf{A}_i\mathbf{x}_i^e - \mathbf{b}_i u_i^e}_{\mathbf{a}_i}, \tag{4.42}$$

whereby the Δ-vectors in (4.41) are split into their absolute state vector $\mathbf{x}$, input value u, and the affine part $\mathbf{a}_i$. This allows us to sum up all linear models into the TS model of the form (A.3)

$$\dot{\mathbf{x}} = \sum_{i=1}^{N_r} h_i(\mathbf{z})\left(\mathbf{A}_i\mathbf{x} + \mathbf{b}_i u + \mathbf{a}_i\right), \tag{4.43}$$

where $\mathbf{a}_i = -\mathbf{A}_i\mathbf{x}_i^e - \mathbf{b}_i u_i^e$. In contrast, the sector-nonlinearity (SL) approach provides an exact description of a given nonlinear system in a predefined operating range (applied in Chapter 3), the membership functions h_i of the approximation method are not obtained from the given nonlinear model, e.g. (4.36). Instead, piecewise linear local membership functions are used to blend the set of locally valid models (4.43) as a function of the premise variables $\mathbf{z}$ into a single TS model (4.43). Thc h_i-functions[5] for TS generation by Taylor linearization (A.16) are determined by

$$\sum_{i=1}^{N_r} h_i(\mathbf{z}) = \prod_{j=1}^{N_l}\sum_{k=1}^{N_j} w_{k,j}(z_j) = 1 \quad \text{with} \quad N_r = \prod_{j=1}^{N_l} N_j \tag{4.44}$$

5 Note that for TS model generation by the SL approach, just two weighting functions are included per premise variable coordinate, see (A.9):

$$\sum_{k=1}^{N_r} h_i(\mathbf{z}) = \prod_{j=1}^{N_l}\sum_{k=1}^{2} w_{k,j}(\cdot), \quad N_r = 2^{N_l}$$

with $\mathbf{z} = (z_1, \dots, z_j, \dots, z_{N_l})^T$, where N_l denotes the number of premise variables and N_j the number of points respectively weighting functions per grid dimension. For the considered problem, $N_l = 3$ premise variables can be specified as follows

$$z_1 = x_1, \qquad z_2 = S, \qquad z_3 = T_c. \tag{4.45}$$

Why is this reasonable? Note, the assignment of the working points (with respect to a fixed S and T_c) is determined uniquely by $x_1 := v_{pv}$ or $d := u_1$, because in the steady state $x_1\, d = v_{dc} \equiv$ const. is fulfilled, see (4.36) for $\dot{x}_2 = 0$. To avoid algebraic loops, the controller output must not be directly dependent on the premise variable. Therefore $x_1 := v_{pv}$ is selected in (4.45) as the premise variable z_1. Furthermore, $z_{2,3}$ are directly related to the time-varying quantities S and T_c and determine $x_{2,c}$ by $x_{2,c} \cdot d = f_{pv}(x_{1,c}, S, T_c) = i_{pv,c}$, see (4.36) for $\dot{x}_1 = 0$. The appropriate selection in (4.45) allows to define all equilibrium points, which are located on a N_l-dimensional grid. In this case for $N_l = 3$ defined by

$$\{z^e_{1,1}, z^e_{1,2}, \dots, z^e_{1,N_1}\} \times \{z^e_{2,1}, z^e_{2,2}, \dots, z^e_{2,N_2}\} \times \{z^e_{3,1}, z^e_{3,2}, \dots, z^e_{3,N_3}\} \tag{4.46}$$

with the total number of models with $N_r = N_1\, N_2\, N_3$ consistent to (4.44). The membership functions h_i result from the product of the weighting functions

$$h_i(\mathbf{z}) = w_{k_1,1}(z_2)\, w_{k_2,2}(z_2)\, w_{k_3,j}(z_3) \quad \text{for} \quad k_j = 1, \dots, N_j, \tag{4.47}$$

which satisfy the convexity condition

$$\sum_{k=1}^{N_j} w_{k,j}(z_j) = 1, \qquad w_{k,j}(z_j) \geq 0. \tag{4.48}$$

The relationship between the model index i (see left-hand side of (4.47)) and the indices of the weighing functions k_j of the right-hand side (4.47) is determined as follows

$$i = k_1 + (k_2 + (k_3 - 1)\, N_2 - 1)\, N_1 \quad \text{for} \quad k_j = 1, \dots, N_j. \tag{4.49}$$

The functions $w_{k,j}$ used in this investigation are constructed as simple as possible. Each linearized submodel is weighted by 1 (i.e. $h_i(\mathbf{z}^e_i) = 1$) at the associated linearization respectively equilibrium point $\mathbf{z}^e_i = (z^e_{1,i}, z^e_{2,i}, \dots, z^e_{N_l,i})^T$ and thus all functions of the product (4.47) also have the value one, whereas all other models are then not active $h_j(\mathbf{z}^e_i) = 0$ for $j \neq i$. To the right and left of equilibrium points, the weighting function decreases until it reaches the value 0. At the edges of the operating range, due to the convexity condition (4.48), the function values increase to one and then remain constant. This results in three different functions for each premise variable z_j, $j = 1, \dots, N_l$. Note for the sake of clarity, k is used instead of k_j and z_k^e instead of $z_{j,k}^e$ with $j = 1, \dots, N_l$:

- The lower boundary function for the left edge with $k = 1$

$$w_{1,j}(z_j) = \begin{cases} 1, & \text{if} \quad z_j < z_1^e, \\ 1 - \frac{z_j - z_1^e}{z_2^e - z_1^e}, & \text{if} \quad z_1^e \leq z_j \leq z_2^e, \\ 0, & \text{otherwise.} \end{cases} \tag{4.50}$$

- The triangle function for $k = 2, \dots, N_j - 1$.

$$w_{k,j}(z_j) = \begin{cases} \frac{z_j - z_{k-1}^e}{z_k^e - z_{k-1}^e}, & \text{if} \quad z_{k-1}^e \leq z_j < z_k^e, \\ 1 - \frac{z_j - z_k^e}{z_{k+1}^e - z_k^e}, & \text{if} \quad z_k^e \leq z_j \leq z_{k+1}^e, \\ 0, & \text{otherwise.} \end{cases} \tag{4.51}$$

- And the upper boundary function for the right edge with $k = N_j$

$$w_{N_j,j}(z_j) = \begin{cases} 1, & \text{if} \quad z_j > z_{N_j}^e, \\ \frac{z_j - z_{N_j-1}^e}{z_{N_j}^e - z_{N_j-1}^e}, & \text{if} \quad z_{N_{j-1}}^e \leq z_j \leq z_{N_j}^e, \\ 0, & \text{otherwise.} \end{cases} \tag{4.52}$$

Figure 4.15 shows the function curves in the composite with the function on the left (4.50) and right edges (4.52) and the triangle functions (4.51) in between. In the literature, other shapes such as Gaussian, trapezoidal, and skew-symmetric functions are also given [Schulte, 2005a]. However, it has been shown that these functions have no or only a slight influence on the approximation accuracy of the TS models [Nelles, 1999].

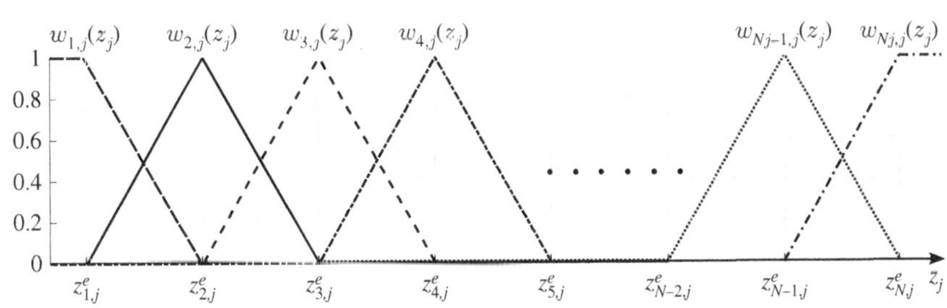

Figure 4.15 Piecewise linear weighting functions of type "lower boundary function" (left), "triangle functions" (in the middle), and "upper boundary function" (right).

However, a major influence on the approximation accuracy of the model in TS form (4.43) compared to the original model $\dot{\mathbf{x}} = \mathbf{f}(\mathbf{x}, \mathbf{u})$ is the number of sub-models N_r. In general, a measure of the local nonlinearity of the right-hand side is the Hessian matrix for the largest nonlinearity in $\mathbf{f}(\mathbf{x}, \mathbf{u})$. Assuming equidistant distances of $w_{k,j}$ under specification of a maximum permissible error of the right sides, the number N_r can be calculated by determining the maximum eigenvalue of the Hessian matrix for the largest nonlinearity in $\mathbf{f}(\mathbf{x}, \mathbf{u})$, see Schulte [2005b]. However, since this is a worst-case estimation, from the user's point of view it is also possible to choose a priori grid of equilibrium points related by (4.38)–(4.40) to the values of premise variables

$$\begin{aligned} v_{pv}^e &= x_1^e = \{0.9,\ 0.925,\ 0.95,\ 0.975,\ 1.0\} \cdot v_{pv,MPP}, \\ S^e &= \{200,\ 333,\ 467,\ 600,\ 733,\ 867,\ 1000\}, \\ T_c^e &= \{273,\ 298,\ 323\}, \end{aligned} \tag{4.53}$$

where $v_{pv,MPP}$ denotes the voltage at MPP related to STC, whereby the set of sub-models $\{\mathbf{A}_i, \mathbf{b}_i, \mathbf{a}_i\}$ for $i = 1, \ldots, N_r$ with $N_r = 5 \cdot 7 \cdot 3 = 105$ can be calculated according to (4.41). Subsequently, it is verified whether the model error is acceptable for the controller design. If not, the mesh is refined step-wise until the maximum error is within the allowable bound.

To demonstrate the quality of the TS model with regard to the original model, both are excited with the same dynamic inputs in the large-signal range. For this purpose, S and T_c profiles were selected to represent the entire workspace of interest. Here, the large operating range is covered by a staircase-shaped change in irradiation S, aggregate cell temperature T_c, and a jump in duty cycle of $\Delta d = -0.1$ at $t = 0.3$ seconds. The simulation results are shown in Figure 4.16. It can be seen that the external excitation is mainly within the model's range of validity defined by the operating points (4.53). Beyond the set of equilibrium points, where the irradiation increases to $S = 1020\ \mathrm{W/m^2}$ in the period from $t = 1.05$ seconds to $t = 1.4$ seconds, the model error $e_{i_L} = i_L - i_{L,TS}$ increases significantly. This model error increase can also be seen in the simulated voltage $e_{v_{pv}} = v_{pv} - v_{pv,TS}$. However, the error is less than 5% in the valid working range. If necessary, a new set of equilibrium points can easily adjust the model when the expected working range is significantly larger. Thus, (4.43) is validated with the selected model coefficients and can be used for the state-feedback design.

4.4.2 State-Feedback Design

In this section, the necessary design steps for the inner-loop voltage-tracking control problem are presented. The proposed structure consists of a state feedback

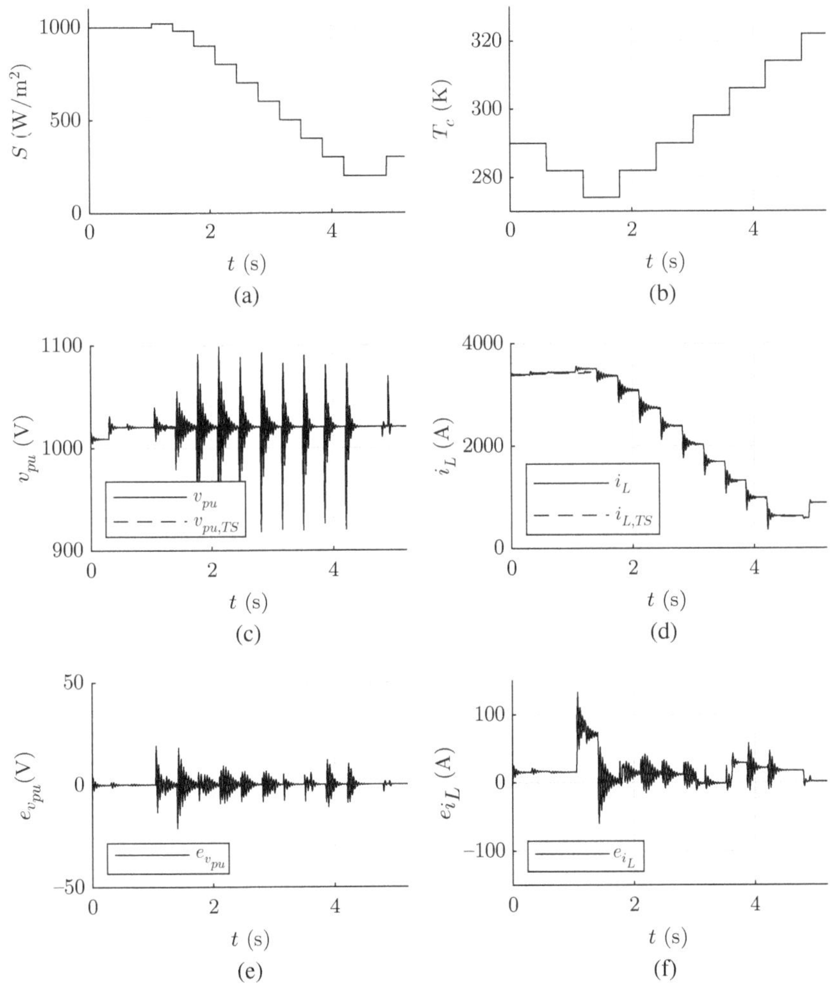

Figure 4.16 TS model with $N_r = 105$ compared to the original nonlinear system: validation over a large input and state-space range.

of the measurable variables v_{pv} and i_L and an integration of the tracking control error $e = v_{pv,ref} - v_{pv}$ shown in Figure 4.17. Alternatively, instead of integrating the control error, the reference value could also be gained with a pre-filter (adjustment of the reference value) and superimposed to the state-feedback output. However, this approach has the disadvantage that the control error is not regulated to zero in a steady-state manner due to model uncertainties. For this reason, the pre-filter is not used. With the integral term, on the other hand, a stationary exact tracking behavior can be achieved.

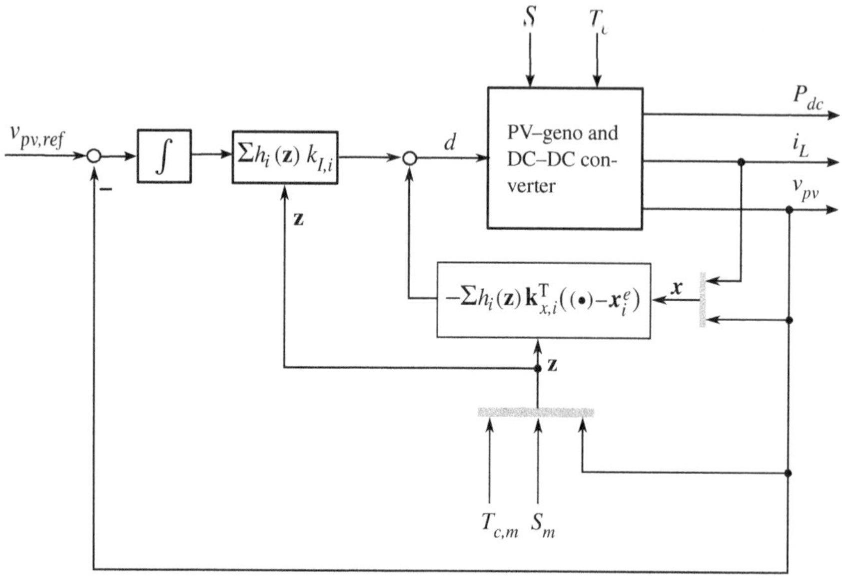

Figure 4.17 Inner-loop control scheme of PV generator voltage controller.

The control objective is to follow a reference voltage $v_{pv,ref}$ calculated by MPPT or demand power tracking (DPT). The DPT is discussed afterward in Section 4.5. The control action here is the duty cycle d, the measurable states are feedback via the $h_i(\mathbf{z})$-weighted sum of controller gains $\mathbf{k}^T_{x,i} \in \mathbb{R}^{1\times 2}$ plus the integration of the control error gained by $\sum_i h_i(\mathbf{z})\, k_{I,i}$. The premise variable vector $\mathbf{z}$ (4.45) results from the measurable variables $S = S_m$, $T_c = T_{c,m}$, and v_{pv} assuming that the irradiation S_m is the same for all PV modules of the plant and that the measurable variables correspond to the actual values without any mismatch. Robustness studies with a defined mismatch between S and S_m is performed in Section 4.5.3. Large areas can be partially shaded, measured by several distributed irradiation meters, e.g. pyranometers, distributed over the entire PV area. For the MPPT or DPT, decentralized voltage regulators for each module string (modules connected in series) or a larger cluster of several strings are applied.

As for the control of wind turbines in Chapter 3, we utilize the integral state-feedback controller in parallel distributed compensator (PDC) form

$$u = -\sum_{i=1}^{N_r} h_i(\mathbf{z})\, \mathbf{k}^T_{x,i}(\mathbf{x} - \mathbf{x}^e_i) + \sum_{i=1}^{N_r} h_i(\mathbf{z})\, k_{I,i} \int_0^t \left(v_{pv,ref} - v_{pv}\right) d\tau, \tag{4.54}$$

see Figure 4.17, where

$$\mathbf{x}_i^e = \begin{pmatrix} v_{pv,i}^e \\ i_{L,i}^e \end{pmatrix}, \qquad x_I := \int_0^t \left(v_{pv,ref} - v_{pv}\right) d\tau = \int_0^t \left(y_{ref} - y\right) d\tau \tag{4.55}$$

with the output equation

$$y = v_{pv} = \mathbf{c}^T \mathbf{x} = (1 \;\; 0)\, \mathbf{x}, \qquad y_{ref} = v_{pv,ref}. \tag{4.56}$$

By introducing the augmented state vector

$$\tilde{\mathbf{x}} = \begin{pmatrix} \mathbf{x} \\ \mathbf{x}_I \end{pmatrix}, \quad \Delta\tilde{\mathbf{x}} = \begin{pmatrix} \mathbf{x} - \mathbf{x}_i^e \\ x_I - x_{I,i}^e \end{pmatrix} \quad \text{with} \quad x_{I,i}^e = 0, \tag{4.57}$$

the control law (4.54) yields the compact form

$$u = -\sum_{i=1}^{N_r} h_i(\mathbf{z}) \left(\mathbf{k}_{x,i}^T,\, k_{I,i}\right) \Delta\tilde{\mathbf{x}} = -\sum_{i=1}^{N_r} h_i(\mathbf{z}) \tilde{\mathbf{k}}_i^T \,\Delta\tilde{\mathbf{x}}, \tag{4.58}$$

where $\tilde{\mathbf{k}}_i^T = \left(\mathbf{k}_{x,i}^T,\, k_{I,i}\right)$. Note these controller coefficients are combined via a convex weighted sum. Since the weights with h_i correspond to the membership functions of the plant model proposed in Section 4.4.1 (4.44), the structure of the used control law is called parallel distributed compensator (PDC), see Niemann et al. [1999]. The parallel distribution refers to the weighted summation of the local controller coefficients. Instead of the h_i functions from the plant model, it would be feasible to use other membership functions in (4.58), e.g. with an additional design parameter $h_i^k(\mathbf{z}, \theta) \neq h_i(\mathbf{z})$. The additional degree of freedom θ in the premise variables could be used to integrate operating point-dependent adaptation of the controller performance into the design. However, this approach is not pursued further here due to the additional design effort and lack of need.

After introducing the control law (4.54), the calculation of the controller coefficients by specifying the desired closed-loop pole region is discussed. For this purpose, LMI criteria for local synthesis and global stability analysis are given to achieve a model-based design. Let us first analyze the closed-loop system dynamics:

Using (4.43), (4.55), and (4.56), the state-space model related to the augmented vector (4.57) results in

$$\dot{\tilde{\mathbf{x}}} = \sum_{i=1}^{N_r} h_i(\mathbf{z}) \left(\tilde{\mathbf{A}}_i \tilde{\mathbf{x}} + \tilde{\mathbf{b}}_i u - \tilde{\mathbf{A}}_i \tilde{\mathbf{x}}_i^e - \tilde{\mathbf{b}}_i u_i^e\right) + (0,\, 0,\, 1\,)^T x_{1,ref} \tag{4.59}$$

with $\tilde{\mathbf{x}} = (\mathbf{x}^T, x_I)^T$ and $x_{1,ref} = v_{pv,ref}$, where

$$\tilde{\mathbf{A}}_i = \begin{pmatrix} \mathbf{A}_i & \mathbf{0}_{2\times1} \\ -\mathbf{c}^T & 0 \end{pmatrix}, \quad \tilde{\mathbf{b}} = \begin{pmatrix} \mathbf{b}_i \\ 0 \end{pmatrix}. \tag{4.60}$$

The closed-loop model with the purpose of LMI controller design is obtained by substituting u in (4.59) with (4.58):

$$\dot{\tilde{\mathbf{x}}} = \sum_{i=1}^{N_r} h_i(\mathbf{z})(\tilde{\mathbf{A}}_i\,\tilde{\mathbf{x}} + \tilde{\mathbf{b}}_i \left(-\sum_{j=1}^{N_r} h_j(\mathbf{z})\,\tilde{\mathbf{k}}_j^T\,\Delta\tilde{\mathbf{x}} \right)$$
$$- \tilde{\mathbf{A}}_i\,\tilde{\mathbf{x}}_i^e - \tilde{\mathbf{b}}_i\,u_i^e) + (0,\,0,\,1)^T y_{ref}$$

Taking advantage of the convexity property of $\sum_{j=1}^{N_r} h_j(\mathbf{z}) = 1$, see (4.44), we get

$$\dot{\tilde{\mathbf{x}}} = \sum_{i=1}^{N_r}\sum_{j=1}^{N_r} h_i(\mathbf{z})h_j(\mathbf{z}) \left(\left(\tilde{\mathbf{A}}_i - \tilde{\mathbf{b}}_i\tilde{\mathbf{k}}_j^T \right) \tilde{\mathbf{x}} + \mathbf{a}_{ij} \right) + \mathbf{e}_3\,y_{ref} \tag{4.61}$$

with

$$\mathbf{a}_{ij} = \left(-\tilde{\mathbf{A}}_i\,\tilde{\mathbf{x}}_i^e - \tilde{\mathbf{b}}_i u_i^e + \tilde{\mathbf{b}}_i\tilde{\mathbf{k}}_j^T\tilde{\mathbf{x}}_j^e \right), \qquad \mathbf{e}_3 = (0,\,0,\,1)^T. \tag{4.62}$$

A quadratic Lyapunov function candidate

$$V(\mathbf{x}) = \tilde{\mathbf{x}}^T\,\mathbf{P}\,\tilde{\mathbf{x}} > 0, \quad \forall\tilde{\mathbf{x}} \neq \mathbf{0} \tag{4.63}$$

is now applied to study the input-to-state stability (ISS) of (4.61). The concept of ISS is based on Sontag and Wang [1995], was introduced in the TS system framework by Lendek et al. [2011], and applied to mechanical speed governor of wind turbines in Schulte and Gauterin [2014], see additional material in the Appendix B.5.3.2. The following shows that $\dot{V}(\tilde{\mathbf{x}})$ satisfies the condition for ISS

$$\dot{V}(\tilde{\mathbf{x}}) \leq -\gamma_1(\|\,\tilde{\mathbf{x}}\,\|) + \gamma_2(|y_{ref}|) \tag{4.64}$$

for all $\tilde{\mathbf{x}}$ and a bounded y_{ref}, where $\gamma_1(\|\,\tilde{\mathbf{x}}\,\|)$ and $\gamma_2(|y_{ref}|)$ satisfy the properties of a $\mathcal{K}^\infty$ function [Sontag and Wang, 1995]. A continuous function $\gamma(r) : [0, a) \to [0, \infty)$ is said to belong to class $\mathcal{K}^\infty$ if it is strictly increasing, $\gamma(0) = 0$ and if $a := \infty$ then $\gamma(r) \to \infty$ for $r \to \infty$ [Khalil, 2002].

From the time derivative of the Lyapunov function candidate (4.63) with (4.61) follows

$$\dot{V}(\tilde{\mathbf{x}}) = \dot{\tilde{\mathbf{x}}}^T\,\mathbf{P}\,\tilde{\mathbf{x}} + \tilde{\mathbf{x}}^T\,\mathbf{P}\,\dot{\tilde{\mathbf{x}}}$$
$$= \tilde{\mathbf{x}}^T \sum_{i=1}^{N_r}\sum_{j=1}^{N_r} h_i(\mathbf{z})h_j(\mathbf{z}) \left(\left(\tilde{\mathbf{A}}_i - \tilde{\mathbf{b}}_i\tilde{\mathbf{k}}_j^T \right)^T \mathbf{P} + \mathbf{P} \left(\tilde{\mathbf{A}}_i - \tilde{\mathbf{b}}_i\tilde{\mathbf{k}}_j^T \right) \right) \tilde{\mathbf{x}}$$
$$+ 2\sum_{i=1}^{N_r}\sum_{j=1}^{N_r} h_i(\mathbf{z})h_j(\mathbf{z})\,\tilde{\mathbf{x}}^T\,\mathbf{P}\,\mathbf{a}_{ij} + 2\tilde{\mathbf{x}}^T\,\mathbf{P}\,\mathbf{e}_3\,y_{ref}.$$

Due to convexity, an upper bound is given for the second double sum term

$$\sum_{i=1}^{N_r}\sum_{j=1}^{N_r} h_i(\mathbf{z})h_j(\mathbf{z})\,\tilde{\mathbf{x}}^T\,\mathbf{P}\,\mathbf{a}_{ij} \leq \tilde{\mathbf{x}}^T\,\mathbf{P}\max_{ij}\mathbf{a}_{ij}.$$

The eigenmotion is exponentially stable with a decay rate of α if a sufficient LMI criterion related to the double sum term is fulfilled:

$$\sum_{i=1}^{N_r}\sum_{j=1}^{N_r} h_i(\mathbf{z})h_j(\mathbf{z}) \left(\left(\tilde{\mathbf{A}}_i - \tilde{\mathbf{b}}_i \tilde{\mathbf{k}}_j^T \right)^T \mathbf{P} + \mathbf{P} \left(\tilde{\mathbf{A}}_i - \tilde{\mathbf{b}}_i \tilde{\mathbf{k}}_j^T \right) \right) \prec -2\alpha \mathbf{P}, \tag{4.65}$$

for $\mathbf{P} \succ 0$ [Tanaka and Wang, 2001]. Thereby (4.65) is fulfilled if a common matrix $\mathbf{P}$ exists such that the LMI problem

$$\tilde{\mathbf{A}}_i^T \mathbf{P} + \mathbf{P}\tilde{\mathbf{A}}_i - \tilde{\mathbf{k}}_j \tilde{\mathbf{b}}_i^T \mathbf{P} - \mathbf{P}\tilde{\mathbf{b}}_i \tilde{\mathbf{k}}_j^T + 2\alpha \mathbf{P} \prec 0, \qquad \mathbf{P} \succ 0 \tag{4.66}$$

holds for all $i, j = 1, \dots, N_r$. This results in

$$\begin{aligned}
\dot{V}(\tilde{\mathbf{x}}) &\le -2\alpha \tilde{\mathbf{x}}^T \mathbf{P} \tilde{\mathbf{x}} + 2\tilde{\mathbf{x}}^T \mathbf{P} \max_{ij} \mathbf{a}_{ij} + 2\tilde{\mathbf{x}}^T \mathbf{P} \mathbf{e}_3 y_{ref} \\
&\le -2\alpha \lambda_{\min}(\mathbf{P}) \| \mathbf{x} \|^2 + 2 \| \mathbf{x} \| \| \mathbf{P} \max_{ij} \mathbf{a}_{ij} \| + 2 \| \mathbf{x} \| \| \mathbf{P} \mathbf{e}_3 \| \; |y_{ref}| \\
&\le -\lambda \| \mathbf{x} \|^2 + 2 \| \mathbf{x} \| \| \mathbf{P} \| \| \max_{ij} \mathbf{a}_{ij} \| + 2 \| \mathbf{x} \| \| \mathbf{P} \| \; |y_{ref}| \\
&\le -\lambda \| \mathbf{x} \|^2 + 2 \| \mathbf{x} \| \| \mathbf{P} \| \left(\frac{\| \max_{ij} \mathbf{a}_{ij} \|}{|y_{ref}|_{\min}} + 1 \right) |y_{ref}| \\
&= -\lambda \| \mathbf{x} \|^2 + 2\delta |y_{ref}| \, \| \mathbf{x} \| \le -\frac{\lambda}{2} \| \mathbf{x} \|^2 + 2\frac{\delta^2}{\lambda} |y_{ref}|^2
\end{aligned}$$

with $\lambda := 2\alpha\lambda_{\min}(\mathbf{P})$, $\delta := \| \mathbf{P} \| \left(\frac{\|\max_{ij} \mathbf{a}_{ij}\|}{|y_{ref}|_{\min}} + 1 \right)$ and $|y_{ref}|_{\min} \le |y_{ref}| \le |y_{ref}|_{\max}$, where $\lambda_{\min}(\mathbf{P})$ denotes the smallest eigenvalue of $\mathbf{P}$. Finally with $\gamma_1(\| \tilde{\mathbf{x}} \|) := \frac{\lambda}{2} \| \mathbf{x} \|^2$ and $\gamma_2(|y_{ref}|) := 2\frac{\delta^2}{\lambda} |y_{ref}|^2$ the TS closed-loop system (4.61) fulfills the ISS property related to Sontag and Wang [1995]. The above analysis shows that the eigenmotion of (4.61) without the reference value ($y_{ref} = 0$) and affine vector ($\mathbf{a}_{ij} = \mathbf{0}$) must be exponentially stable so that the entire system with reference values and affine components is ISS. The proof of exponential stability results from a feasible solution of (4.66).

Global Controller Synthesis

As previously noted, the PDC controller (4.58) with setpoint specification is now designed. We call this approach "global controller synthesis" since the $\tilde{\mathbf{k}}_i$ coefficients are directly determined via a global closed-loop stability criterion based on a common Lyapunov function using LMI constraints from Theorem B.6. The controller design is to be based on the previously derived and validated TS model (see Figure 4.16) with $N_r = 105$ local sub-models proposed in Section 4.4.1. The result of the design represented by the comparison of open-loop and calculated

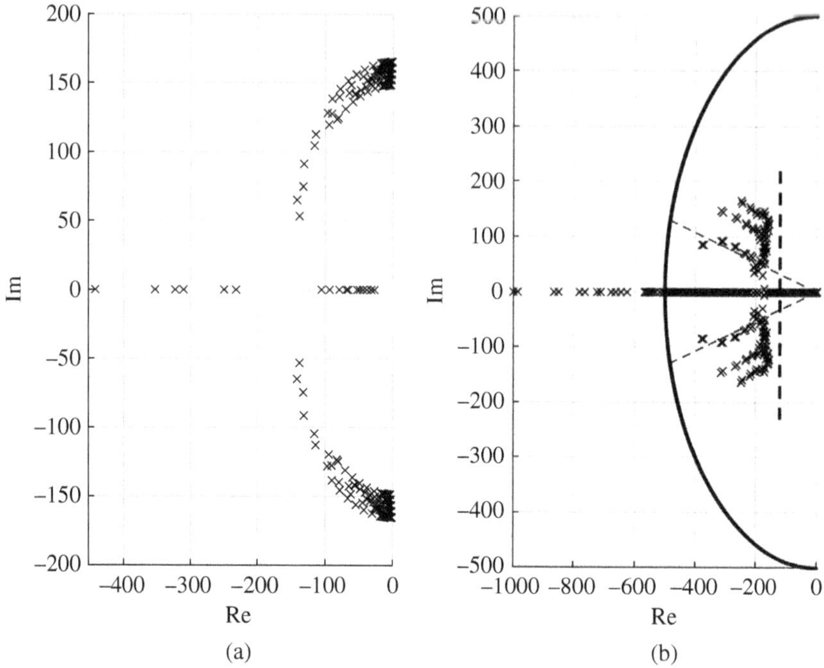

Figure 4.18 Pole locations in the complex plane of the open-loop system (a) vs. closed-loop system (b) using LMI-based global feedback synthesis with $N_r = 105$ sub-models.

closed-loop pole location is shown in Figure 4.18. For the design, the following region constraints were specified

$$\alpha = 120, \qquad r = 300, \qquad \theta = 15^\circ, \tag{4.67}$$

which should guarantee a fast decay of the eigenmotion with good damping in the overall working range of the PV Generator DC–DC system. However, *no feasible set is found for that large number of local models*, which is also evident from the pole distribution in Figure 4.18 of the closed-loop system (4.61).

It is clearly seen that the boundaries of desired pole region are not fulfilled. Some eigenvalues even lie in the right half plane. Therefore, this approach is not suitable for the given plant model and the closed-loop requirements. For this reason, we now use a two-step approach that has already been successfully in Schulte and Hahn [2004] and Pöschke et al. [2022].

Local Synthesis and Global Stability Analysis

In the two-step approach for TS controller synthesis used now, the associated sub-controller is first designed for each ith sub-model. That means the $\tilde{\mathbf{k}}_i$ controller gains in (4.58) are determined individually. In the second step, whether the whole system is stable is checked. For this purpose, a single quadratic Lyapunov function (LF) candidate of the closed loop (with the previously computed $\tilde{\mathbf{k}}_i$ coefficient) can be given, and an LMI-based feasibility criterion can be applied so that the finding of an LF is reduced to the optimization problem of finding a common positive $\mathbf{P}$ matrix for the constraint (B.11), see Theorem B.1, Appendix B.

In the first step, the LMI conditions are given related to the pole-region specification $\mathcal{S}(\alpha, r, \theta)$ for *local controller design*. The closed-loop system with the eigenvalues $\lambda_i = \text{eig}\left(\tilde{\mathbf{A}}_i - \tilde{\mathbf{b}}_i\,\tilde{\mathbf{k}}_i^T\right)$ with (4.58) and (4.59) are all located in the given pole region $\mathcal{S}(\alpha, r, \theta)$ (illustrated in Figure B.3) if there exist $\mathbf{X}_i = \mathbf{X}_i^T \succ 0$ and $\mathbf{m}_i^T := \tilde{\mathbf{k}}_i^T\mathbf{X}_i$, $i = 1, 2, \dots, N_r$ such that the LMI problem

$$
\begin{aligned}
\boldsymbol{\Gamma}_i^1 &= \tilde{\mathbf{A}}_i\mathbf{X}_i + \mathbf{X}_i\tilde{\mathbf{A}}_i^T - \tilde{\mathbf{b}}_i\,\mathbf{m}_i^T - \mathbf{m}_i\tilde{\mathbf{b}}_i^T + 2\alpha\mathbf{X}_i,\\
\boldsymbol{\Gamma}_i^2 &= \begin{pmatrix} (\tilde{\mathbf{A}}_i\mathbf{X}_i + \mathbf{X}_i\tilde{\mathbf{A}}_i^T - \tilde{\mathbf{b}}_i\mathbf{m}_i^T - \mathbf{m}_i\tilde{\mathbf{b}}_i^T)\sin\theta & (\tilde{\mathbf{A}}_i\mathbf{X}_i - \mathbf{X}_i\tilde{\mathbf{A}}_i^T - \tilde{\mathbf{b}}_i\mathbf{m}_i^T + \mathbf{m}_i\tilde{\mathbf{b}}_i^T)\cos\theta \\ (\mathbf{X}_i\tilde{\mathbf{A}}_i^T - \tilde{\mathbf{A}}_i\mathbf{X}_i - \mathbf{m}_i\tilde{\mathbf{b}}_i^T + \tilde{\mathbf{b}}_i\mathbf{m}_i^T)\cos\theta & (\tilde{\mathbf{A}}_i\mathbf{X}_i + \mathbf{X}_i\tilde{\mathbf{A}}_i^T - \tilde{\mathbf{b}}_i\mathbf{m}_i^T - \mathbf{m}_i\tilde{\mathbf{b}}_i^T)\sin\theta \end{pmatrix},\\
\boldsymbol{\Gamma}_i^3 &= \begin{pmatrix} -r\mathbf{X}_i & \tilde{\mathbf{A}}_i\mathbf{X}_i - \tilde{\mathbf{b}}_i\,\mathbf{m}_i^T \\ \mathbf{X}_i\tilde{\mathbf{A}}_i^T - \mathbf{m}_i\tilde{\mathbf{b}}_i^T & -r\mathbf{X}_i \end{pmatrix}
\end{aligned}
\tag{4.68}
$$

with

$$\boldsymbol{\Gamma}_i^k(\mathbf{X}_i, \mathbf{m}_i) \prec 0, \quad k = 1, 2, 3 \quad \text{for all} \quad i = 1, 2, \dots, N_r$$

is feasible. Based on the definition of $\mathbf{m}_i$, the control coefficient of the voltage controller (4.54) is finally calculated according to

$$\tilde{\mathbf{k}}_i^T = \mathbf{m}_i^T\,\mathbf{X}_i^{-1}, \quad i = 1, 2, \dots, N_r. \tag{4.69}$$

For sufficient damping, short rise time, and the bounded controller output of $0 \leq d \leq 1$ the closed-loop pole region

$$\alpha = 120, \qquad r = 150, \qquad \theta = 15^\circ \tag{4.70}$$

is chosen. For the calculation of the convex LMI constrained problem (4.68), as in Chapter 3, the LMI terms (4.68) are implemented with the help of the toolbox proposed in Löfberg [2004] and the "semidefinite programming solver" (SEDUMI). Results for the desired pole region and related model parameters of Table D.4 are illustrated in Figure 4.19b by the closed-loop eigenvalue distribution $\text{eig}\left(\tilde{\mathbf{A}}_i - \tilde{\mathbf{b}}_i\tilde{\mathbf{k}}_j^T\right)$ for all i,j. It can be clearly seen that all eigenvalues of the closed-loop systems of the plant model/controller combinations are in the desired region. Furthermore, compared with the open-loop pole distribution of Figure 4.19a , one can see the effect of state feedback. Note that in order to avoid imposing a higher natural frequency on the system, which increases the control effort, the radius $r = 150$ given in (4.70) of the closed-loop pole region corresponds to the position of the poles of the open-loop system, see Figure 4.19.

Finally, the ISS of the closed-loop TS system (4.61) is verified with the previously calculated coefficients $\tilde{\mathbf{k}}_i$ by computing a joint $\mathbf{P}$ matrix with (4.66) for the working

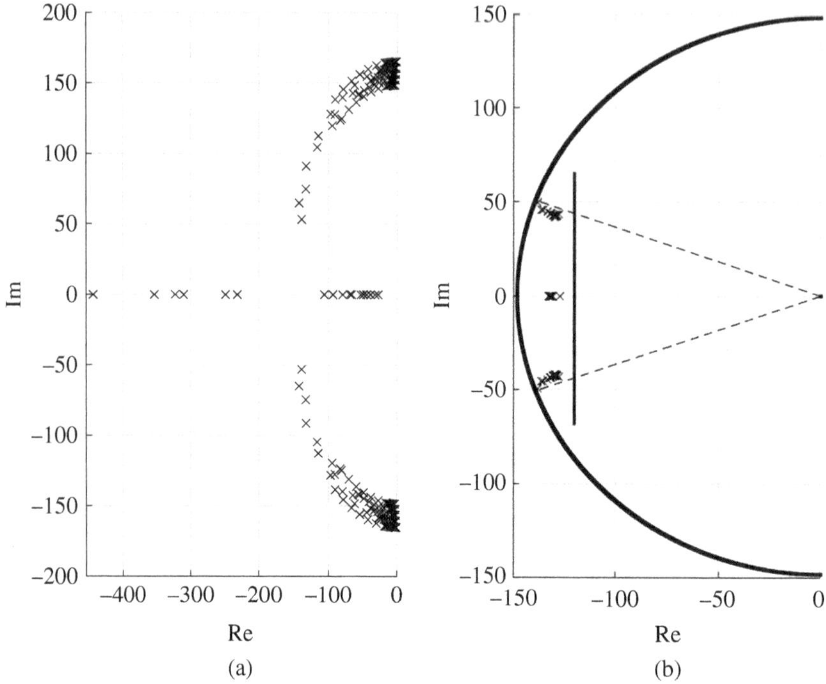

Figure 4.19 Pole locations in the complex plane of the open-loop system (a) vs. closed-loop system (b) using LMI-based local feedback design with $N_r = 105$ sub-models.

range of interest (4.53) with the restriction that only temperature and irradiation fluctuations between two neighboring grid points have been assumed.

Simulation

After the design, the control law (4.58) including the calculated $\tilde{\mathbf{k}}_j$ coefficients for all $j = 1, \ldots, 105$ is now validated by simulation. The set up corresponds to the control scheme in Figure 4.17. To investigate the large-signal behavior, the entire possible reference voltage range is covered via step changes (see the Figure 4.20a and 4.21b) and by the change of irradiation S and cell temperature T_c (see Figure 4.20c and 4.21c). To demonstrate the advantages of model-based controller design in the TS framework using the control law (4.54), a simulation with a linear control law is also presented. The linear controller results from the "freezing of weighting functions" in the TS controller (4.54). In this case only state feedback of $\tilde{\mathbf{k}}_{70}$ is used, i.e. $h_{70} = 1$, where all others h_i with $i \neq 70$ are zero. The index $i = 70$ is chosen because the simulation starts at the operating point $i = 70$. After $t = 0.5$ seconds this working point is left, which leads to other models becoming valid as well, and thus other $\tilde{\mathbf{k}}_i$ enter the TS control law through h_i unequal weights ($h_i > 0$), see diagram

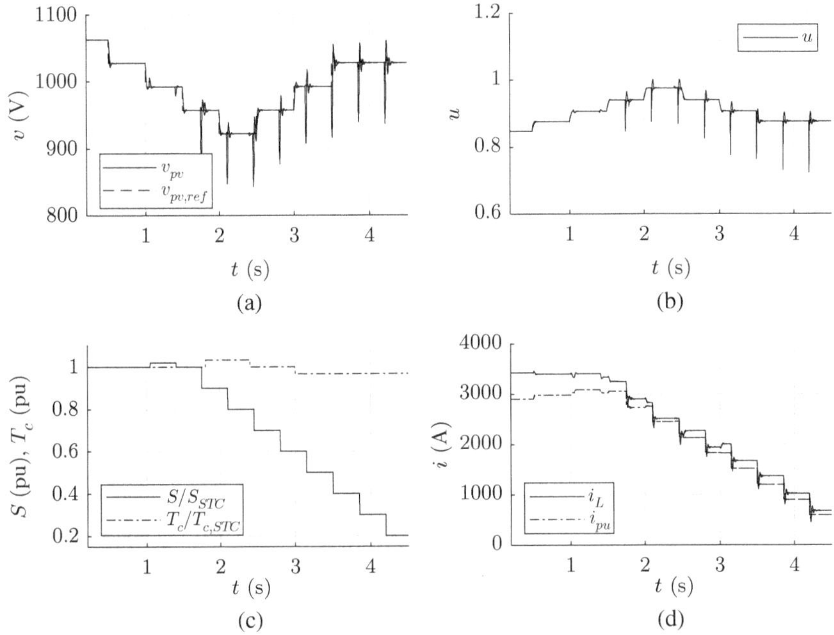

Figure 4.20 Validation of the linear controller designed with one model $i = 70$ where $u = d$ as duty cycle.

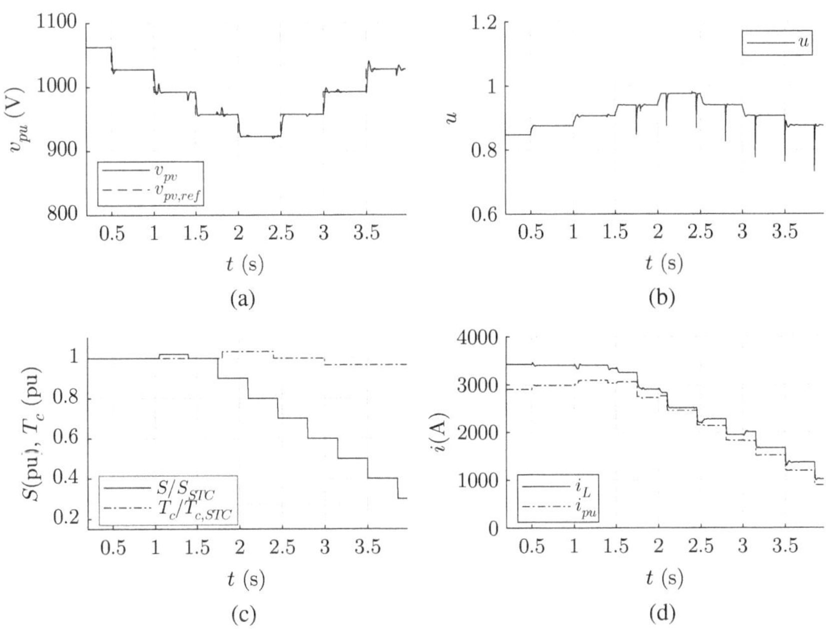

Figure 4.21 Validation of the Takagi–Sugeno PDC controller designed with $i = 1, \ldots, 105$ models based on the control law (4.58).

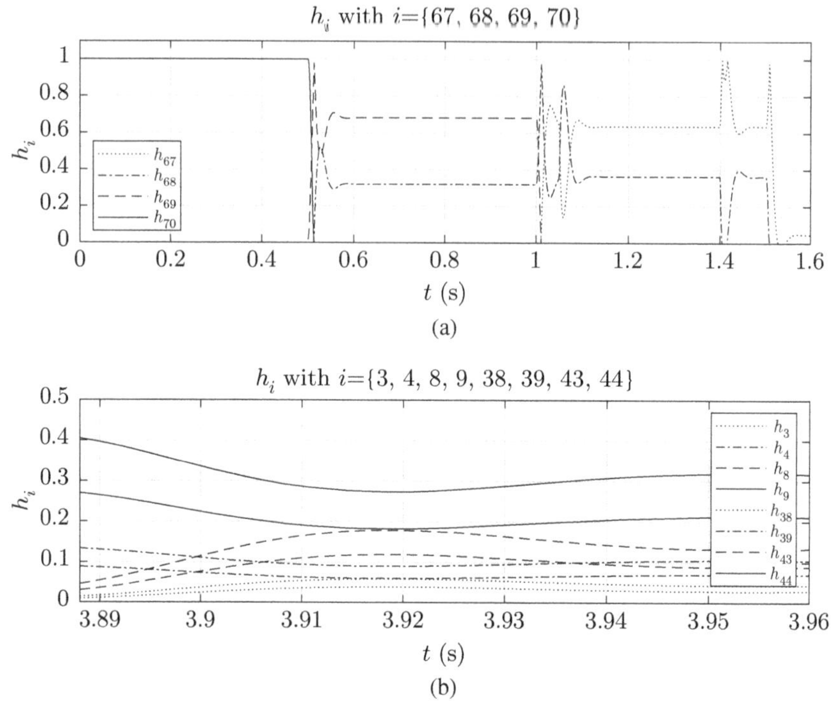

Figure 4.22 Variation of the membership function values h_i of the controller law (4.58) in time period $t \in$ [0, 1.6 seconds] and $t \in$ [3.89 seconds, 3.96 seconds] related to simulation shown in Figure 4.21.

of Figure 4.22a. By direct comparison of the regulated variable v_{pv} in Figure 4.21 with Figure 4.20 it is clearly seen that the TS controller significantly outperforms the linear one in its control performance. Here, the PV voltage v_{pv} follows the reference value well at all irradiation and temperature changes. On the other hand using the linear controller, large overshoots occur. In order to better understand the inner mechanism of the $\tilde{\mathbf{k}}_i$ weighting by h_i, which cause time-varying state feedback and integral gain, an illustration of the h_i curves unequal zero is shown in Figure 4.22b.

For a detailed comparison, cutouts from Figures 4.20 and 4.21 are shown in Figure 4.23. It is evident that the performance of the linear controller significantly degrades as the distance to the design region of the linear controller increases. In Figure 4.23a, the performance of the linear controller is almost identical to that of the TS controller, where the command responses (reference step at $t = 1$ second) and disturbance rejection (S step at $t = 1.05$ seconds) are sufficient for both the linear and TS controller. This can be explained simply by the fact that the linear

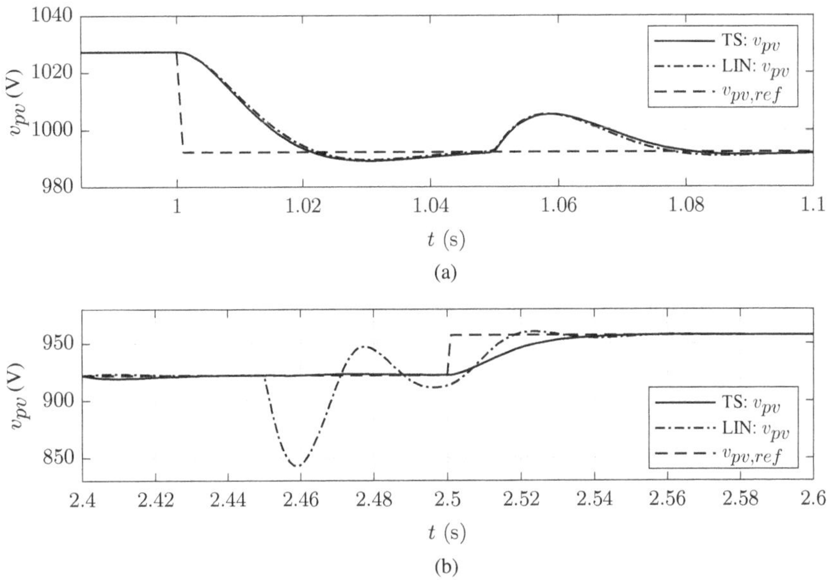

Figure 4.23 Comparison between the Takagi–Sugeno PDC controller (TS) and a linear controller with the gain $\tilde{\mathbf{k}}_{70}$.

regulator operates during the period of $t \in [0.9, 1.1]$ seconds in a range for which it was designed. However, in Figure 4.23b, the performance of the linear regulator is insufficient. The disturbance, caused by an S step at $t = 2.45$ seconds (see Figure 4.20), cannot be rejected by the linear controller, but it can be by the TS controller.

4.5 Demanded Power Tracking

So far, the studies have focused on controlling the voltage of PV generators. In preparation for demanded power point-tracking functionality, an operating range has been considered for which a model-based TS controller is proposed using the buck converter as a control element. The design approach can also be applied to the other types like boost and buck–boost converters, whose generic state-space average models has been specified in Section 4.3.2.

For the purpose of DC power control, a demanded power-tracking (DPT) or simply power-tracking[6] control scheme is presented as an extension of the MPPT. The conventional control objective is when after leaving the point of maximum

6 This is the term from Section 3.4.4 that denotes the same functionality in wind turbines.

power, e.g. due to a sudden module shading or temperature change, the point of maximum power is reached again by MPPT. In contrast, the DPT algorithm determines the reference voltage to provide the requested, not the maximum, power. Of course, only if the requested power is lower than the maximum available power. Indeed, this is already known from the power tracking of wind turbines; see Section 3.4.4. Otherwise, storage elements must be installed to provide additional power.

In this section, the general controller structure is first described, then the DPT control scheme is introduced, including the previously developed voltage controller as a low-level function. Therefore, two different mechanisms for power tracking with and without prior model knowledge of the P–V curve are presented.

4.5.1 Overall Control Structure

Overall structure of the control loops is illustrated in Figure 4.24 where $\Delta P_{dc,ref}$ denotes the desired power change and P_{dc} the measured power related to the DC bus. The notations S_m and $T_{c,m}$ are introduced to distinguish between the measured (index m) and physical values. The proposed method assumes that the actual values are known; therefore, $S_m \approx S$, $T_{c,m} \approx T_c$ is valid. Figure 4.24 clearly shows the inner control loop with the state feedback of v_{pv} and i_L for controlling the PV voltage by the duty cycle d of the DC–DC converter. The outer loop with the power-tracking block provides the set point $v_{pv,ref}$ for the lower-level voltage control. If the reference is set to $\Delta P_{dc,ref} = 0$, the controller function follows the MPPT algorithm. However, when $\Delta P_{dc,ref} \neq 0$ is set, the optimal operating point is left and the function denoted as demand power tracking (DPT) in the following is activated. For the development of a DPT strategy, a distinction is made between a model-based approach and a model-free approach. For this purpose, a control scheme for each of the two contrary approaches is presented. Before proceeding,

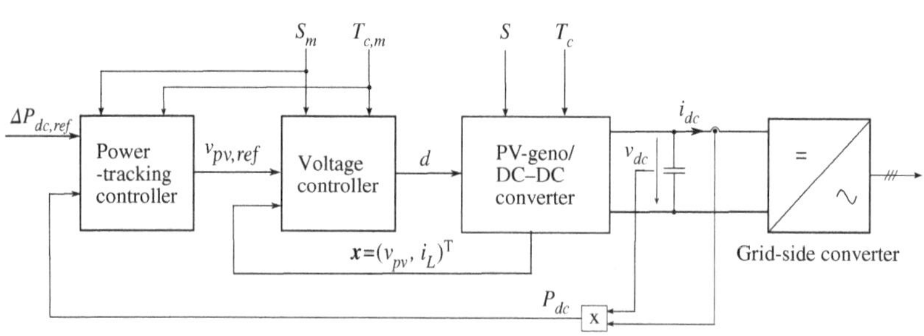

Figure 4.24 PV power plant controller of the primary conversion with inner voltage and outer power-tracking controller.

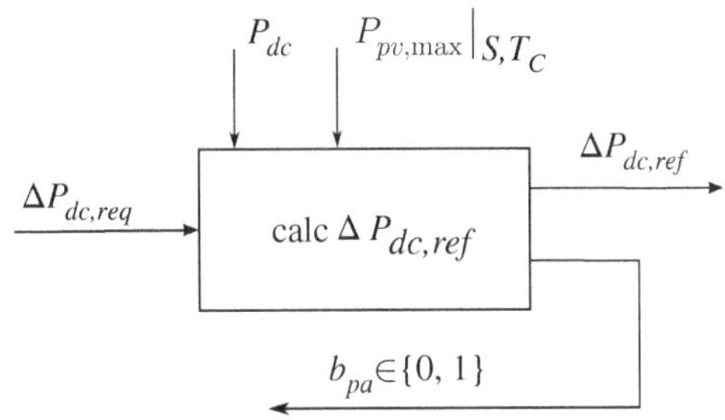

Figure 4.25 Calculation of the difference reference power $\Delta P_{dc,ref}$ as a function of the difference requested power $\Delta P_{dc,req}$. A confirmation bit b_{pa} is returned to the higher-level grid controller.

we will first discuss the calculation of the reference value as an input to the DPT strategy.

Usually, higher-level grid controllers request a differential power respectively power change denoted as $\Delta P_{dc,req}$, so it must be checked locally at the power plant level whether such a change can be supplied instantly.[7] Analogous to the wind turbine control in (3.85), the following distinction is made:

$$\Delta P_{dc,ref} = \begin{cases} \Delta P_{dc,req}, & P_{dc,av} + \Delta P_{dc,req} \leq P_{pv,max}\big|_{S,T_c} \\ P_{pv,max}\big|_{S,T_c} - P_{dc,av}, & P_{dc,av} + \Delta P_{dc,req} > P_{pv,max}\big|_{S,T_c} \end{cases} \tag{4.71}$$

with

$$P_{dc,av}(t) = \frac{1}{N} \sum_{i=0}^{N-1} P_{dc}(t - i\,T_s), \tag{4.72}$$

where T_s denotes a constant sampling time and $P_{pv,max}\big|_{T_c,S}$ is the maximum achievable power related to the cell temperature and irradiation. The corresponding block diagram for the integration of (4.71) into control scheme in Figure 4.24 is illustrated in Figure 4.25.

4.5.2 Model-Based Power Tracking

A model-based approach to power tracking is applicable if the PV generator model (4.9) describes sufficiently well the primary conversion from the irradiation S(×irradiated area) to i_{pv}(×v_{pv}). Under this assumption, a model-based feedforward respectively open-loop control scheme is presented for the classical MPPT and power-tracking task by exploiting the PV generator's generic current source model (4.9).

7 How to quantify "instantly" is subject of this study and will be examined in closer detail in Chapter 6, taking into account the most important influencing factors.

Therefore, let us start with the MPPT scheme. To reach the MPP, a voltage denoted as $v_{pv,opt}$ has to be set by the DC–DC converter that satisfies

$$\left.\frac{\partial P_{pv}(v_{pv})}{\partial v_{pv}}\right|_{S,T_c} \overset{!}{=} 0 \quad \rightarrow \quad v_{pv,opt}(S, Tc), \tag{4.73}$$

where $P_{pv} = i_{pv}(v_{pv})\, v_{pv}$ with (4.9) so that

$$\left.P_{pv,max}\right|_{S,T_c} = P_{pv}(v_{pv,opt}, S, T_c). \tag{4.74}$$

This approach is based on the circumstance that the P–V curve, as shown in Figures 4.7 and 4.8, has only one extreme value each for S and T_c, respectively. The derivation in (4.73) is calculated as follows

$$\frac{\partial P_{pv}(v_{pv})}{\partial v_{pv}} = \frac{\partial i_{pv}}{\partial v_{pv}} v_{pv} + i_{pv}.$$

Hence, after a short calculation using (4.9) we get

$$\left.\frac{\partial P_{pv}(v_{pv})}{\partial v_{pv}}\right|_{S,T_c} = N_p\, i_{ph} - N_p\, i_s \left[e^{\frac{v_{pv}}{N_s A_n v_{T,STC}}} \left(\frac{v_{pv}}{N_s A_n v_{T,STC}} + 1 \right) - 1 \right] - 2 v_{pv} \frac{N_p}{N_s R_h} \overset{!}{=} 0. \tag{4.75}$$

The solution of the implicit function (4.75) over v_{pv} for the instantaneous values S and T_c (measured) denoted as $v_{pv,opt}(S, T_c)$ is calculated numerically by successive combination of interpolation methods according to Dekker [1969] implemented in the Matlab function `fzero`. With the online calculation of $v_{pv,opt}(S, T_c)$, a solution is found after a few iterations, the reference value $v_{pv,ref}$ can be passed to the voltage controller resulting in a fast regulation of the maximum power, see Figure 4.24. The performance of this approach will be shown later by simulation results.

We now extend the MPPT function to add power-tracking capability. For this purpose, the intersection of the P–V curve and the reference power $P_{pv,ref}$, where

$$P_{pv,ref} = \Delta P_{dc,ref} + P_{dc,av} \tag{4.76}$$

applies with the use of (4.71) and (4.72). Note that if the second case in (4.71) is valid, in other words, the desired power cannot be delivered completely, only up to the maximum available power it corresponds to the MPPT functionality. However, let's consider the first case in (4.71), where the requested is less than the maximum achievable power. For clarification, this circumstance is shown in Figure 4.26. It can be seen that to generate the requested power by the PV generator with the assumption that S and T_c do not change, two different voltages $v_{pv,I}$ and $v_{pv,II}$ with the relation $v_{pv,I} \leq v_{pv,opt} \leq v_{pv,II}$ can be set. To control the power in the linear range

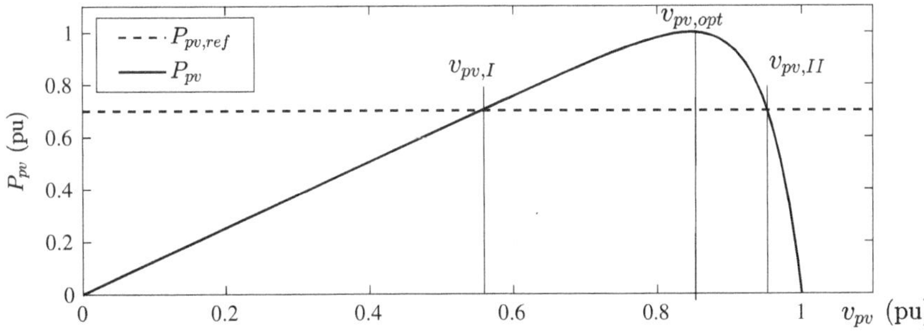

Figure 4.26 Intersections of the generic power curve with constant reference power.

with almost constant current i_{pv} (see the I–V curve in Figures 4.7 and 4.8), the voltage value left from the maximum, see Figure 4.26, with

$$v_{pv,ref} = v_{pv,DPT} := v_{pv,I} \tag{4.77}$$

should be used as the reference value for the underlying voltage controller. This value results for a given $P_{pv,ref}$, as in the MPPT case (4.73), from calculating a zero of the following function of v_{pv}

$$P_{pv,ref} - P_{pv}(v_{pv})\Big|_{S,T_c} \overset{!}{=} 0 \quad \text{with} \quad 0 \leq v_{pv} \leq v_{pv,opt} \quad \rightarrow \quad v_{pv,DPT}(S, Tc). \tag{4.78}$$

Since only solutions to the left of maximum power $P_{pv}(v_{pv,opt})$ are determined, the solution space is limited. Figure 4.27 shows the necessary calculation steps as a block diagram to illustrate the DPT scheme using (4.73), (4.75), and (4.78). It can be seen that the scheme also determines the value $P_{pv,max}\Big|_{S,T_c}$ for testing with (4.71), i.e. whether the requested power value is available from the generating plant.

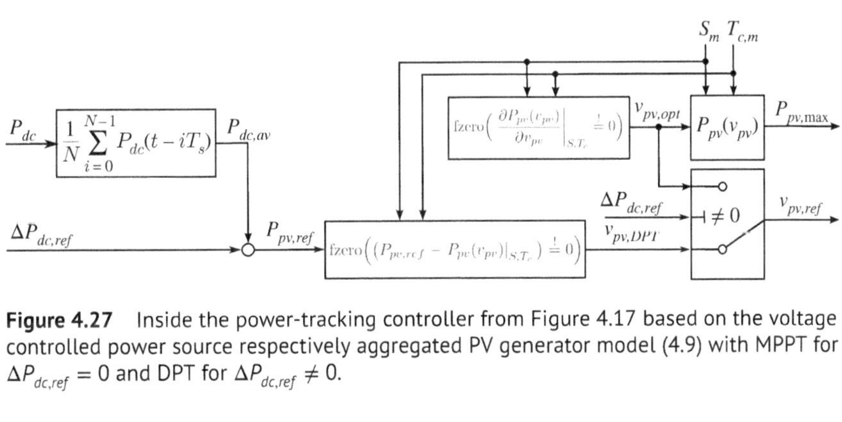

Figure 4.27 Inside the power-tracking controller from Figure 4.17 based on the voltage controlled power source respectively aggregated PV generator model (4.9) with MPPT for $\Delta P_{dc,ref} = 0$ and DPT for $\Delta P_{dc,ref} \neq 0$.

4.5.2.1 Performance and Robustness Analysis of the Cascaded Controller

Finally, simulation results are presented combining the proposed DPT scheme (outer-loop) with the inner-loop voltage control from Section 4.4.2. To investigate the transient dynamics and robust performance of the power-tracking control, various simulations were performed. Figure 4.28 shows an example of power demand with a staircase-shaped decrease and a step-shaped increase. In Figure 4.28a we compare the produced PV generator power to the demanded reference power. Here, the non-minimum phase behavior is clearly visible. This is due to the fact that when v_{pv} is reduced by the inner-loop controller, the current through the induction $i_L = i_{pv}$ in the considered buck converter initially increases for a short time and thus the power in the DC link increases as well. However, the effect disappears after a short time by a further drop in PV voltage, so that the real power follows the required power after about 100 ms. The dynamics of fast power-tracking are determined by the dynamics of the voltage regulation and current rate limitation by the converter inductance.

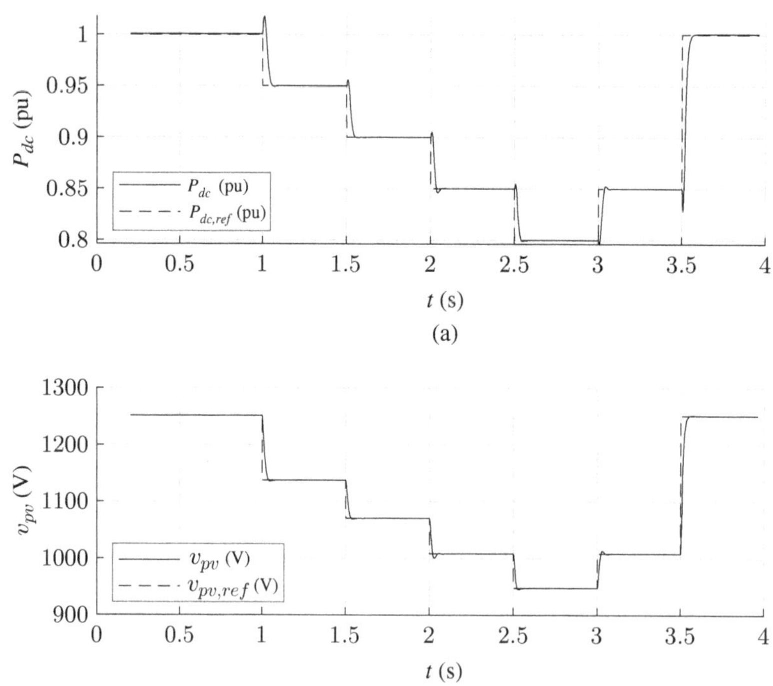

Figure 4.28 Power-tracking performance with decreasing and increasing power demand at constant irradiation for $S_m = S$ and cell temperature $T_{c,m} = T_c$.

In Section 4.5.1 we have assumed that the actual irradiation and the cell temperature are known so that $S_m = S$ and $T_{c,m} = T_c$ are satisfied. For the robustness analyses of control scheme, we now consider a mismatch between the measured (m-index) and actual (no index) values with the variations:

$$T_{c,m} = T_c + \Delta T_{c,m}, \qquad \Delta T_{c,m} \in \{-5, -10, -15\}. \tag{4.79}$$

$$S_m = S + \Delta S_m, \qquad \Delta S_m \in \{5, 10, 15, 20, 25\}. \tag{4.80}$$

The effect of measurement uncertainties in cell temperature during power tracking is shown in Figure 4.29. It can be seen that even with a relatively high-temperature deviation of $\Delta T_c = 15°$, the deviation in the adjusted power is relatively small. A larger influence on the power-tracking performance comes from the uncertainties in the irradiation measurements shown in Figure 4.30. However, the question of the requirement can only be defined for a specific power plant requirement. An additional external control loop with an integral component could achieve accuracy.

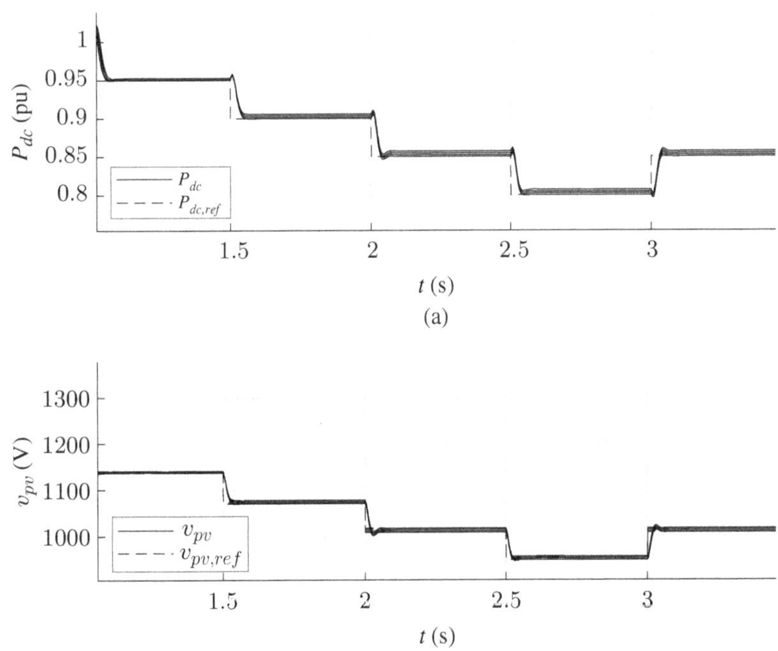

(a)

(b)

Figure 4.29 Power-tracking performance with mismatch (4.79) between the measured $T_{c,m}$ and the actual cell temperature T_c.

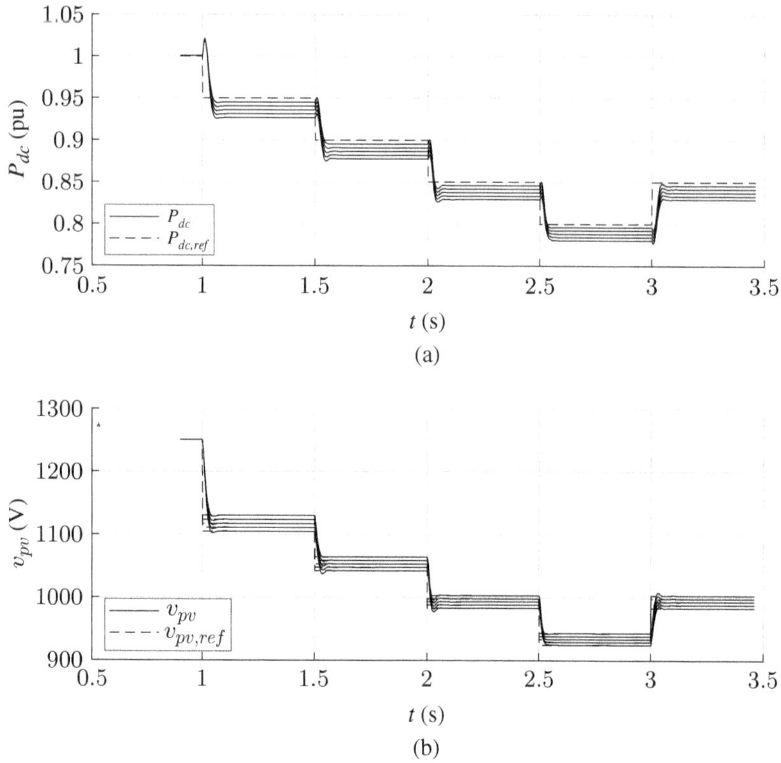

Figure 4.30 Power-tracking performance with mismatch (4.80) between the measured S_m and the actual irradiation S.

4.5.3 Model-Free Power Tracking

Opposite to the previous model-based power tracking for PV systems with the assumption that the P–V diagram is available, a model-free approach, the extremum-seeking control (ESC) [Benosman, 2016] is introduced. In addition to these fundamentally different paradigms of model-free and model-based, other methods related to MPPT, but not for DPT, have been proposed in recent years that can be placed methodologically between them. Procedures with a data-based online estimation of a polynomial P–V model using recursive least squares methods proposed in Xiao et al. [2006] are worth mentioning. Alternatively, other rather model-free methods like the hill-climbing (HC) method, commonly referred to as the perturbation and observation (P&O) method [Xiao, 2017] frequently discussed in the literature, are used in practice.

HC-based methods continuously perturb the system, waiting until a steady state is reached where the new set power is compared to the previous power.

By comparing, it is determined whether the reference voltage of the internal controller (see Figure 4.24) must be further increased or decreased. In contrast to the HC algorithm, the ESC used here excites the system with a sinusoidal signal. In general ESC as an online optimization method can be applied to systems where the objective optimization function has a local optimum. This is fulfilled when finding the MPPT and can be used to a limited extent in the DPT optimization problem if the permissible solution space bounded and left from the MPP optimum as illustrated in Figure 4.26.

The first concept of ESC can be found in Leblanc [1922], while the algorithm was described in detail for the first time in the 1950s [Draper and Li, 1951]. Due to the stability proof, the ESC experienced a renaissance at the beginning of this century [Krstić and Wang, 2000]. As a result, the method found the appropriate reputation in the automatic control community. With application to MPPT in PV power plants [Brunton et al., 2010] and microwind turbines [Dietrich et al., 2021].

For a concrete study of the application in PV power plants, let us start with the control scheme for ESC shown in Figure 4.31. It contains all the necessary elements for an adaptive ESC, like the generation of a sinusoidal perturbation signal

$$u_p(t) = a_p \sin(\omega_p t), \tag{4.81}$$

the high-pass filter HP[·]

$$G_h(s) = \frac{s}{s + \omega_h}, \tag{4.82}$$

and the multiplicative of the high-pass filtered signal with the perturbation signal (4.81). The integration of the result and the subsequent superposition (addition) with the disturbance signal (4.81) result in the input of the plant

$$v_{pv,ref} = a_p \sin(\omega_p t) + \int_0^t \mathrm{HP}[P^*_{dc}] \cdot \sin(\omega_p t) d\tau. \tag{4.83}$$

In this application, it is the reference voltage for the inner voltage control loop; see Figure 4.17. If no power change is requested from a higher-level system ($\Delta P_{dc,ref} = 0$), the ESC operates in MPPT mode. This means that the switch in Figure 4.31 is closed in the upper position; otherwise for $\Delta P_{dc,ref} \neq 0$, as shown, the lower position is closed, and instead of the measured power, a transformed power value is used in the DPT mode. The determination of P^*_{dc} is thus as follows

$$P^*_{dc} = \begin{cases} P_{dc}, & \Delta P_{dc,ref} = 0, \\ \frac{P^2_{dc,ref} - (P_{dc,ref} - P_{dc})^2}{P^2_{dc,ref}}, & \Delta P_{dc,ref} \neq 0. \end{cases} \tag{4.84}$$

The result of calculating the steady-state P^*–V curve for the DPT case is shown in Figures 4.32b and 4.33b. It is obvious that the function (4.84) for $\Delta P_{dc,ref} \neq 0$,

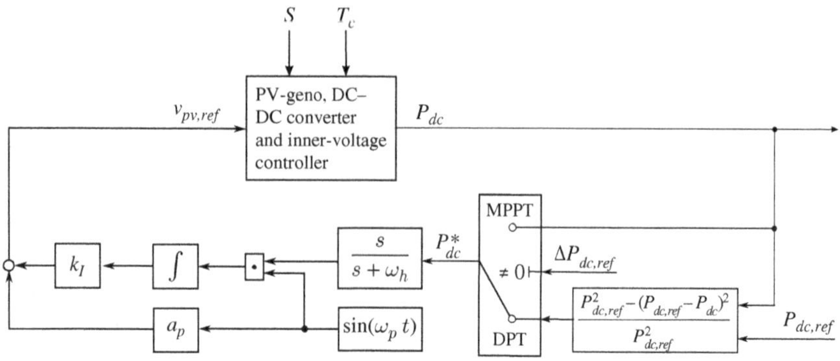

Figure 4.31 Switchable ESC for MPPT and DPT (demanded power tracking).

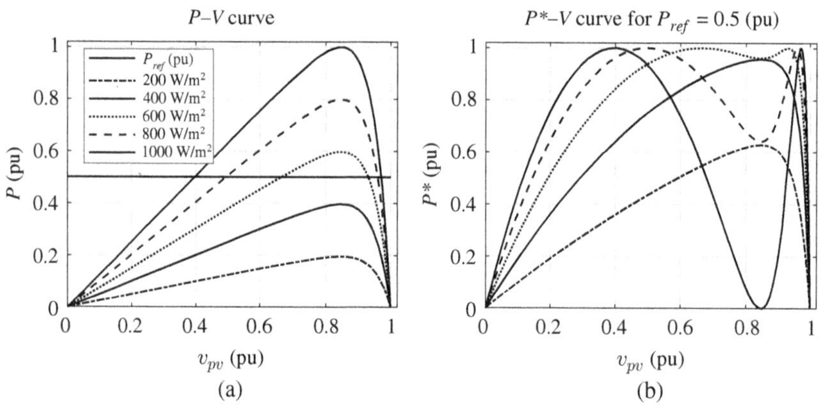

Figure 4.32 P–V (a) vs. transformed P^*-V (b) curve for model-free DPT.

reaches maximum values at $\Delta P_{dc,ref} = P_{dc}$. In the classic MPPT mode, however, shown in Figures 4.32a and 4.33a, the MPP forms the maximum value as usual.

The dynamic behavior of the ESC is determined by four design parameters. The cut-off frequency of the high-pass filter denoted as ω_h, the frequency and amplitude of perturbation signal (4.81) denoted as ω_p respectively a_p, and the integrator gain denoted as k_I. Note that the frequency ω_p is an order of magnitude above the cut-off frequency of the high-pass filter. The upper limit of ω_p is determined by the cut-off frequency of the controlled system denoted as ω_c. To ensure that the controlled system, in this case the system with the lower voltage control loop, follows the reference value well, the perturbation frequency must be three to four times smaller than ω_c. Further, the perturbation amplitude a_p in (4.81) should be as small as possible. It must excite the system sufficiently to reach the maximum point quickly, but may not create fatigue loads in the system due

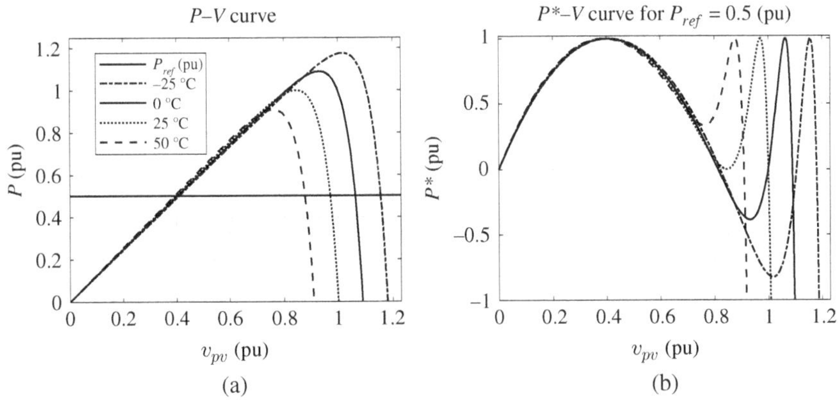

Figure 4.33 P–V (a) vs. transformed P^*-V (b) curve for model-free DPT.

to the superimposed frequencies at the input. In summary, the design parameter setup criteria are as follows

$$\begin{aligned} &\omega_h \ll \omega_p < \omega_c, \\ &a_p \approx 0.001, \dots, 0.01\, u_{\max}, \end{aligned} \tag{4.85}$$

where u_{abs} denotes the maximum permissible input, i.e. the value that does not lead to input saturation.

After this preliminary work, the entire system will be examined based on the previous example. For an appropriate parameter setting of the ESC, the control loop dynamics of the PV power plant from Section 4.4.2 will first be converted into a suitable form. The distribution of the closed-loop poles caused by the LMI-based design of the state controller shown in Figure 4.18, together with the output equation, determines the dynamics of the controlled system. Assuming that the PV generator is operated in the linear range, i.e. to the left of the power optimum of the P–V curve, and that the current is therefore not dependent on the output voltage and constant for a given S and T_c with the associated value $i_{pv,c}(S,T_c)$, the output equation is linear in the states. In a good approximation, the pole distribution in Figure 4.19b (closed loop), can be aggregated by three poles with

$$s_{p,1} = -120, \qquad s_{p,2} = -120 + j45, \qquad s_{p,3} = -120 - j45. \tag{4.86}$$

with which the plant dynamics containing the PV generator, DC–DC converter, and inner-voltage control shown in Figure 4.31 can be described by the third-order linear time invariant (LTI) system

$$G_p(s) = i_{pv,c}(S, T_c)\frac{s_{p,1}\, s_{p,2}\, s_{p,3}}{(s - s_{p,1})(s - s_{p,2})(s - s_{p,3})} \tag{4.87}$$

with the poles determined in (4.86). The transfer function (4.87) has a cut-off frequency of $\omega_c = 68$ rad/s at the −3 db limit. Thus, the following ESC parameter setting with the exception of k_I is based on the previous comments and design guideline (4.85)

$$\begin{aligned}
\omega_p &= 2\pi f_p = 22 \text{ rad/s} < \omega_c, \qquad f_p = 3.5\, 1/\text{s}\\
\omega_h &= 0.1\,\omega_p\\
a_p &= 0.004\, V_{oc,STC}\, N_s = 5\,\text{V}\\
k_I &= 0.009
\end{aligned}$$

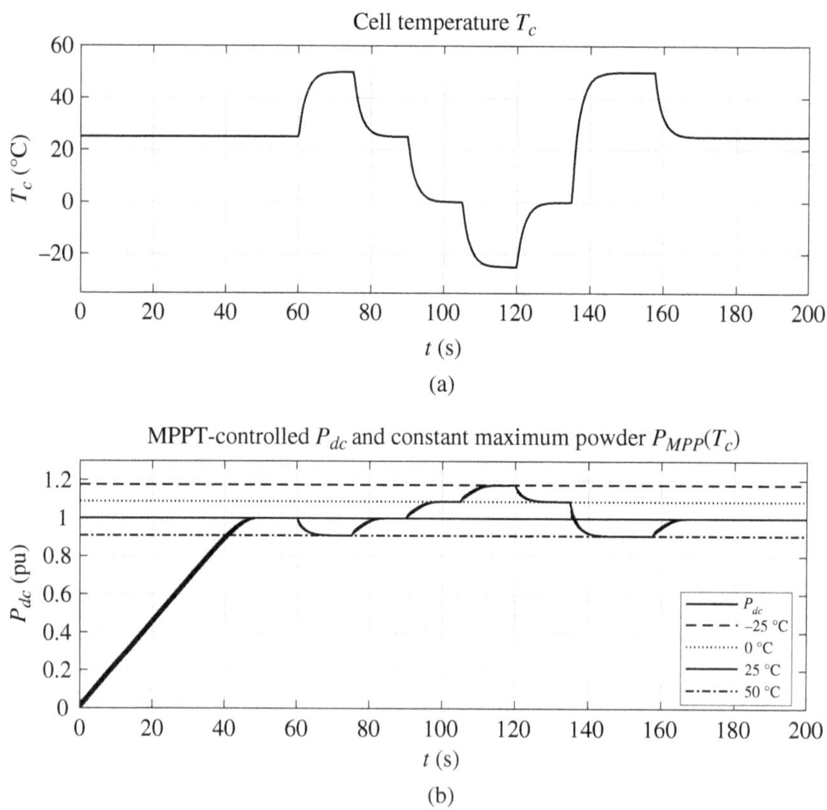

Figure 4.34 Maximum power point tracking with changing cell temperature.

The used PV plant parameters $V_{oc,STC}$ and N_s are listed in Table D.4. The integrator gain is set experimentally using simulation runs. At this point, an analyzable and derivable design rule for the k_I could be determined. A possible approach is presented in Section 4.7 (Problems).

Finally, the simulation results with the model-free MPPT and DPT power controller are discussed. First, for the MPPT controller validation, as illustrated in Figure 4.34a, the cell temperature is varied from $t = 60$ seconds within a range of 50 to −25 °C. The MPP is shifted within 20 seconds by rapid temperature change, which is not realistic but is well suited as a test scenario for evaluating the convergence of the ES algorithm. The simulation results show that the point of maximum power is reached again in about 10 seconds after a temperature change. The horizontal lines represent the maximum power value corresponding to the steady-state cell temperature {−25, 0, 25, 50 °C }. These values are taken from the P–V curve in Figure 4.33. In addition to the tracking behavior, the simulation shows the initiation and ramp-up phase of the extremum-seeking controller from $t = 0$ to 60 seconds. It is good to see how the integral behavior and the superimposed oscillation (the ripples in the signal) lead to a slow interaction to the maximum point without prior knowledge, i.e. the initialization with a reference value of zero.

Second, the simulation result in Figure 4.35 using the model-free DPT controller shows a good convergence and thus a good tracking behavior. The power reference value is set to a stepped decrease and increase between 0.8 and 1 pu. Due to the nonlinear characteristic, the step response depends on the range and step direction. Due to the small-scale range used in Figure 4.35 compared to Figure 4.34, the ripples in the DC power at the primary converter output of the PV system are clearly visible. Therefore, Chapter 6 examines how this affects the AC power fed

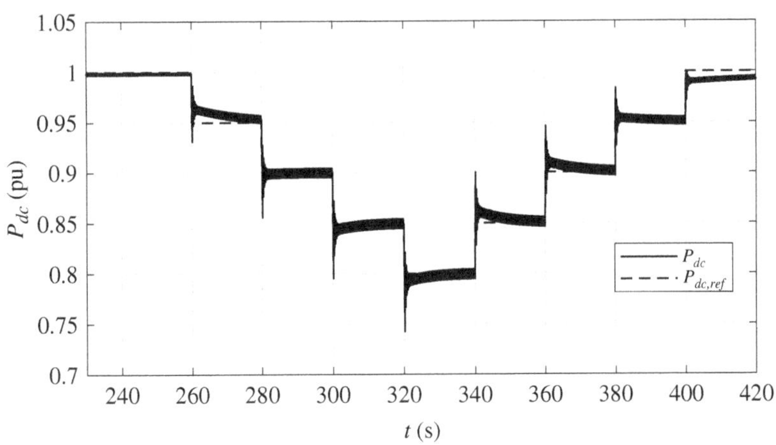

Figure 4.35 Demanded power tracking with step-change reference power $P_{dc,ref}$.

into the grid. Even if the tracking behavior is generally good, a rclatively large overshoot can be seen. This compromise had to be made in order to achieve a fast response and short response time. However, due to the lack of model knowledge, the step response is not as fast as in the model-based method from Section 4.5.2. Furthermore, it can be seen that at the end of simulation from $t = 400$ seconds (see Figure 4.35) the final value is no longer reached in 20 seconds as before. There is still potential for improvement.

4.6 Summary

A generic PV power plant model for primary conversion was presented. It consists of an aggregated PV generator model and a switching DC–DC converter to feed the DC bus. Three standard second-order DC–DC converters were considered and mathematically described by average second-order state-space models. Based on the PV generator represented by a voltage-controlled current source equivalent and the DC–DC converter model, a state-feedback structure in Takagi–Sugeno form was designed to control the voltage in the inner loop in a wide operating range also far from the MPP. The inner-loop control is overlaid by an optimal power control where either the MPP or a specific reference power (DPT) can be selected. Model-based and adaptive model-free control schemes have been proposed to implement the MPPT and DPT functions. It has been shown that PV power plants can provide positive and negative control power, depending on the desired higher-level control, e.g. for fast frequency control.

4.7 Problems

4.1 As an alternative to the simplified single-diode model (4.6) of a PV cell related to the equivalent circuit from Figure 4.4, specify two alternative equivalent circuits. What is the advantage of the model (4.6) used here and what is not considered or neglected?

4.2 The aggregation of all PV cells into a PV power plant is based on which assumption? What is not considered in the model and how can it be extended?

4.3 Design a voltage controller for the boost converter in Figure 4.10 described by the state-space average model (4.30) analogous to the LMI-based controller design in Section 4.4 of the buck converter (4.29).

4.4 Explain the adaptive behavior of the ESC shown in Figure 4.31 using the simulation results of Figure 4.34. Draw the signal curve before the high-pass filter, after multiplication with the perturbation signal and after integration. Analyze this for the case where the power is to the left, to the right of the optimum, or at the optimum.

4.5 Perform a stability analysis of the ESC control scheme in Figure 4.31 depending on the integrator gain for the MPPT mode. Based on the small-signal behavior, investigate the pole shift of the closed loop as the k_I increases.

4.8 Bibliography

Other recent works on PV generator power control are now briefly presented. First, to improve the classical MPPT algorithms related to robustness and dynamics, several new control concepts have been developed in recent years. In Esram and Chapman [2007], 19 distinct methods for MPPT of PV arrays are compared and discussed. However, the methods studied do not include approaches for operating PV systems below their maximum operating point. For reasons as outlined in the second part of this book chapter, recent research has considered on PV power plant designs that can provide fast time variation of the injected power.

An overview of frequency control challenges and possible countermeasures in future low-inertia power systems with renewable energy sources is given in Shazon et al. [2022]. One focus is the frequency control by deloading and power curtailment. A comprehensive review of so-called flexible maximum power-point-tracking (FMPPT) algorithms for grid-connected PV systems is given in Sharma et al. [2023]. The term FMPPT is used to summarize all the algorithms that control the PV generator power to a specific value, rather than continuously to the MPP. For this purpose, a detailed comparative study of different control strategies is performed in terms of current-, voltage-, and power-based tracking capabilities. In addition, there are currently several approaches to the subject: The main contribution of [Gomez-Merchan et al., 2021] is to propose a model-free binary-search-based algorithm for FMPPT. The binary search (BNS) is a searching algorithm that performs a logarithmic search within a sorted data array until the target voltage value is found. In Messasma et al. [2023b] a power reserve control (PRC) technique based on PV voltage variation is proposed. It has a two-mode cycle, with the first mode being used to calculate and reach the MPPT point and the second mode being used to provide the appropriate power reserve. The approach is applied to offer primary frequency control and virtual inertia. In Messasma et al. [2023a], a strategy with a flexible maximum power-tracking controller using an artificial neural network (ANN) is proposed. The maximum available power is estimated using an ANN model in the first step. In the second step, the current source of the PV generator is controlled by a proportional–integral (PI) voltage regulator to calculate the requested power considering the estimated maximum available power. A so-called pseudo maximum power point tracking (PMPPT) is presented in Kumar et al. [2020] which makes it possible to operate PV power plants with reserve power capacity without employing a battery for storage. In Zarina et al. [2014] control strategies for frequency support by hybrid power plants consisting of PV and rotating generators without battery energy storage have been formulated. In particular, two modes of operation for deloaded PV are suggested, where a cost analysis shows that the proposed concept is economical compared to battery usage for frequency control. Studies of interaction and stability issues of PV power plants with fast frequency response based on a small-signal analysis connected to the grid are presented in Salazar-Chiralt et al. [2022].

In this book chapter, we have assumed an aggregated system of primary conversion. That means the PV generator consists of single cells connected in parallel and series, exposed to the same irradiation and temperature. In real plants, especially in large-scale PV plants in the 10 and 100 MW range, the PV park controller is the higher-level controller of the individual string or central inverters. In Bullich-Massagué et al. [2016] the entire design process, from requirement definitions to the controller implementation, is described by means of a 9.4 MW PV power plant. The controller is designed to fulfill the requirements of the most demanding grid codes and combines the utilization of the PV inverters, fixed switched capacitors, and STATCOMs. A review of components as photovoltaic panels, converters, and transformers utilized in large-scale photovoltaic power plants are addressed in Cabrera-Tobar et al. [2016].

5

Modeling and Control of Voltage Source Converters

5.1 Introduction

The power electronics (PE)-based voltage source converter (VSC) is the interface between primary power conversion (wind turbine rotor-generator and PV-generator) and the electrical grid, see Figure 1.3. Generally, VSC is used to convert power from direct current into alternating quantities and the other way around. This conversion process is conducted by semiconductor-switching scheme to effectively provide a desired voltage at the converter terminal. Depending on the control concept, different metrics as injected current (current source) or voltage (voltage source) may guide the possibly cascaded closed-loop structure. The topology of a generic two-level three-phase VSC is illustrated in Figure 5.1. An example with the purpose for experimental studies and rapid-control prototyping, i.e. validation of novel control concepts like PLL-free grid-forming (GFM) control is shown in Figure 5.2.

In addition to electronic switches, voltage-fed converters require passive elements such as capacitors and inductors on the DC and AC sides. The capacitor C_{dc} on the DC side is the most important storage element. Moreover, only a small part, usually 5% of the energy, is stored on the AC side. Therefore, the primary function of the capacitor is to smooth the DC voltage. The low-pass LCL filter, consisting of $L_{f_{1,2}}$ and C_f, provides attenuation of the pulse-width modulation (PWM) carrier and sideband harmonics. The passive elements L_g and R_g as lumped inductance and resistance determine the impedance of the power line [Kundur, 1994]. The interface to the generated power on the primary power conversion side is modeled by a controlled current source i_u and a capacitor C_{dc} connected in parallel, see Figure 5.1. Note that most existing work on VSC modeling and control assumes that the DC source is ideal, and thus the primary and secondary power conversion are decoupled [Rocabert et al., 2012; Pan et al., 2020; Peng et al., 2018b]. Decoupling assumption is insufficient for the design of grid-supporting inverter-based power plants, especially for fast frequency

Advanced Control of Grid-Integrated Renewable Energy Power Plants: LMI based Design in the Takagi-Sugeno Framework, First Edition. Horst Schulte.

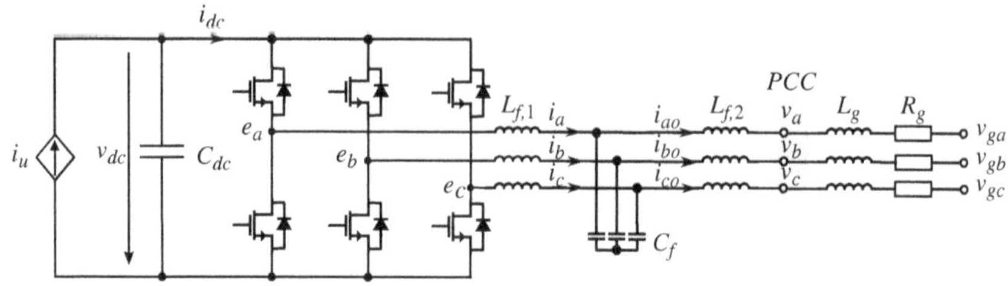

Figure 5.1 Topology of three-phase voltage source converter with PCC (point of common coupling) and simplified grid model (L_g, R_g) with a single line element and three-phase voltage source $v_{g,abc}$ at the end

Figure 5.2 Voltage source converter with controller board and LCL filter (HTW Berlin), "Windkraftwerk," Advanced Control Lab for Grid Integration Studies. Source: Kisser et al. [2017]/HTW Berlin

responses without intermediate large energy storage, since the coupling of primary and secondary side must be considered in the controller design for grid support.

Control concepts of power electronic inverters have been divided into grid-following (GFL) and GFM inverters in recent years. The concept of GFM converters was originally introduced in the field of microgrid and islanding applications [Lasseter, 2002; Rocabert et al., 2012] and has been widely recognized as a flexible solution to improve the system stability and resilience of larger interconnected power systems with a high proportion of power electronic-based generation [Rosso et al., 2020; Rosso et al., 2021]. In this regard, the GFL and GFM control concepts differ as follows: GFL converter includes a synchronization unit to synchronize the generated voltage to a given line voltage, often implemented using phase-locked loops (PLLs). The current injection is controlled by the suitable adjustment of the angular position of the inverter voltage with respect to the grid voltage at a defined measuring point. Inverters with a GFL control concept behave like controlled current sources with a high parallel impedance.

In contrast, GFM inverters are not synchronized to a specific line voltage. The power is controlled directly by regulating the output terminal voltage of the VSC. The control loop is closed by feeding back the calculated power at the PCC (calculated from the measured current $\mathbf{i}_{abc}$ and voltage $\mathbf{v}_{abc}$) and comparing them with the reference power. Such an inverter operates like a voltage source coupled to the grid via a low-series coupling impedance. GFM inverter maintains an internal voltage phasor that is constant or nearly constant in the sub-transient to transient time frame. This allows an immediately response[1] to changes in the external system and maintain stability during challenging network conditions.

5.2 Three-Phase Signals and Systems

For the reduced-order VSC modeling and power system behavior analysis in the upcoming Section 5.3 the three-phase signal is first introduced

$$\mathbf{x}_{abc}(t) = \left(x_a(t), x_b(t), x_c(t) \right)^T \in \mathbb{R}^3, \tag{5.1}$$

which generally describes the voltages and currents at the nodes and branches of the circuit. We assume that (5.1)

1. is a periodic signal with zero average: $x_i(t) = x_i(t+T)$ for all $i \in \{a, b, c\}$, where $T > 0$ denotes the common period in second;
2. describes a balanced three-phase system

$$\mathbf{x}_{abc}(t) = \begin{pmatrix} x_a(t) \\ x_b(t) \\ x_c(t) \end{pmatrix} = X_m \begin{pmatrix} \cos(\omega t + \phi) \\ \cos\left(\omega t + \phi - \frac{2\pi}{3}\right) \\ \cos\left(\omega t + \phi + \frac{2\pi}{3}\right) \end{pmatrix}, \tag{5.2}$$

1 According to Weise et al. [2019], a response is required within a quarter period (4.2 or 5 ms) of the AC system with 60 or 50 Hz.

where X_m denotes the signal amplitude with $X_m = \sqrt{2}X$ for all $i \in \{a, b, c\}$ with X as the RMS value of x_i. The time-variable RMS value is defined by the integral over the last period.

$$X(t) = \sqrt{\frac{1}{T}\int_{t-T}^{t} x_i^2(\tau)d\tau}, \quad i \in \{a, b, c\}. \tag{5.3}$$

Note that $X_m = \sqrt{2}X$ is not valid for sinusoidal signals whose amplitude X_m changes significantly within one period. For an arbitrary common angle $\phi(t)$ and amplitude $X_m > 0$

$$x_a(t) + x_b(t) + x_c(t) = 0, \tag{5.4}$$

holds for all $t \geq 0$.

3. (5.2) is synchronous in the steady state with constant frequency ω and amplitude X_m.

Three-phase quantities such as (5.2) are often expressed using the phasor notation. It is defined by the parameters of the magnitude and phase ϕ. The magnitude of the phasor denoted as X_i is the root mean square (RMS) value of $x_i(t)$, $i \in \{a, b, c\}$, where the phase ϕ_i corresponds to the angle of the voltage $x_i(t)$. The three-phase signal (5.2) can be assigned to three phasors implicitly related to a fixed frequency ω

$$\begin{aligned} \underline{X_a} &= X_a\, e^{j\phi}, \\ \underline{X_b} &= X_b\, e^{j\left(-\frac{2\pi}{3}+\phi\right)}, \\ \underline{X_c} &= X_c\, e^{j\left(+\frac{2\pi}{3}+\phi\right)} \end{aligned} \tag{5.5}$$

with the common RMS values $X = X_a = X_b = X_c$ for a balanced system. Note the limitation of the phasor representation (5.5) is that a known frequency must be assumed and cannot be used to represent signals containing more than one frequency component, such as signals with harmonics. Each voltage phasor in (5.5) can be converted to a three-phase time function using Euler's relation with

$$x_i(t) = \mathrm{Re}\left\{\sqrt{2}\underline{X_i}\, e^{j\omega t}\right\}, \quad i \in \{a, b, c\}. \tag{5.6}$$

To study three-phase systems (5.2), (5.6) it is very helpful to work with coordinate transformations. In this and the next Chapter 6, especially for controlling currents, voltages, and instantaneous powers in the secondary conversion unit of RE power plants, we use the synchronous $dq0$-reference frame generated by the power-invariant Park transform of the stationary reference frame $\alpha\beta$. First, the coordinates $\alpha\beta 0$ are calculated using the Clarke transform $\mathbf{T}_{\alpha\beta 0} : \mathbf{x}_{abc} \to \mathbf{x}_{\alpha\beta 0}$ from the measurable quantities (5.2):

$$\mathbf{x}_{\alpha\beta 0} = \mathbf{T}_{\alpha\beta 0}\,\mathbf{x}_{abc}\,, \qquad \mathbf{T}_{\alpha\beta 0} = \sqrt{\frac{2}{3}}\left(\begin{array}{ccc} 1 & -\frac{1}{2} & -\frac{1}{2} \\ 0 & \frac{\sqrt{3}}{2} & -\frac{\sqrt{3}}{2} \\ \hline \frac{1}{\sqrt{2}} & \frac{1}{\sqrt{2}} & \frac{1}{\sqrt{2}} \end{array}\right), \tag{5.7}$$

where $\mathbf{T}_{\alpha\beta 0}$ denotes the power-invariant Clarke transform. For the three-phase system (5.2), we get the related stationary reference frame $\alpha\beta 0$:

$$\mathbf{x}_{\alpha\beta 0} = \left(\begin{array}{c} x_\alpha \\ x_\beta \\ \hline x_0 \end{array}\right) = \mathbf{T}_{\alpha\beta 0}\,\mathbf{x}_{abc} = \sqrt{3}X\left(\begin{array}{c} \cos(\omega t + \phi) \\ \sin(\omega t + \phi) \\ \hline 0 \end{array}\right), \tag{5.8}$$

where x_α and x_β are the two-phase orthogonal components in the stationary $\alpha\beta$-reference frame. The two components uniquely define the balanced three-phase system (5.2) and can be represented by the space vector $\vec{\mathbf{x}} \in \mathbb{C}$ in a complex plane, which is defined at any time

$$\vec{\mathbf{x}}_{\alpha\beta} = x_\alpha + jx_\beta = \sqrt{3}X\cos(\omega t + \phi) + j\sqrt{3}X\sin(\omega t + \phi). \tag{5.9}$$

The matrix $\mathbf{T}_{\alpha\beta 0}$ is orthogonal, that is the inverse is equal to the transpose with $\mathbf{T}_{\alpha\beta 0}^{-1} = \mathbf{T}_{\alpha\beta 0}^{T} : \mathbf{x}_{\alpha\beta 0} \to \mathbf{x}_{abc}$

$$\mathbf{x}_{abc} = \mathbf{T}_{\alpha\beta 0}^{-1}\,\mathbf{x}_{\alpha\beta 0}\,, \qquad \mathbf{T}_{\alpha\beta 0}^{-1} = \mathbf{T}_{\alpha\beta 0}^{T} = \sqrt{\frac{2}{3}}\left(\begin{array}{cc|c} 1 & 0 & \frac{1}{\sqrt{2}} \\ -\frac{1}{2} & \frac{\sqrt{3}}{2} & \frac{1}{\sqrt{2}} \\ -\frac{1}{2} & -\frac{\sqrt{3}}{2} & \frac{1}{\sqrt{2}} \end{array}\right). \tag{5.10}$$

Next, the power-invariant Park transform $\mathbf{T}_{dq0} : \mathbf{x}_{\alpha\beta 0} \to \mathbf{x}_{dq0}$ is introduced for the case when the d-axis is aligned with the α-axis:

$$\mathbf{x}_{dq0} = \mathbf{T}_{dq0}(\theta)\,\mathbf{x}_{\alpha\beta 0}\,, \qquad \mathbf{T}_{dq0}(\theta) = \left(\begin{array}{cc|c} \cos(\theta) & \sin(\theta) & 0 \\ -\sin(\theta) & \cos(\theta) & 0 \\ \hline 0 & 0 & 1 \end{array}\right). \tag{5.11}$$

This transforms the stationary $\alpha\beta 0$-reference frame (5.8) into the rotating $dq0$-reference frame with the angle θ:

$$\mathbf{x}_{dq0} = \left(\begin{array}{c} x_d \\ x_q \\ x_0 \end{array}\right) = \sqrt{3}X\left(\begin{array}{c} \cos(\omega t + \phi - \theta) \\ \sin(\omega t + \phi - \theta) \\ \hline 0 \end{array}\right). \tag{5.12}$$

If $\theta = \omega t$ it is obtained

$$\mathbf{x}_{dq0} = \begin{pmatrix} x_d \\ x_q \\ \hline x_0 \end{pmatrix} = \sqrt{3}X \begin{pmatrix} \cos(\phi) \\ \sin(\phi) \\ \hline 0 \end{pmatrix}, \tag{5.13}$$

i.e. the $dq0$ coordinate system rotates synchronously with the vector $\mathbf{x}_{\alpha\beta 0}$ in the $\alpha\beta 0$ coordinate system. As for the stationary reference frame with (5.9), a space vector is also specified

$$\vec{\mathbf{x}}_{dq} = x_d + jx_q = \sqrt{3}X\cos(\phi) + j\sqrt{3}X\sin(\phi). \tag{5.14}$$

The inverse Park transform results from the orthogonality $\mathbf{T}_{dq0}^{T}(\theta) = \mathbf{T}_{dq0}^{-1}(\theta), \forall\theta$:

$$\mathbf{x}_{\alpha\beta 0} = \mathbf{T}_{dq0}^{-1}(\theta)\mathbf{x}_{dq0}, \qquad \mathbf{T}_{dq0}^{-1}(\theta) = \left(\begin{array}{cc|c} \cos(\theta) & -\sin(\theta) & 0 \\ \sin(\theta) & \cos(\theta) & 0 \\ \hline 0 & 0 & 1 \end{array}\right). \tag{5.15}$$

For the balanced three-phase system applied to the proposed VSC control concepts in Section 5.7, the component x_0 is always zero (5.4) and use the coordinate frames $\alpha\beta$ and dq with the transform matrix $\mathbf{T}_{\alpha\beta} : \mathbf{x}_{abc} \rightarrow \mathbf{x}_{\alpha\beta}$

$$\mathbf{x}_{\alpha\beta} = \mathbf{T}_{\alpha\beta}\mathbf{x}_{abc}, \quad \mathbf{T}_{\alpha\beta} = \sqrt{\frac{2}{3}} \begin{pmatrix} 1 & -\frac{1}{2} & -\frac{1}{2} \\ 0 & \frac{\sqrt{3}}{2} & -\frac{\sqrt{3}}{2} \end{pmatrix}, \tag{5.16}$$

and the inverse transformation

$$\mathbf{x}_{abc} = \mathbf{T}_{\alpha\beta}^{T}\mathbf{x}_{\alpha\beta}, \quad \mathbf{T}_{\alpha\beta}^{T} = \sqrt{\frac{2}{3}} \begin{pmatrix} 1 & 0 \\ -\frac{1}{2} & \frac{\sqrt{3}}{2} \\ -\frac{1}{2} & -\frac{\sqrt{3}}{2} \end{pmatrix}. \tag{5.17}$$

And also $\mathbf{T}_{dq} : \mathbf{x}_{\alpha\beta} \rightarrow \mathbf{x}_{dq}$

$$\mathbf{x}_{dq} = \mathbf{T}_{dq}(\theta)\mathbf{x}_{\alpha\beta}, \quad \mathbf{T}_{dq}(\theta) = \begin{pmatrix} \cos(\theta) & \sin(\theta) \\ -\sin(\theta) & \cos(\theta) \end{pmatrix}, \tag{5.18}$$

with the related inverse

$$\mathbf{x}_{\alpha\beta} = \mathbf{T}_{dq}^{T}(\theta)\mathbf{x}_{dq}, \quad \mathbf{T}_{dq}^{T} = \begin{pmatrix} \cos(\theta) & -\sin(\theta) \\ \sin(\theta) & \cos(\theta) \end{pmatrix}. \tag{5.19}$$

The combination of (5.16) and (5.18) gives the dq-components $\mathbf{x}_{dq}$ directly from the instantaneous abc three-phase system $\mathbf{x}_{abc}$:

$$\mathbf{x}_{dq} = \tilde{\mathbf{T}}_{dq}(\theta)\mathbf{x}_{abc}, \quad \tilde{\mathbf{T}}_{dq}(\theta) := \mathbf{T}_{dq}(\theta)\mathbf{T}_{\alpha\beta}$$

$$\tilde{\mathbf{T}}_{dq}(\theta) = \sqrt{\frac{2}{3}} \begin{pmatrix} \cos(\theta) & \cos(\theta - \frac{2\pi}{3}) & \cos(\theta + \frac{2\pi}{3}) \\ -\sin(\theta) & -\sin(\theta - \frac{2\pi}{3}) & -\sin(\theta + \frac{2\pi}{3}) \end{pmatrix}. \tag{5.20}$$

The inversion of the transformation is based on the exploitation of the orthogonality with $\tilde{\mathbf{T}}_{dq}^T(\theta) = (\mathbf{T}_{dq}\,\mathbf{T}_{\alpha\beta})^T = \mathbf{T}_{\alpha\beta}^T\,\mathbf{T}_{dq}^T$:

$$\mathbf{x}_{abc} = \tilde{\mathbf{T}}_{dq}^T(\theta)\,\mathbf{x}_{dq}\,, \quad \tilde{\mathbf{T}}_{dq}^T(\theta) = \sqrt{\frac{2}{3}}\begin{pmatrix} \cos(\theta) & -\sin(\theta) \\ \cos(\theta - \frac{2\pi}{3}) & -\sin(\theta - \frac{2\pi}{3}) \\ \cos(\theta + \frac{2\pi}{3}) & -\sin(\theta + \frac{2\pi}{3}) \end{pmatrix}. \tag{5.21}$$

5.3 Average and Instantaneous Power

After presenting the necessary coordinate transformations, we deal with the essentials of the instantaneous power theory for the three-phase system (5.2) in $\alpha\beta$ and dq-coordinates. Further details can be found in Akagi et al. [2007]. Based on the previous assumptions (5.2), (5.4) (sinusoidal and balanced system) the three-phase voltage and line current are given by

$$\mathbf{v}_{abc}(t) = \begin{pmatrix} v_a(t) \\ v_b(t) \\ v_c(t) \end{pmatrix} = \sqrt{2}V\begin{pmatrix} \cos(\omega t + \phi_v) \\ \cos(\omega t + \phi_v - \frac{2\pi}{3}) \\ \cos(\omega t + \phi_v + \frac{2\pi}{3}) \end{pmatrix}, \tag{5.22}$$

$$\mathbf{i}_{abc}(t) = \begin{pmatrix} i_a(t) \\ i_b(t) \\ i_c(t) \end{pmatrix} = \sqrt{2}I\begin{pmatrix} \cos(\omega t + \phi_I) \\ \cos(\omega t + \phi_I - \frac{2\pi}{3}) \\ \cos(\omega t + \phi_I + \frac{2\pi}{3}) \end{pmatrix}, \tag{5.23}$$

with the RMS values V and I where ϕ_v and ϕ_I denote the phase angle of the voltage and current, which are to be distinguished from each other. For a three-phase system, the instantaneous active power denoted as $p_{3\phi}(t)$ describes the total energy flow per time unit being transferred between two subsystems [Akagi et al., 2007], and it is given by

$$p_{3\phi}(t) = \mathbf{v}_{abc}^T(t)\,\mathbf{i}_{abc}(t) = v_a(t)\,i_a(t) + v_b(t)\,i_b(t) + v_c(t)\,i_c(t). \tag{5.24}$$

Substituting (5.22) and (5.23) in (5.24) and applying the harmonic addition theorem

$$\cos(\alpha)\cos(\beta) = \frac{1}{2}\left(\cos(\alpha - \beta) + \cos(\alpha + \beta)\right), \tag{5.25}$$

yields

$$p_{3\phi}(t) = 3\,VI\cos(\phi_v - \phi_I) + \underbrace{VI\left[\cos(\delta(t)) + \cos\left(\delta(t) - \frac{2\pi}{3}\right) + \cos\left(\delta(t) + \frac{2\pi}{3}\right)\right]}_{=0}, \tag{5.26}$$

where $\delta(t) = 2\omega t + \phi_v + \phi_I$. Since the term in (5.26) with time-dependent $\delta(t)$ is always equal to zero, the three-phase power is time-invariant and depends only on the phase angle difference $\phi_v - \phi_I$ and RMS values V and I

$$p_{3\phi}(t) = 3\,VI\cos(\phi_v - \phi_I) = P_{3\phi} = 3P_{\phi}, \tag{5.27}$$

where $P_{3\phi}$ is the average active power of a three-phase system corresponding to the three times power of a single-phase system denoted as P_{ϕ}. For the three-balanced system, the three-phase power can also be defined from the voltage and current in phasor form (5.5), similar to the definition of a single-phase system under sinusoidal conditions. Three-phase apparent power is defined as three times single-phase apparent power

$$\begin{aligned} S_{3\phi} = 3\underline{V}\,\underline{I}^* &= 3\,Ve^{j\phi_v}\,Ie^{-j\phi_I} \\ &= 3VI\cos(\phi_v - \phi_I) + j3VI\sin(\phi_v - \phi_I) \\ &= P_{3\phi} + j\,Q_{3\phi}, \end{aligned} \tag{5.28}$$

which results in the three-phase reactive power

$$Q_{3\phi} = 3\,VI\sin(\phi_v - \phi_I). \tag{5.29}$$

The power definitions described above can also be defined in the stationary $\alpha\beta$-reference frame and synchronous rotating dq-reference frame. The three-phase voltage $\mathbf{v}_{abc}$ (5.22) and current $\mathbf{i}_{abc}$ (5.23) transformed by (5.16) into $\alpha\beta$ is given

$$\begin{aligned} v_\alpha &= \sqrt{3}V\cos(\omega t + \phi_v), \qquad & i_\alpha &= \sqrt{3}I\cos(\omega t + \phi_I) \\ v_\beta &= \sqrt{3}V\sin(\omega t + \phi_v), \qquad & i_\beta &= \sqrt{3}I\sin(\omega t + \phi_I). \end{aligned} \tag{5.30}$$

This allows a voltage and current vector to be derived in the complex plane with $\mathbf{v}, \mathbf{i} \in \mathbb{C}$:

$$\begin{aligned} \mathbf{v} = v_\alpha + j v_\beta &= \sqrt{3}V\left(\cos(\omega t + \phi_v) + j\sin(\omega t + \phi_v)\right) \\ &= \sqrt{3}V\,e^{j(\omega t + \phi_v)}, \end{aligned} \tag{5.31}$$

$$\begin{aligned} \mathbf{i} = i_\alpha + j i_\beta &= \sqrt{3}I\left(\cos(\omega t + \phi_I) + j\sin(\omega t + \phi_I)\right) \\ &= \sqrt{3}I\,e^{j(\omega t + \phi_I)}, \end{aligned} \tag{5.32}$$

Analogous to the concept of complex power defined in (5.28) an instantaneous complex power using the voltage $\mathbf{v}$ and current vector $\mathbf{i}$ is introduced [Akagi et al., 2007]

$$\mathbf{s} = \mathbf{v}\mathbf{i}^* = (v_\alpha + j v_\beta)(i_\alpha - j i_\beta) = \underbrace{(v_\alpha i_\alpha + v_\beta i_\beta)}_{p_{3\phi}} + j\underbrace{(v_\beta i_\alpha - v_\alpha i_\beta)}_{q_{3\phi}}. \tag{5.33}$$

The definition of instantaneous power $p_{3\phi}$ and $q_{3\phi}$ can be given in a compact matrix notation

$$\begin{pmatrix} p_{3\phi} \\ q_{3\phi} \end{pmatrix} = \begin{pmatrix} v_\alpha & v_\beta \\ v_\beta & -v_\alpha \end{pmatrix} \begin{pmatrix} i_\alpha \\ i_\beta \end{pmatrix}. \tag{5.34}$$

The instantaneous active power related to the rotating dq-frame results formally from the substitution of the $\alpha\beta$ components in $p_{3\phi} = v_\alpha\, i_\alpha + v_\beta\, i_\beta$ determined with the inverse Park transform (5.15) for the current and voltage components

$$\begin{aligned} v_\alpha &= \cos(\theta)v_d - \sin(\theta)v_q, & i_\alpha &= \cos(\theta)i_d - \sin(\theta)i_q \\ v_\beta &= \sin(\theta)v_d + \cos(\theta)v_q, & i_\beta &= \sin(\theta)i_d + \cos(\theta)i_q, \end{aligned} \tag{5.35}$$

follows

$$\begin{aligned} p_{3\phi} &= \underbrace{(\cos(\theta)\, v_d - \sin(\theta)\, v_q)}_{v_\alpha} \underbrace{(\cos(\theta)\, i_d - \sin(\theta)\, i_q)}_{i_\alpha} + \\ &\quad \underbrace{(\sin(\theta)\, v_d + \cos(\theta)\, v_q)}_{v_\beta} \underbrace{(\sin(\theta)\, i_d + \cos(\theta)\, i_q)}_{i_\beta} \\ p_{3\phi} &= v_d\, i_d + v_q\, i_q. \end{aligned} \tag{5.36}$$

In the same manner one obtains for $q_{3\phi} = v_\beta\, i_\alpha - v_\alpha\, i_\beta$ with (5.35)

$$q_{3\phi} = v_q\, i_d - v_d\, i_q. \tag{5.37}$$

Since the power values $p_{3\phi}$ and $q_{3\phi}$ depend only on the angle between current and voltage, but not on the absolute angular position of the space vector, the power relationship is independent of the $\alpha\beta$- or dq-reference frame. Finally, the compact form of the power in dq-coordinates in matrix notation

$$\begin{pmatrix} p_{3\phi} \\ q_{3\phi} \end{pmatrix} = \begin{pmatrix} v_d & v_q \\ v_q & -v_d \end{pmatrix} \begin{pmatrix} i_d \\ i_q \end{pmatrix}. \tag{5.38}$$

Notes on the Amplitude-Invariant Clarke Transform

In the original work of Clarke [Duesterhoeft et al., 1951], instead of the power-invariant transformation, an amplitude-invariant transformation was presented. It was introduced with the following equations[2]

$$x_a = x_\alpha + x_0$$

$$x_b = -\frac{1}{2} + x_0 + \frac{\sqrt{3}}{2} x_\beta$$

2 In Duesterhoeft et al. [1951] the equations are related to the current; here, we formulate it in general for $x \in \{i, v\}$.

$$x_c = -\frac{1}{2} + x_0 - \frac{\sqrt{3}}{2} x_\beta$$

and is equivalent to the matrix notation

$$\begin{pmatrix} x_a \\ x_b \\ x_c \end{pmatrix} = \underbrace{\begin{pmatrix} 1 & 0 & 1 \\ -\frac{1}{2} & \frac{\sqrt{3}}{2} & 1 \\ -\frac{1}{2} & -\frac{\sqrt{3}}{2} & 1 \end{pmatrix}}_{\tilde{\mathbf{T}}_{\alpha\beta 0}^{-1}} \begin{pmatrix} x_\alpha \\ x_\beta \\ x_0 \end{pmatrix}. \tag{5.39}$$

The reverse transformation is

$$\begin{pmatrix} x_\alpha \\ x_\beta \\ x_0 \end{pmatrix} = \frac{2}{3} \underbrace{\begin{pmatrix} 1 & -\frac{1}{2} & -\frac{1}{2} \\ 0 & \frac{\sqrt{3}}{2} & -\frac{\sqrt{3}}{2} \\ \frac{1}{2} & \frac{1}{2} & \frac{1}{2} \end{pmatrix}}_{\tilde{\mathbf{T}}_{\alpha\beta 0}} \begin{pmatrix} x_a \\ x_b \\ x_c \end{pmatrix}. \tag{5.40}$$

It is evident that the transformation matrix $\tilde{\mathbf{T}}_{\alpha\beta 0}$ is not orthogonal. However, due to the transformation, the length of the phase vector $\mathbf{x}_{\alpha\beta}$ remains unchanged compared to the *abc*-three-phase system (5.2)

$$\begin{pmatrix} x_\alpha \\ x_\beta \end{pmatrix} = \frac{2}{3} \underbrace{\begin{pmatrix} 1 & -\frac{1}{2} & -\frac{1}{2} \\ 0 & \frac{\sqrt{3}}{2} & -\frac{\sqrt{3}}{2} \end{pmatrix}}_{\tilde{\mathbf{T}}_{\alpha\beta}} \begin{pmatrix} x_a \\ x_b \\ x_c \end{pmatrix} = \sqrt{2} X \begin{pmatrix} \cos(\omega t + \phi) \\ \sin(\omega t + \phi) \\ \hline 0 \end{pmatrix}. \tag{5.41}$$

The power in the rotating reference frame with the transformation $\tilde{\mathbf{T}}_{\alpha\beta}$ is determined with

$$\begin{pmatrix} p_{3\phi} \\ q_{3\phi} \end{pmatrix} = \frac{3}{2} \begin{pmatrix} v_\alpha & v_\beta \\ v_\beta & -v_\alpha \end{pmatrix} \begin{pmatrix} i_\alpha \\ i_\beta \end{pmatrix}. \tag{5.42}$$

As a result, the power values in the synchronous rotating reference frame are also provided with the same factor of 3/2:

$$\begin{pmatrix} p_{3\phi} \\ q_{3\phi} \end{pmatrix} = \frac{3}{2} \begin{pmatrix} v_d & v_q \\ v_q & -v_d \end{pmatrix} \begin{pmatrix} i_d \\ i_q \end{pmatrix}. \tag{5.43}$$

Therefore, to avoid this scaling factor, we use the previously described power-invariant Clarke transform (5.16), (5.17) together with the Park transform (5.18), (5.19) in the following studies.

5.4 Reduced-Order VSC Model

For control-oriented modeling, the following simplifications are established. First, we assume that the terminal voltage of the VSC has only a fundamental frequency based on a balanced three-phase system and is described by $\mathbf{e}_{abc}$ (5.2), $\mathbf{e}_{\alpha\beta}$ (5.16), or $\mathbf{e}_{dq}$ (5.18). The harmonics due to the switching semiconductors are not taken into account. Second, only symmetrical faults in the lines are treated, so that the balanced system (5.2) is preserved. Third, the line filter (LCL) shown in Figure 5.1 is simplified to a resulting longitudinal impedance. The influence of the filter capacitor C_f on the grid stability can be neglected for most issues [Klaes et al., 2020]. Especially if only the fundamental frequency is examined without harmonics [Teodorescu et al., 2011]. That allows us to sum up the filter inductances to

$$L_f = L_{f,1} + L_{f,2}. \tag{5.44}$$

For the application case where the PCC resp. measurement point is located behind the feed-in transformer[3] as illustrated in Figure 1.3, L_f is extended by the longitudinal transformer inductance L_t

$$L_{ft} = L_{f,1} + L_{f,2} + L_t. \tag{5.45}$$

Fourth, the power losses of the semiconductor switches, in the magnetic components of the LCL filter and the transformer, and of the internal lines are concentrated in one linear loss function proportional to the current. Fifth, in the case of significant grid disturbance events, the VSC must react with an adequate fault current contribution (if hardware limits allow) as protection systems require. This fault ride-through (FRT) capability is only considered for the symmetrical fault case, so asymmetric faults are not considered. Sixth, the power injection laws are an elementary part of the control law in Section 5.7 and are derived in the following. Based on all six assumptions, an equivalent circuit of the VSC under consideration is obtained, shown in Figure 5.3.

The corresponding mathematical model of the VSC is derived first for the AC side: With Kirchhoff's voltage law related to the equivalent circuit (Figure 5.3), we obtain with the definition of (5.45)

$$-\mathbf{e}_{abc} + L_{ft}\frac{d\mathbf{i}_{abc}}{dt} + \mathbf{v}_{abc} = \mathbf{0}, \tag{5.46}$$

that describes the voltage/current dynamics with three balanced phases (5.2). It is transformed into the time-dependent rotating dq-reference frame with

3 The feed-in transformer corresponds to the machine transformer of conventional power stations.

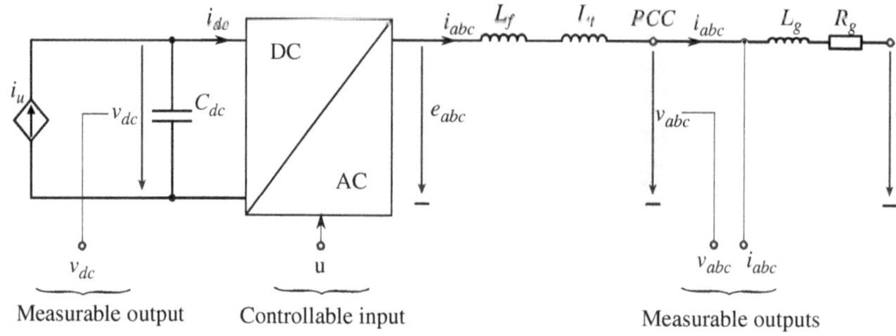

Figure 5.3 Scheme of the reduced-order voltage source converter model

$\mathbf{x}_{abc} = \tilde{\mathbf{T}}_{dq}^{T}(\theta)\,\mathbf{x}_{dq}$ (5.21), where the rotation angle is set to

$$\theta(t) \overset{!}{=} \omega t, \tag{5.47}$$

with the grid frequency ω at the PCC. Note that this step is also common in synchronous machine modeling[4] Sauer et al. [2018]. The frequency ω is determined by the directly measurable shaft speed $\omega_{shaft} = \frac{2}{p}\,\omega$ of the rotating machine, where p denotes the number of magnetic poles. In contrast to the synchronous machine, the phase in VSC must be estimated by means of PLL methods if the control law is formulated in dq-coordinates (see the control concept of GFL VSC in Section 5.7.2). Substituting the abc- by dq-variables with (5.21) results in

$$\begin{aligned} & -\tilde{\mathbf{T}}_{dq}^{T}(\theta)\,\mathbf{e}_{dq} + L_{ft}\frac{d}{dt}(\tilde{\mathbf{T}}_{dq}^{T}(\theta)\,\mathbf{i}_{dq}) + \tilde{\mathbf{T}}_{dq}^{T}(\theta)\,\mathbf{v}_{dq} = \mathbf{0} \mid \tilde{\mathbf{T}}_{dq}(\theta)\cdot \\ \Rightarrow \quad & -\mathbf{e}_{dq} + L_{ft}\,\tilde{\mathbf{T}}_{dq}(\theta)\frac{d}{dt}(\tilde{\mathbf{T}}_{dq}^{T}(\theta)\,\mathbf{i}_{dq}) + \mathbf{v}_{dq} = \mathbf{0}. \end{aligned} \tag{5.48}$$

The calculation of

$$\begin{aligned} \tilde{\mathbf{T}}_{dq}(\theta)\frac{d}{dt}(\tilde{\mathbf{T}}_{dq}^{T}(\theta)\,\mathbf{i}_{dq}) &= \tilde{\mathbf{T}}_{dq}(\theta)\left(\tilde{\mathbf{T}}_{dq}^{T}(\theta)\frac{d\mathbf{i}_{dq}}{dt} + \frac{d\tilde{\mathbf{T}}_{dq}^{T}(\theta)}{d\theta}\underbrace{\frac{d\theta}{dt}}_{\omega}\,\mathbf{i}_{dq}\right) \\ &= \frac{d\mathbf{i}_{dq}}{dt} + \tilde{\mathbf{T}}_{dq}(\theta)\frac{d\tilde{\mathbf{T}}_{dq}^{T}(\theta)}{d\theta}\omega\mathbf{i}_{dq} \\ &= \frac{d\mathbf{i}_{dq}}{dt} + \omega\begin{pmatrix} 0 & -1 \\ 1 & 0 \end{pmatrix}\mathbf{i}_{dq} \\ &= \begin{pmatrix} \frac{di_d}{dt} \\ \frac{di_d}{dt} \end{pmatrix} + \begin{pmatrix} -\omega i_q \\ \omega i_d \end{pmatrix} \end{aligned}$$

4 Balanced symmetrical machines modeling approach in [Sauer et al., 2018], Section 3.3.

and replacing in (5.48) yields

$$-\begin{pmatrix} e_d \\ e_q \end{pmatrix} + L_{ft}\begin{pmatrix} \frac{di_d}{dt} \\ \frac{di_q}{dt} \end{pmatrix} + L_{ft}\begin{pmatrix} -\omega\, i_q \\ \omega\, i_d \end{pmatrix} + \begin{pmatrix} v_d \\ v_q \end{pmatrix} = \begin{pmatrix} 0 \\ 0 \end{pmatrix}.$$

From that follows finally the reduced state space model of the VSC represented in the dq-reference frame rotating with $\theta = \omega t$ (synchronize condition)

$$\dot{\mathbf{x}} = \begin{pmatrix} 0 & \omega \\ -\omega & 0 \end{pmatrix}\mathbf{x} + \frac{1}{L_{ft}}\begin{pmatrix} e_d \\ e_q \end{pmatrix} - \frac{1}{L_{ft}}\begin{pmatrix} v_d \\ v_q \end{pmatrix}, \tag{5.49}$$

$$\mathbf{x} = \begin{pmatrix} i_d \\ i_q \end{pmatrix}, \quad \mathbf{v} = \begin{pmatrix} v_d \\ v_q \end{pmatrix}, \quad \mathbf{e} = \begin{pmatrix} e_d \\ e_q \end{pmatrix}, \tag{5.50}$$

where $\mathbf{x}$ denotes the state vector, $\mathbf{e}$ is the controllable input from the VSC, and $\mathbf{v}$ denotes the voltage at the PCC as an uncontrollable input vector.

The DC side shown in Figure 5.3 with the current source and capacitor C_{dc} can be modeled using Kirchhoff's current law:

$$\frac{dv_{dc}}{dt} = \frac{1}{C_{dc}}(i_u - i_{dc}), \tag{5.51}$$

Extended by the voltage, this gives the power balance equation between the primary power conversion $P_{pp} \in \{P_g, P_{pv}\}$ and secondary power conversion $P_{sp} = P_{dc}$ (see Figure 1.3)

$$\frac{dv_{dc}}{dt}v_{dc} = \frac{1}{C_{dc}}(\underbrace{i_u\, v_{dc}}_{P_{pp}} - \underbrace{i_{dc}\, v_{dc}}_{P_{sp}}). \tag{5.52}$$

Assuming a lossless converter with $P_{ac} = P_{dc}$, then

$$\frac{dv_{dc}}{dt}v_{dc} = \frac{1}{C_{dc}}(P_{pp} - P_{ac}), \tag{5.53}$$

holds. Considering the switching losses in the electronic switches, the previous equation can still be extended by a current-dependent power loss with

$$\frac{dv_{dc}}{dt}v_{dc} = \frac{1}{C_{dc}}(P_{pp} - P_{ac} - i_{dc}^2 R_L), \tag{5.54}$$

where R_L denotes an equivalent VSC loss resistor.

5.5 Power Injection in the Steady State

Power injection at the point-of-common coupling (PCC) by the VSC is controlled by $\mathbf{e}_{abc}$, see Figure 5.1. Here, the power flow (active and reactive power) is a function of $\mathbf{e}_{abc}$ and the uncontrolled voltage v_{abc} at the PCC. Without loss of generality,

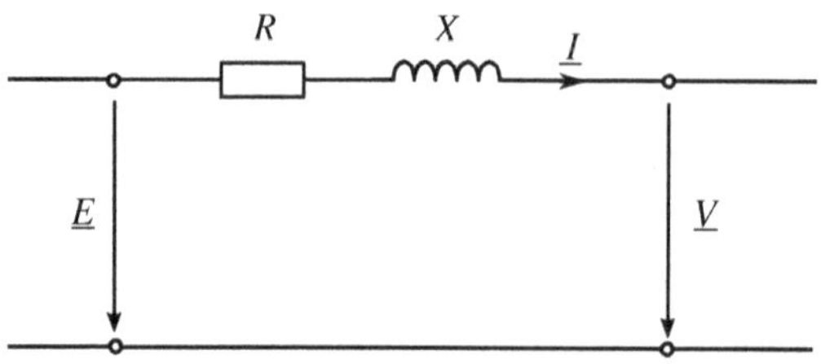

Figure 5.4 Single-phase equivalent circuit of a line with R and $X = \omega L$

we can derive for a balanced three-phase system (5.2) the mathematical equation from Kirchhoff's laws by phasor calculus approach. For this purpose, let us consider an equivalent circuit of a single line between the injector and PCC with a resistance R and reactance $X = \omega L$ in series illustrated in Figure 5.4. The reference voltage $\underline{V}$ at the PCC in Figure 5.4 is represented by a phasor with a pure real value. In contrast, $\underline{E}$ is variable in angle and amplitude

$$\underline{V} = V e^{j0^\circ} = V, \qquad \underline{E} = E e^{j\delta} = E(\cos\delta + j\sin\delta). \tag{5.55}$$

where δ denotes the so-called power angle. From Kirchhoff's loop rule, it follows that

$$-\underline{E} + \underline{V} + (R + jX)\underline{I} = 0, \tag{5.56}$$

where $X = \omega L$. With (5.55) one obtains after a short calculation

$$\underline{I} = \frac{R(E\cos\delta - V) + XE\sin\delta}{R^2 + X^2} - j\frac{X(E\cos\delta - V) - RE\sin\delta}{R^2 + X^2}, \tag{5.57}$$

With the definition of three-phase apparent power (5.28)

$$S_{3\phi} = 3\underline{V}\,\underline{I}^* = P_{3\phi} + jQ_{3\phi},$$

and by substituting $\underline{I}^*$ by the conjugate complex expression of (5.57), we get

$$\begin{aligned} P_{3\phi} &= \frac{3V}{R^2 + X^2}\,[R(E\cos\delta - V) + XE\sin\delta\,], \\ Q_{3\phi} &= \frac{3V}{R^2 + X^2}\,[XE\cos\delta - RE\sin\delta - XV\,]. \end{aligned} \tag{5.58}$$

With the special case that $X \gg R$, the resistance R is neglected, where

$$P_{3\phi} = \frac{3EV\sin\delta}{X}, \qquad Q_{3\phi} = 3\,\frac{EV\cos\delta - V^2}{X}, \tag{5.59}$$

is obtained.

5.6 Voltage Drop in the Steady State

In addition to the steady-state power flow calculation, the characterization of voltage drop on the lines must also be understood. That allows a suitable injection of (reactive) power by the VSC to control the voltage level at a nearby point in the

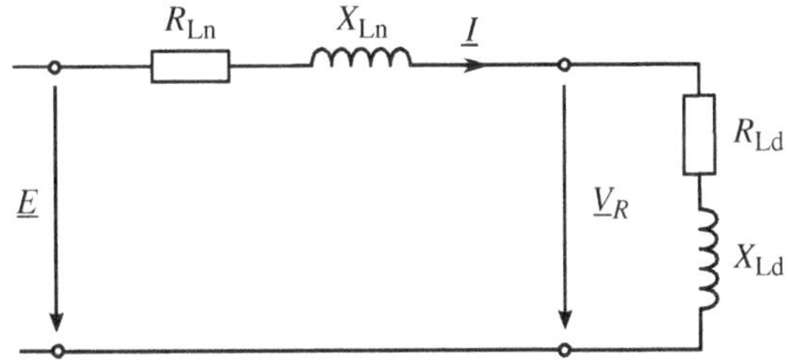

Figure 5.5 Single-phase equivalent circuit of a line terminated with a load impedance

grid. The equivalent circuit diagram in Figure 5.5 shows the single-phase scenario to be investigated. From the equivalent circuit in Figure 5.5 we obtain the voltage equations

$$-\underline{V}_R + \underline{Z}_{Ld}\underline{I} = 0, \tag{5.60}$$

$$-\underline{E} + \underline{Z}_{Ln}\underline{I} + \underline{V}_R = 0. \tag{5.61}$$

The phasor notation of the voltage and current variables

$$\underline{E} = E\,e^{j\delta}, \qquad \underline{V}_R = V_R\,e^{j0^\circ} = V_R, \qquad \underline{I} = I_d + jI_q, \tag{5.62}$$

and the line impedance

$$\underline{Z}_{Ln} = R_{Ln} + jX_{Ln} = Z_{Ln}\cos\varphi_{Ln} + jZ_{Ln}\sin\varphi_{Ln}, \tag{5.63}$$

and load impedance

$$\underline{Z}_{Ld} = R_{Ld} + jX_{Ld} = Z_{Ld}\cos\varphi_{Ld} + jZ_{Ld}\sin\varphi_{Ld}, \tag{5.64}$$

where δ is the power angle and φ_{Ld} the power factor related to the load with $\underline{Z}_x = \sqrt{R_x^2 + X_x^2}$ and $X_x = \omega L_x$, $x \in \{Ln, Ld\}$ are used for the following calculations. Note that the d-axis and q-axis of the current phasor of a single phase do not correspond to the dq-coordinates of the rotating three-phase dq-reference frame in Section 5.2. To avoid confusion, the dq-components of the three-phase rotating system are usually written in lower case and the single-phase phasor calculus in upper case. By substituting $\underline{Z}_{Ld}$ and $\underline{I}$ in (5.60) with the expressions from (5.62) and (5.63), we obtain the equations separated into a real and an imaginary part and compactly represent it in the matrix notation:

$$\begin{pmatrix} Z_{Ld}\cos\varphi_{Ld} & -Z_{Ld}\sin\varphi_{Ld} \\ Z_{Ld}\sin\varphi_{Ld} & Z_{Ld}\cos\varphi_{Ld} \end{pmatrix}\begin{pmatrix} I_d \\ I_q \end{pmatrix} = \begin{pmatrix} V_R \\ 0 \end{pmatrix}.$$

Using the transpose of the orthogonal rotation matrix we obtain the current components

$$\begin{pmatrix} I_d \\ I_q \end{pmatrix} = \begin{pmatrix} \cos\varphi_{Ld} & \sin\varphi_{Ld} \\ -\sin\varphi_{Ld} & \cos\varphi_{Ld} \end{pmatrix}\begin{pmatrix} \frac{V_R}{Z_{Ld}} \\ 0 \end{pmatrix} = \begin{pmatrix} \frac{V_R}{Z_{Ld}}\cos\varphi_{Ld} \\ -\frac{V_R}{Z_{Ld}}\sin\varphi_{Ld} \end{pmatrix}, \tag{5.65}$$

that are still used later. In the next step the second voltage equation (5.61) will be addressed. Analogous as before, $\underline{I}$ and $\underline{Z}_{Ln}$ are replaced by (5.62) and (5.63) and one gets

$$\underline{E} - \underline{V}_R = (R_{Ln} I_d - X_{Ln} I_q) + j(X_{Ln} I_d + R_{Ln} I_q).$$

With the definition

$$\Delta\underline{V} := \underline{E} - \underline{V}_R = \Delta V_d + j\Delta V_q, \tag{5.66}$$

where $\Delta\underline{V}_d$ denotes the direct-axis voltage drop and $\Delta\underline{V}_q$ the quadrature-axis voltage drop, we get

$$\Delta V_d = R_{Ln} I_d - X_{Ln} I_q,$$
$$\Delta V_q = X_{Ln} I_d + R_{Ln} I_q.$$

Replacing the currents $I_{d,q}$ with the previously calculated load-dependent terms (5.73) results in

$$\Delta V_d = \frac{V_R}{Z_{Ld}} \left(R_{Ln} \cos\varphi_{Ld} + X_{Ln} \sin\varphi_{Ld}\right), \tag{5.67}$$

$$\Delta V_q = \frac{V_R}{Z_{Ld}} \left(X_{Ln} \cos\varphi_{Ld} - R_{Ln} \sin\varphi_{Ld}\right). \tag{5.68}$$

For the case of a purely inductive line with R_{Ln}=0 the load- and line-dependent voltage drop $\Delta\underline{V} = \Delta V_d + j\Delta V_q$ is illustrated in Figure 5.6. And the matrix notation can also help for a better understanding:

$$\begin{pmatrix} \Delta V_d \\ \Delta V_q \end{pmatrix} = \frac{V_R}{Z_{Ld}} \begin{pmatrix} \cos\varphi_{Ld} & \sin\varphi_{Ld} \\ -\sin\varphi_{Ld} & \cos\varphi_{Ld} \end{pmatrix} \begin{pmatrix} R_{Ln} \\ X_{Ln} \end{pmatrix}. \tag{5.69}$$

Finally, let us consider the important case, where both the line Z_{Ln} and load Z_{Ld} are *purely inductive*. Given

$$R_{Ln} = 0,\ X_{Ln} > 0, \qquad R_{Ld} = 0,\ X_{Ld} > 0 \Rightarrow Z_{Ld} = X_{Ld},\ \varphi_{Ld} = \pi/2,$$

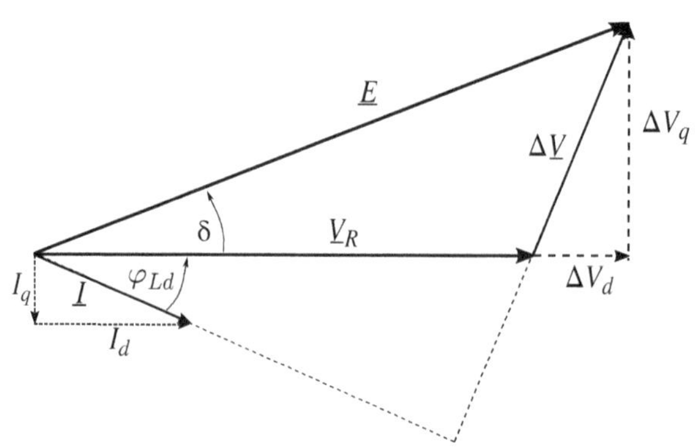

Figure 5.6 Phasor diagram of a single line for $Z_{Ln} = X_{Ln}$, $R_{Ln} = 0$. The direct- axis voltage drop ΔV_d and quadrature-axis voltage drop ΔV_q are a function of I_d, I_q, respectively, the power factor angle φ_{Ld}

then one obtains for (5.69)

$$\begin{pmatrix} \Delta V_d \\ \Delta V_q \end{pmatrix} = \frac{V_R}{X_{Ld}} \begin{pmatrix} 0 & 1 \\ -1 & 0 \end{pmatrix} \begin{pmatrix} 0 \\ X_{Ln} \end{pmatrix} = V_R \begin{pmatrix} \frac{X_{Ln}}{X_{Ld}} \\ 0 \end{pmatrix}, \tag{5.70}$$

with the related (reactive) current

$$\begin{pmatrix} I_d \\ I_q \end{pmatrix} = \begin{pmatrix} 0 \\ -\frac{V_R}{X_{Ld}} \end{pmatrix}. \tag{5.71}$$

This means that if the inductive load due to a short circuit decrease ($X_{Ld} \downarrow$), the resulting increase in the voltage drop ($\Delta V_d \uparrow$) must be compensated by a reactive current opposite to I_q. Thus, reactive power Q must be injected by the VSC. The practically irrelevant but still interesting case[5] is the *purely resistive* load. Given

$$R_{Ln} > 0,\ X_{Ln} = 0, \qquad R_{Ld} > 0,\ X_{Ld} = 0 \ \Rightarrow\ Z_{Ld} = R_{Ld},\ \varphi_{Ld} = 0,$$

then one obtains for (5.69)

$$\begin{pmatrix} \Delta V_d \\ \Delta V_q \end{pmatrix} = \frac{V_R}{R_{Ld}} \begin{pmatrix} 1 & 0 \\ 0 & 1 \end{pmatrix} \begin{pmatrix} R_{Ln} \\ 0 \end{pmatrix} = V_R \begin{pmatrix} \frac{R_{Ln}}{R_{Ld}} \\ 0 \end{pmatrix}. \tag{5.72}$$

with the related (active) current

$$\begin{pmatrix} I_d \\ I_q \end{pmatrix} = \begin{pmatrix} \frac{V_R}{R_{Ld}} \\ 0 \end{pmatrix}. \tag{5.73}$$

This corresponds to the conditions for a DC voltage divider.

5.7 Selected Control Concepts

5.7.1 Control and Functional Requirements

To determine the objectives of VSC control, it is first necessary to decide what is required from the perspective of grid operation. That means which control functions are recommended to provide robust dynamic support. After that, a matrix is set up that helps to indicate which combinations of the primary and secondary power conversion control can lead to the desired results. Finally, test scenarios are described that are used for the overall system validation in

5 In fact, the resistive component is 5 to 25 times smaller (increasing with increasing voltage level) for cables and overhead lines, see the examples of overhead line and cable parameters in Kundur [1994].

Chapter 6. For the operation of primary power conversion in Chapters 3 and 4, we have distinguished between the standard maximum power tracking (MPPT) respective power limitation (PL) for wind turbines in the full-load region and the advanced active/demanded power tracking (DPT) functionality. Now, the proposed advanced controller functionality of the primary converter side for solar PV and wind power systems will be suitably combined with the VSCs as the main element of the secondary power converter side (see Figure 1.3). Inverter-based regenerative energy systems like wind turbines and solar PV power plants have to support the grid in different kinds. As already mentioned in the introduction, Chapter 1, the requirements for inverter-based regenerative power plants, mainly in the supply of ancillary services, are increasing worldwide (see Figure 1.2). Soon, power plants must be able to support the grid not only in the transmission grid but also in the distribution grid at the medium-voltage level. In other words, many more inverter-based generation plants (up to three times as many), even in the power class between 10 and 100 MW, must actively contribute to the grid's stability. For this purpose, fully automatized generation assets are needed to provide robust dynamic support to the grid including the following functional requirements[6]

1. Grid feeding (without ancillary service)
2. Grid frequency stabilization
3. Grid voltage stabilization
4. Operating in low system strength condition
5. Re-synchronization capability to restore and reconnect to the grid events
6. System restoration and black-start capability by grid-leading
7. FRT for large grid disturbance

In this study, we focus on the providing of the function nos. 1–4 including primary power conversion (wind and PV generator control proposed in Chapters 3 and 4). In particular, nos. 5 and 6 require extensive system testing with complex multi-machine scenarios, which would exceed the scope of this book. Furthermore, we consider the normal grid operation, i.e. the balanced three-phase system, proposed in Section 5.2, can describe the instantaneous grid state. The FRT functionality in no. 7, especially in the case of asymmetric faults using symmetric components, is not examined either. This would go beyond the scope of the book.

---

6 In systems engineering, a functional requirement defines a function of a system, which describes a specified behavior between inputs and outputs. In this case, the measurement at the PCC is the input and the actuation comes from the current or voltage source feed by a variable renewable energy resource.

Table 5.1 Matrix of primary and secondary power conversion control scheme

	Conventional primary control: MPPT/power limitation	Advanced primary control: DPT/Active power control
GFL VSC	1. Grid feeding	2. Grid freq. stabilization
GFM VSC	3. Grid voltage stabilization	2. Grid freq. stabilization 3. Grid voltage stabilization 4. Operating in low system strength condition

As an overview, Table 5.1 shows the functions to be implemented as a suitable combination of the previously developed controller of the primary power conversion by MPPT or DPT and the secondary by GFL or GFM VSC. The minimum requirement for grid feeding without support is fulfilled with the conventional primary controller and the GFL power injection. On the other hand, a GFM VSC is required to stabilize the grid voltage. In addition, for the primary side a demanded power control can be used, but it is not necessary for voltage stabilization, as almost only reactive power is injected. For the grid frequency stability shown experimentally in Andrejewski et al. [2023], this can also be accomplished with GFL inverters if the frequency response is within the time range of 4–6 seconds, see the upper right matrix entry in Table 5.1. However, if the frequency of power systems is to be controlled for weak networks, the behavior as a voltage source is sufficient. In order to be able to make reliable statements about this, scenarios are defined for the overall system validation in Chapter 6.

5.7.2 Control Scheme of Grid-Following VSC

The control scheme of the classic GFL VSC control system for grid parallel operation of converter-based generators is illustrated in Figure 5.7. The LCL filter of the VSC is simplified and modeled by an impedance L_f (5.44), and the transformer by a series impedance L_t. The scheme consists of two proportional integral (PI) inner-loop current control[7] related to the synchronous reference frame (5.13) and a slower outer loop for power control. For this purpose, the instantaneous active and reactive power at the PCC are typically calculated by the measurement at the feed-in transformer on the medium voltage-level side. The calculation is done in dq-coordinates with (5.36) and (5.37). For plausibility check, the active power in the original abc-coordinates with (5.24) can be calculated additionally. The synchronization with the electrical grid (terminal voltage) is done by a PLL. Several

7 Is also called voltage-oriented vector current control (DQCC) [Teodorescu et al., 2007].

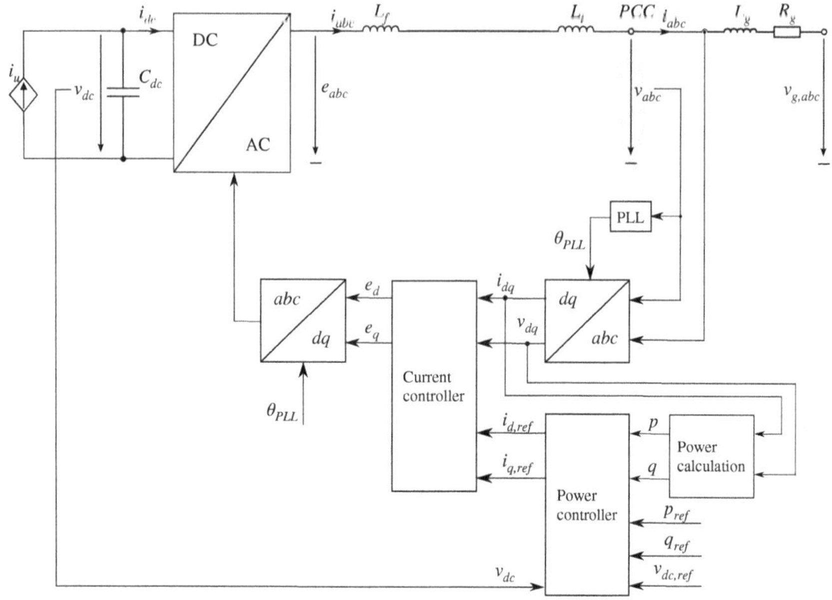

Figure 5.7 Control scheme of voltage source converter for grid-following (GFL)

PLL implementations have been proposed in recent years [Golestan et al., 2017]. In this study, the conventional synchronous dq-reference frame PLL (SRF-PLL) [Golestan et al., 2017] is used and aligned with v_d such that $v_q = 0$. The PLL estimated angle θ is used for the park and inverse park transform, see Figure 5.7. Reference current $i_{dq,ref}$ in d and q coordinates are calculated with the power controller. The current controller output provides an instantaneous voltage (angle and absolute value), converted by space vector modulation (SVM) into pulse patterns for the electronic switches. In addition to the control loops shown above, a third loop can be created around the two inner circles, where the reference values can be calculated from the droop characteristic $P_{ref} = P(f)$ and $Q_{ref} = Q(v)$ to obtain a demanded active and reactive power. However, this is not the control objective here because we assume an external input of ΔQ_{ref} and ΔP_{ref} by a higher-level controller as defined in the general concept in Figure 1.3. The higher-level controller can be provided, e.g. by a park or hybrid power plant controller. These are already commercially available. Or methods for a more optimal distributed network operation are used, such as the concept of dynamic virtual power plants (DVPP) [Marinescu et al., 2022]. Let us now turn to the description of the conventional GFL control algorithms consisting of the power (P, Q) and the internal current control (i_d, i_q).

5.7.2.1 PQ Control of Grid-Following VSC

The power controller block illustrated in Figure 5.7 contains two PI controllers[8]

$$\begin{aligned} i_{d,ref} &= k_{p,p}\, e_{pf} + k_{I,p} \int_0^t e_{pf}(\tau)d\tau + u_{dc,ref}, \\ i_{q,ref} &= k_{p,q}\, e_{qf} + k_{I,q} \int_0^t e_{qf}(\tau)d\tau, \end{aligned} \tag{5.74}$$

with the related power signal filters

$$\dot{e}_{pf} = -\frac{1}{\tau_{pf}} e_{pf} + \frac{1}{\tau_{pf}} e_p, \qquad \dot{e}_{qf} = -\frac{1}{\tau_{qf}} e_{qf} + \frac{1}{\tau_{pf}} e_q, \tag{5.75}$$

where τ_{pf} and τ_{qf} denote filter time constants, which provide additional design freedom. A third PI controller regulates the DC-bus voltage fluctuations caused by $p_{dc} \neq p_{3\phi}$

$$u_{dc,ref} = k_{p,v_{dc}} (v_{dc} - v_{dc,ref}) + k_{I,v_{dc}} \int_0^t (v_{dc} - v_{dc,ref})d\tau, \tag{5.76}$$

by overriding the $i_{d,ref}$ value of the active power controller in (5.74) by $u_{dc,ref}$, which leads to the balancing of the power flow $p_{dc} = p_{3\phi}$ from the primary to secondary side. The power errors in (5.74)

$$e_{pf} = p_{3\phi,ref} - p_{3\phi}, \qquad e_{qf} = q_{3\phi,ref} - q_{3\phi}, \tag{5.77}$$

are determined from the reference values and the calculated power in the synchronous dq-reference frame (5.38)

$$\begin{pmatrix} p_{3\phi} \\ q_{3\phi} \end{pmatrix} = \begin{pmatrix} v_d & v_q \\ v_q & -v_d \end{pmatrix} \begin{pmatrix} i_d \\ i_q \end{pmatrix}.$$

The underlying control approach is based on the assumption that the d-axis of rotating reference frame is perfectly aligned with the grid voltage so that $v_q = 0$. Therefore with

$$p_{3\phi} = v_d\, i_d, \qquad q_{3\phi} = -v_d\, i_q, \tag{5.78}$$

the active power can be adjusted in that ideal case via the d-current and the reactive power via the q-current. With the assumption of constant voltage at the PCC the active power and reactive power are proportional to i_d, respectively, i_q. Note using the definition of the PI controller states

$$x_{I,p} := \int_0^t e_{pf}(\tau)d\tau, \qquad x_{I,q} := \int_0^t e_{qf}(\tau)d\tau, \tag{5.79}$$

8 The PQ controller and the inner-loop i_d,i_q controller with the decoupling network follows the structure in Teodorescu et al. [2011], Figure 9.19.

allows the PQ regulator to be formulated in the state space with the differential equation

$$\begin{aligned}
\dot{e}_{pf} &= -\frac{1}{\tau_{pf}} e_{pf} + \frac{1}{\tau_{pf}} (p_{3\phi,ref} - v_d\, i_d - v_q\, i_q), \\
\dot{e}_{qf} &= -\frac{1}{\tau_{qf}} e_{qf} + \frac{1}{\tau_{qf}} (q_{3\phi,ref} - v_q\, i_d + v_d\, i_q), \\
\dot{x}_{I,p} &= e_{pf}, \\
\dot{x}_{I,q} &= e_{qf},
\end{aligned} \tag{5.80}$$

and the output equation of the controller

$$\begin{aligned}
i_{d,ref} &= k_{p,p}\, e_{pf} + k_{I,p}\, x_{I,p}, \\
i_{q,ref} &= k_{p,q}\, e_{qf} + k_{I,q}\, x_{I,q},
\end{aligned} \tag{5.81}$$

where

$$\begin{aligned}
&\mathbf{x} = (e_{pf}\,,\, e_{pf}\,,\, x_{I,p}\,,\, x_{I,q}\,)^T, \\
&\mathbf{u} = (p_{3\phi,ref}\,,\, q_{3\phi,ref}\,)^T, \qquad \mathbf{y} = (\, i_{d,ref}\,,\, i_{q,ref}\,)^T.
\end{aligned} \tag{5.82}$$

The quantities $i_{d,q}$ in (5.80) are determined by (5.87) and the voltage coordinates in $v_{d,q}$ at the PCC are disturbance variables from the control point of view.

5.7.2.2 Current Control of GFL Voltage Source Converter

The current controller, which is part of the overall regulator scheme in Figure 5.7, results from the VSC model. Based on VSC model (5.49) according to (5.49)

$$\begin{aligned}
L_{ft} \frac{di_d}{dt} &= -v_d + \omega L_{ft}\, i_q + e_d, \\
L_{ft} \frac{di_q}{dt} &= -v_q - \omega L_{ft}\, i_d + e_q,
\end{aligned} \tag{5.83}$$

an appropriate controller law is obtained

$$\begin{aligned}
e_d &= \underbrace{v_d - \omega L_{ft}\, i_q}_{\text{decoupling term}} + k_{p,i_d} (i_{d,ref} - i_d) + k_{I,i_d} \int_0^t (i_{d,ref} - i_d)\, d\tau, \\
e_q &= \underbrace{v_q + \omega L_{ft}\, i_q}_{\text{decoupling term}} + k_{p,i_q} (i_{q,ref} - i_q) + k_{I,i_q} \int_0^t (i_{q,ref} - i_q)\, d\tau.
\end{aligned} \tag{5.84}$$

If we substitute $e_{d,q}$ in (5.83) by the control law (5.84), we get an augmented state space model with

$$x_{I,i_d} := \int_0^t (i_{d,ref} - i_d) d\tau\,, \qquad x_{I,i_q} := \int_0^t (i_{q,ref} - i_q) d\tau, \tag{5.85}$$

for

$$\mathbf{x} = (i_d, x_{I,i_d}, i_q, x_{I,i_q})^T, \qquad \mathbf{u} = (i_{d,ref}, i_{q,ref})^T, \tag{5.86}$$

that represents a closed-loop system with two decoupled subsystems:

$$\begin{aligned}
\frac{di_d}{dt} &= -\frac{k_{p,i_d}}{L_{ft}} i_d + \frac{k_{I,i_d}}{L_{ft}} x_{I,i_d} + \frac{k_{p,i_d}}{L_{ft}} i_{d,ref} \\
\frac{dx_{I,i_d}}{dt} &= -i_d + i_{d,ref} \\
\frac{di_d}{dt} &= -\frac{k_{p,i_q}}{L_{ft}} i_q + \frac{k_{I,i_q}}{L_{ft}} x_{I,i_q} + \frac{k_{p,i_q}}{L_{ft}} i_{q,ref} \\
\frac{dx_{I,i_q}}{dt} &= -i_q + i_{q,ref},
\end{aligned} \tag{5.87}$$

Thus, the appropriate choice of the feed-forward terms in (5.84) causes a decoupling of the d and q components, and in the ideal case, two independent systems to be controlled are obtained. However, if the synchronization conditions (5.47) and/or the balancing condition (5.4) are no longer fulfilled, the GFL network operation can lead to instabilities. Also for this reason in addition to the already mentioned, research in the field of GFM converters has increased in recent years. Two controller structures first presented in Klaes et al. [2020] and Kusche et al. [2022] are now presented.

5.7.3 Control Scheme of Grid-Forming VSC

The control scheme of the GFM VSC control proposed in Klaes et al. [2020] is shown in Figure 5.8. Here, it is immediately noticeable that no PLL is used. Optionally, a frequency lock loop (FLL) is implemented to adapt the droop characteristic in the power controller block to the average grid frequency. That is usually only necessary for weak grids or if the secondary control cannot sufficiently compensate the frequency drop. Due to the lack of PLL, unlike the GFL scheme, the power is calculated in the $\alpha\beta$ and not in the dq-coordinates.

An extension of the concept from Klaes et al. [2020] presented here includes a voltage regulation of the DC link.

To achieve this, a voltage regulator output is superimposed on the active power control by setting the power angle δ. However, the share of an extra $\Delta\delta$ for voltage control should be small compared to the demanded power. Therefore, synchronizing the power flow from the primary power conversion unit (PV-Generator/Wind-Power-Generator) into the secondary power conversion unit with power injection control by the VSC is critical. Further details on active power flow synchronization in the overall system are addressed in Chapter 6.

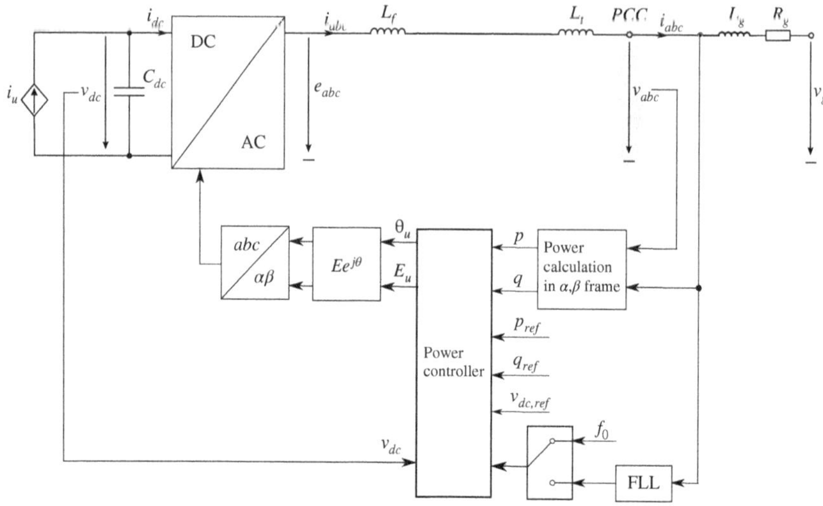

Figure 5.8 Control scheme of voltage source converter for grid-forming (GFM)

5.7.3.1 PQ Control by Direct Droop-Based Voltage Control

The power calculation of the control scheme in Figure 5.8 is done in the stationary $\alpha\beta$-reference frame

$$\mathbf{i}_{\alpha\beta} = \mathbf{T}_{\alpha\beta}\,\mathbf{i}_{abc}, \qquad \mathbf{v}_{\alpha\beta} = \mathbf{T}_{\alpha\beta}\,\mathbf{v}_{abc}, \tag{5.88}$$

using (5.16) with the matrix formula (5.34)

$$\begin{pmatrix} p_{3\phi} \\ q_{3\phi} \end{pmatrix} = \begin{pmatrix} v_\alpha & v_\beta \\ v_\beta & -v_\alpha \end{pmatrix} \begin{pmatrix} i_\alpha \\ i_\beta \end{pmatrix},$$

where $\mathbf{i}_{\alpha\beta} = (i_\alpha, i_\beta)^T$ and $\mathbf{v}_{\alpha\beta} = (v_\alpha, v_\beta)^T$. Along with the reference power p_{ref} and q_{ref}, the power controller sets the necessary power angle δ and the voltage amplitude E of the space vector $\vec{e}(t)$ (5.9). From this, the voltage values in the stationary $\alpha\beta$-reference frame can be calculated with

$$\vec{\mathbf{e}}_{\alpha\beta} = \sqrt{3}E_u e^{j\theta_u} \quad \text{where} \quad \mathbf{e}_{\alpha\beta} = \begin{pmatrix} e_\alpha \\ e_\beta \end{pmatrix} = \begin{pmatrix} \sqrt{3}E_u \cos(\omega t + \theta_u) \\ \sqrt{3}E_u \sin(\omega t + \theta_u) \end{pmatrix}, \tag{5.89}$$

and the reverse transformation (5.17) yields the *abc*-phases of the three-wire system

$$\mathbf{e}_{abc} = \mathbf{T}_{\alpha\beta}^T\,\mathbf{e}_{\alpha\beta}. \tag{5.90}$$

The control law of the power regulator is given as

$$\begin{aligned}\theta_u &= 2\pi \int_0^t \left(k_{I,p}\, e_{p,f}(\tau) + f_0\right) d\tau + k_{p,p}\, e_{p,f},\\ E_u &= k_{I,q} \int_0^t e_{q,f}\, d\tau + k_{p,q}\, e_{q,f} + |\mathbf{v}_{\alpha\beta}|,\end{aligned} \tag{5.91}$$

where $|\mathbf{v}_{\alpha\beta}| = \sqrt{v_\alpha^2 + v_\beta^2}$ using the amplitude-invariant Clarke transform $\mathbf{v}_{\alpha\beta} = \tilde{\mathbf{T}}_{\alpha\beta}\mathbf{v}_{abc}$ with (5.41) and the filtered control error signals of the active and reactive power

$$\begin{aligned}\dot{e}_{pf} &= -\frac{1}{\tau_{pf}} e_{pf} + \frac{1}{\tau_{pf}} \underbrace{(p_{3\phi,ref} - p_{3\phi})}_{e_p},\\ \dot{e}_{qf} &= -\frac{1}{\tau_{qf}} e_{qf} + \frac{1}{\tau_{qf}} \underbrace{(q_{3\phi,ref} - q_{3\phi})}_{e_q}.\end{aligned} \tag{5.92}$$

The instantaneous active power $p_{3\phi}$ and reactive $q_{3\phi}$ are calculated by (5.34) using the measurements v_{abc} and i_{abc} at the PCC transformed with (5.88). The signal processing of the power error signal follows the procedure of the GFL control (5.80).

Using the control law (5.95) for GFM VSC proposed in Klaes et al. [2020], the power angle is determined by a droop factor $k_{I,p}$ as the gain of the integral part as well as directly by a proportional controller part by $k_{p,p}$. According to Weise et al. [2019] the provision of grid support in terms of power/frequency requires a fast response. That can be achieved with this controller structure. However, fast frequency response behavior alone does not ensure stable grid operation, i.e. additional slower control measures such as power signal filters (5.92) are required to achieve the fulfillment of classical system voltage and frequency stability. To fulfill requirements in the time domain approximately between 20 and 60 seconds, higher-level control such as classical secondary control or new approaches like dynamic virtual power plants [Marinescu et al., 2022, Ochoa et al., 2023], which cannot be further investigated here, must be applied.

It should be noted that not only the control structure but also the parameterization of the controller is important for a GFM behavior. For controller parameterization, the following experimentally validated settings were proposed in [Klaes et al., 2020]:

$$k_{I,p} = \frac{v_{sc}}{3\,\pi\, f_0\, \tau_{qf}}, \qquad k_{p,p} = 2\,\pi\, k_f\, f_0\, \tau_{qf}, \tag{5.93}$$

where v_{sc} denotes the relative short circuit drop of the LCL filter impedance. In order to be able to control the DC link voltage, the control law (5.95) for the power angle setting is extended by a DC voltage droop

$$\theta_u = 2\pi \int_0^t \left(k_{I,p}\, e_{p,f}(\tau) + f_0\right) d\tau - k_{p,p}\, e_{p,f} + k_{p,v}(v_{dc} - v_{dc,ref}). \tag{5.94}$$

The controller extension is similar to the concept presented in Subotić and Groß [2022], where the difference of squared voltages is used instead of the voltage difference. The overall concept there is called dual droop.

5.7.4 Grid-Forming VSC Control With Inner-Loop Current Regulator

The second type of GFM converter contains an extra inner-loop dq current controller and an outer-loop PQ power controller illustrated in Figure 5.9. Here, the power regulator does not set the voltage directly. Instead, the desired instantaneous voltage vector is adjusted via the current control, similar to the GFL controller. The reference current for the internal current controller is calculated via the voltage difference between the measurable PCC voltage and the desired voltage $\mathbf{e}_{abc}$ to be set and the VSC impedance L_{ft}. The equations of the cascaded controller scheme for GFM VSC are given as follows.

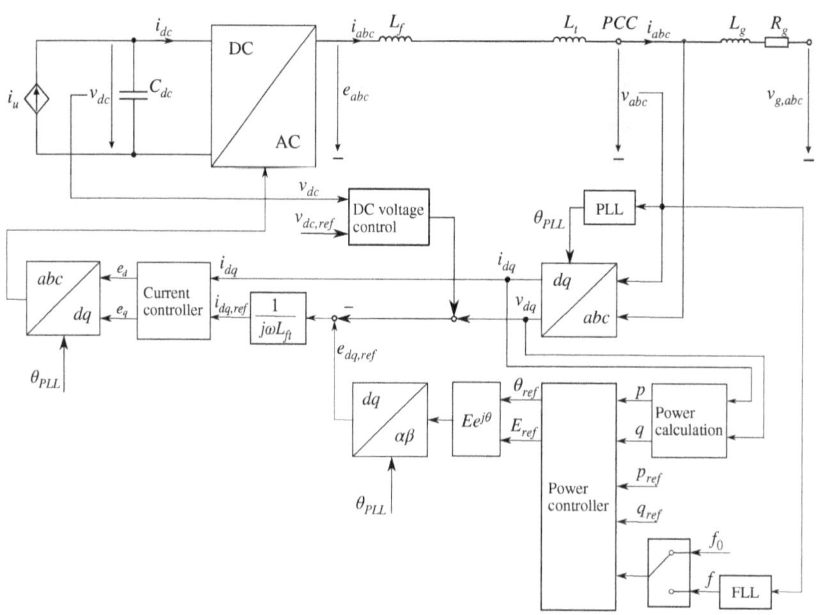

Figure 5.9 Grid-forming VSC scheme with inner-loop current controller

5.7.4.1 PQ Control as Reference Signal Generator for Lower-Level Control

The power calculation is based on the PCC voltages and current measurement values in the rotating dq-reference frame, which corresponds to the GFL procedure in Section 5.7.2 with (5.38). Also, as before in (5.80), the filtered control errors of active and reactive power control are used in the *power controller* block. The power controller is determined by

$$\begin{aligned} \theta_{ref} &= 2\pi \int_0^t \left(k_{I,p}\, e_{p,f}(\tau) + f_0\right) d\tau + k_{p,p}\, e_{p,f}, \\ E_{ref} &= k_{I,q} \int_0^t e_{q,f}\, d\tau + k_{p,q}\, e_{q,f} + |\mathbf{v}_{\alpha\beta}|, \end{aligned} \tag{5.95}$$

where $|\mathbf{v}_{\alpha\beta}| = \sqrt{v_\alpha^2 + v_\beta^2}$ using the Clarke transform (5.41) to get $\mathbf{v}_{\alpha\beta} = \tilde{\mathbf{T}}_{\alpha\beta}\, \mathbf{v}_{abc}$ from the measurements at the PCC. The voltage space vector $\vec{\mathbf{e}}_{ref} = f(E_{ref}, \theta_{ref})$ is then converted first into

$$\mathbf{e}_{\alpha\beta,ref} = \begin{pmatrix} e_{\alpha,ref} \\ e_{\beta,ref} \end{pmatrix} = \begin{pmatrix} \sqrt{3}\, E_{ref} \cos(\omega t + \theta_{ref}) \\ \sqrt{3}\, E_{ref} \sin(\omega t + \theta_{ref}) \end{pmatrix}, \tag{5.96}$$

and finally with the help of Park's transformation (5.18) in

$$\mathbf{e}_{dq,ref} = \mathbf{T}_{dq}(\theta_{PLL})\, \mathbf{e}_{\alpha\beta,ref}. \tag{5.97}$$

The reference values $\mathbf{i}_{dq,ref}$ for the lower-level current controller result from the reduced-order VSC model (5.49) for the steady-state case

$$\mathbf{0} = \begin{pmatrix} 0 & \omega \\ -\omega & 0 \end{pmatrix} \mathbf{x} + \frac{1}{L_{ft}} \begin{pmatrix} e_{d,ref} \\ e_{q,ref} \end{pmatrix} - \frac{1}{L_{ft}} \begin{pmatrix} v_d \\ v_q \end{pmatrix},$$

where $\mathbf{x} = (\, i_{d,ref}\, ,\, i_{q,ref}\,)^T$, which directly results in

$$i_{q,ref} = -\frac{1}{\omega L_{ft}}(e_{d,ref} - v_d), \qquad i_{d,ref} = \frac{1}{\omega L_{ft}}(e_{q,ref} - v_q). \tag{5.98}$$

Optionally, the setpoint calculation of the direct current $i_{d,ref}$ can be extended by a DC voltage controller analogous to the control law (5.74) (d-reference-current) of the GFL converter

$$i_{d,ref} = \frac{1}{\omega L_{ft}}(e_{q,ref} - v_q) + u_{dc,ref}, \tag{5.99}$$

where $u_{dc,ref}$ denotes the DC voltage controller output of (5.76). Note, the formula (5.98) can also be derived with the complex-valued calculation below

$$(i_{d,ref} + j\, i_{q,ref}) = \frac{(e_d + j\, e_q) - (v_d + j\, v_q)}{j\, \omega L_{ft}}, \tag{5.100}$$

which is shown as a block diagram equivalent in Figure 5.9. The inner current control that is fed with the reference values $i_{d,ref}, i_{q,ref}$ exactly corresponds to the PI controller structure (5.84) of the GFL converter in Section 5.7.2.

5.8 Summary

VSCs with GFM control can build the grid voltage and synchronize with other converters or synchronous generators. For this purpose, two different GFM controller concepts were presented: A direct voltage control concept with a PI droop controller without current control and a cascaded controller scheme with an inner current loop and outer power controller to set the desired voltage at the PCC. The latter concept can be considered as a hybrid GFL-GFM converter control scheme. It contains a limited current controller but is also a regulated voltage source. To classify the two GFL concepts presented, a classic GFL converter structure with power control and inner dq current control is explained in detail, too. Chapter 6 considers the overall behavior of the converter concepts and the power controllers' primary resources, i.e. the wind turbine and PV generator, for different scenarios.

5.9 Problems

5.1 Draw the phasor diagram of a single line with a load impedance shown in Figure 5.5 for $R_{Ln} > 0$ and $X_{Ln} > 0$.

5.2 Derive the transfer function of an LCL filter for the system in Figure 5.1. For this purpose, use the components L_{f1}= 5.6 mH, C_f = 3.86 μF, L_{f2} = 5.6 mH. In general, explain and show for the specific example that the filter for the dynamic range of interest can be replaced by an impedance with L_f to (5.44).

5.3 Show with (5.20) and (5.21) that the equality

$$\tilde{\mathbf{T}}_{dq}(\theta)\frac{d\tilde{\mathbf{T}}_{\mathrm{dq}}^{T}(\theta)}{dt} = \begin{pmatrix} 0 & \omega \\ -\omega & 0 \end{pmatrix} \tag{5.101}$$

is valid.

5.4 Simulate the closed-loop dynamics for the GFL VSC according to Section 5.7.2. Therefore, calculate suitable PI controller coefficients for the d- and q-components and choice suitable P and Q filter coefficient.

5.5 Simulate the closed-loop dynamics for the GFM VSC according to Section 5.7.3. Therefore, calculate suitable controller and filter coefficients.

5.6 Describe the advantages and disadvantages of the presented GFM control concepts for VSC of Sections 5.7.3 and 5.7.4.

5.10 Bibliography

For further studies about modeling and control of VSCs, the textbooks Teodorescu et al. [2007] and Yazdani and Iravani [2010] are quite useful. There, among other things, different concepts for realizing PLLs (also for the unbalanced case), different converter topologies with more than two voltage levels, and a more detailed modeling of VSVs are discussed.

The precise, generally applicable definition of the requirements for GFM converters in large power grids is still an open problem. Network operators, converter manufacturers, and research laboratories worldwide are studying that and similar issues. An attempt to define the characteristics and requirements is described in Weise et al. [2019] and NN [2021]. It should be emphasized that different concepts for GFM controls have been recently developed but could not be presented here comprehensively. There are still exciting concepts to be mentioned like the virtual synchronous machine (VISMA)-based VSC control [Beck and Hesse, 2007], universal droop controller [Zhong et al., 2016], direct voltage control [Erlich et al., 2018], GFM with current limitation algorithm [Klaes et al., 2020, Qoria et al., 2020], and dual port control [Subotić and Groß, 2022]. To name a few. Thus, the research on GFM control of VSC for grid support is still ongoing.

6

Advanced Grid Integration of PV and Wind Power Plants

6.1 Introduction

In this chapter, we will now combine the primary power conversion with the secondary power conversion part, see the overview in Figure 1.3. For a reliable analysis, relevant scenarios based on grid events, volatile regenerative resources, and load changes are defined first in Section 6.2. The focus of this study is to evaluate the ancillary services capability of wind power plants (Section 6.3) and photovoltaic power plants (Section 6.4) connected to the grid via power-electronic-based VSC. The investigations are limited to a single-machine system with a grid equivalent. Further considerations like the multi-machine behavior in an x-bus system like the two-area 70-bus system proposed in Mañé et al. [2023] would exceed the scope of this book.

The single-machine system will not be integrated – as many studies do - with an infinity bus. Instead, we use a reduced equivalent grid model based on a *swing equation* and *balanced three-phase voltage source* with time-variable phase angle ϕ_g and RMS voltage magnitude V_g. The swing equation used here is

$$\begin{aligned} \Delta\dot{\omega}_g &= \frac{1}{M_g}(P_{g,S} - P_{L,S} - d_g\,\Delta\omega_g) \quad \text{with} \quad \Delta\omega_g = \omega_g - \omega_s, \\ \dot{\phi}_g &= \Delta\omega_g + \omega_s, \end{aligned} \tag{6.1}$$

where ω_s denotes the electrical synchronous angular frequency, $P_{g,S}$ denotes the total generated power, and $P_{L,S}$ the sum of all loads. The model parameters are d_g as an aggregated damping factor and M_g as the angular moment defined by

$$M_g = \frac{2\,H_g\,S_n}{\omega_s}, \tag{6.2}$$

Advanced Control of Grid-Integrated Renewable Energy Power Plants: LMI-based Design in the Takagi-Sugeno Framework, First Edition. Horst Schulte.

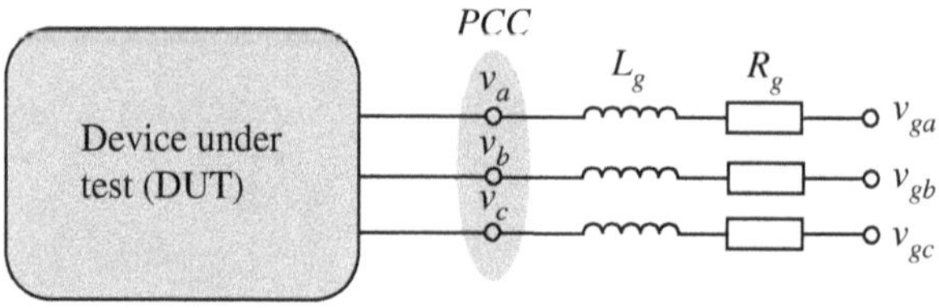

Figure 6.1 Schematic of the test setup with a voltage source given in (6.1), (6.3) represented by grid equivalent

with the aggregated inertia constant H_g (1.2) and rated apparent power S_n of the aggregated system. Balanced three-phase voltage source contains the phase angle $\phi_{gv} = \phi_g$ specified by (6.1)

$$\mathbf{v}_{g,abc}(t) = \begin{pmatrix} v_{ga}(t) \\ v_{gb}(t) \\ v_{gc}(t) \end{pmatrix} = \sqrt{2}V_g(t) \begin{pmatrix} \cos(\omega t + \phi_{gv}(t)) \\ \cos(\omega t + \phi_{gv}(t) - \frac{2\pi}{3}) \\ \cos(\omega t + \phi_{gv}(t) + \frac{2\pi}{3}) \end{pmatrix}, \tag{6.3}$$

which is electrically connected to the PCC via the line impedance $Z = R_g + jX_g$ with $X_g = \omega_g L_g$. To illustrate the generic settings formally described in (6.1) and (6.3), Figure 6.1 shows the terminals of the voltage source with the PCC interface to the device-under-test (DUT). The DUTs are in our case the wind turbine (proposed in Chapters 2 and 3) or the photovoltaic system (proposed in Chapter 4) connected to PCC by the voltage-source converter (proposed in Chapter 5).

6.2 Description of Test Scenarios

The test cases for the DUT are designed to cover various functionalities occurring when participating in a bulk power system. The focus is on studying the systems response to active and reactive power demands, as well as responses to the change in voltage magnitude or phase angle of the equivalent grid (6.1), (6.3). This is to emulate certain grid events, such as phase jumps due to line tripping and voltage drop due to short circuit in the line. Furthermore, system responses from sudden changes in available primary energy sources are also to be analyzed.

Investigation of the response of the DUT to a higher-level control request is equivalent to the command response from the *control engineering point of view*. It is assumed that the request for active power can be fully converted to a reference signal based on the condition for wind power systems described in (3.85), (3.86) with Figure 3.31 and for PV power systems (4.71), (4.72) with Figure 4.25. That means there must always be enough primary power available for the power request. If this is not possible, i.e. local generation reaches its limit, the demand must be dispatched to additional power plants.

The previously mentioned grid events that affect the DUT via the VSC at the PCC are, from a control point of view, disturbances and must be rejected. Disturbance rejection is first investigated on the grid side, where mainly the secondary control of the VSC takes effect. If the power flow needs to be re-regulated, it can also affect the control of the primary power conversion in wind turbines and PV systems.

In contrast to grid events, rapid changes in regenerative resources always affect the primary control. For example, the effects of rapid power change in wind turbines on mechanical fatigue loading have been studied in Clemens et al. [2020]. The control system's performance regarding the disturbance rejection for wind speed, irradiation, and temperature changes has already been discussed in Chapters 3 and 4. The following consideration in this chapter focuses on the effect of the overall system, especially the grid feed-in. Enclosed now is the description of five test scenarios applied to wind power plants in Section 6.3 and photovoltaic systems in Section 6.4.

Case 1: Active power request:

The active power demand is increased step-wise by 0.1 per unit (pu) for testing. Regarding the PV and wind turbine (WT) systems without energy storage, this active power command can only be achieved if the current primary energy source provides sufficient power. For this reason, the RES power plants are operated in a derating operation, where the wind turbine and PV generator operate at a lower power than it could deliver based on the wind resp. irradiation conditions. This ensures that the power plant under investigation operates within and does not exceed the allowable limits quantified by (3.85) and (4.71). The formal description of the test case is

$$\Delta P_{req} = \begin{cases} 0 & t < t_e \\ 0.1\,S_B & t \geq t_e \end{cases}, \tag{6.4}$$

related to the absolute power values

$$P_{req} = \begin{cases} 0.8\,S_B & t < t_e \\ 0.9\,S_B & t \geq t_e \end{cases}, \tag{6.5}$$

where S_B denotes the base power related to three-phase power $S_{3\phi}$ and t_e denotes for all cases time of occurrence of the event. Note the per-unit-value is defined as follows [Machowski et al., 2008]

$$\text{per-unit value} = \frac{\text{actual value}}{\text{base value}}, \tag{6.6}$$

Case 2: Step-wise change of the primary source:

The second case is based on a step-wise change of the primary source: For photovoltaic power plants, the irradiance is changed; for wind power plants, a change in the effective wind speed magnitude is used to excite the wind turbine system. In this test case, the control loops and dynamics of the primary power conversion determine the system's step response. Because the secondary conversion dynamics, containing the VSC with LCL filter, are much faster than the primary dynamics and, therefore, follow the paths given by the wind turbine rotor or PV-generator DC–DC converter dynamics.

Case 3: Reactive power request:

The reactive power demand is varied step-wise by 0.3 pu. This process is mainly governed by the secondary conversion side, where either the current setpoint is recalculated, or the exciter of the synchronous generator is used to provide the desired level of reactive power. A variation of reactive power directly acts on the voltage magnitude of the considered connection point within the electrical grid, thus enhancing voltage stabilization. The changes following a reactive power demand step are governed by either the outer loop power control acting on the current setpoint of the VSC. The closed-loop dynamics of this process are located in the range of several 10 of milliseconds or seconds, respectively, and investigate the reference-tracking dynamics rather than the disturbance dynamics in case of, e.g. voltage drops. The formal description of the test case is

$$Q_{req} = \Delta Q_{req} = \begin{cases} 0 & t < t_e \\ 0.3\, S_B & t \geq t_e \end{cases}. \tag{6.7}$$

Case 4: Voltage magnitude drop (grid event):

In this test case, a voltage magnitude drop to 0.9 pu symmetrically at all three phases determined by (6.3) is performed:

$$V_g = \begin{cases} 1.0\, V_B & t < t_e \\ 0.9\, V_B & t \geq t_e \end{cases}, \tag{6.8}$$

where V_B denotes the base voltage. The drop level is chosen to evaluate the initial disturbance rejection dynamics in normal operation, i.e. without reaching limitations or activating a fault mode due to an identified under-voltage. This relates to small-signal stability. The transfer to a new operating point, which leads back to the feed-in of the power before the event, is performed by the cascaded controller in the voltage source converter.

Case 5: Phase shift (grid event):

In this test case, a step-wise change of the voltage angle of $\Delta\phi = 18°$ is conducted by

$$\Delta\phi = \begin{cases} 0 & t < t_e \\ 18° & t \geq t_e \end{cases}, \tag{6.9}$$

Usually, the phase angle jump occurs in parallel to a variation of the magnitude due to, e.g. a line tripping or a sudden change in loading. For separate analysis, only the phase angle is varied. This test case allows studying the transients of the generation unit, especially considering the synchronization process of the power plants to the grid's voltage.

As we will see in Section 6.3 test scenarios used have different effects on the response of the sub-systems of the considered power plants. This means that different time scales must be considered; thus, the coupling or non-coupling of the primary and secondary sides must be considered. The dynamics of the secondary power conversion side is usually faster than that of the primary, so the phenomena delayed from the primary side have similar dynamics at the PCC and are directly measurable in the injected power. It is investigated and discussed in which test scenarios the transmission characteristics of the VSC are important and when these are negligible. These crucial findings enable justifiable model reduction for advanced system considerations in large test bed networks such as the IEEE 68-bus NETS/NYPS[1] [Singh and Pal, 2013].

Note the results presented in Sections 6.3 and 6.4 were first published in the report [Kusche et al., 2022] and will be analyzed and evaluated in more detail here.

6.3 Integration of Wind Power Plants

Let us now examine the overall system dynamics of the wind power plant consisting the control of turbine rotor by the generator torque and the collective pitch angle proposed as a model-based design concept in Chapter 3 in combination with the automatic controlled VSC for grid-following (GFL) and grid-forming (GFM) operation presented in Chapter 5. Investigations of GFM operation are focused on the second GFM concept with lower-level current controllers proposed in Section 5.7.4.

In the simulation studies the power generated by the wind turbine is the input to dynamic model of the DC link, which captures the DC voltage as an integral state and depending on the size of the capacitance. The power injected into the secondary conversion process is based on the power fed into the DC link model, where the power balance of the generated wind power and the injected power of the converter determines the voltage state of the DC link. It is described by model Eqs. (5.51) to (5.54), which are used in the following simulations.

1 New England Test System/New York Power System.

6.3.1 Response to Active Power Request

Within the simulated Test Case 1 defined in (6.4), a wind speed above the rated wind speed is chosen that can supply 1 pu available power. The power demand is set to 0.8 pu before the request of step-wise variation of the active power. Figure 6.2 shows that the altered power demand is processed by the wind turbine controller adjusting the generator torque and pitch angle to provide the desired power of 0.9 pu. The generator torque can be actuated at a fast scale, where the timescale of the rotational motion is due to rotor inertia in the range of seconds. Caused by the change in active power, the voltage at the PCC varies that results in a transient of the reactive power below 0.01 pu as illustrated in Figure 6.2. The voltage transient at the PCC also has an effect on the estimated frequency of the PLL. The cascaded power and current control loops used in GFM operation of the inverter (see Figure 5.9) regulate the active and reactive power to the setpoints.

Figure 6.3 illustrates the simulated wind turbine states for Test Case 1. The change in power demand results in adjusting both the pitch angle and generator torque to reach the desired power level. While the pitch angle is decreased to allow for an increasing aerodynamical torque produced by the wind turbine rotor, the generator torque increases to match the rotor torque. Due to the selected operating strategy for active power tracking presented in Section 3.4 for wind speeds above the nominal wind speed, the turbine operates at the rated rotor speed.

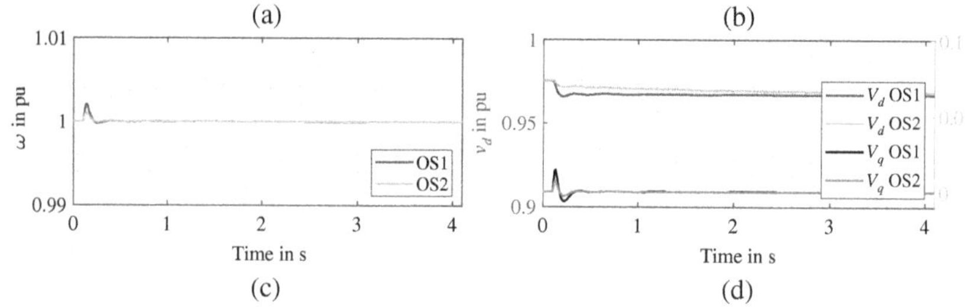

Figure 6.2 Test Case 1: System response after step change of active power request P_{req} at $t_e = 0.1$ s related to the generated active power of the generator P_g and converter power P_{con} are shown in (a). The reactive power Q_{con} is shown in (b), the estimated PLL frequency denoted as ω_{pll} in (c) and the voltage at PCC in dq-coordinates in (d). Note that the control method presented in Section 3.4 corresponds to the operating strategy 1 (OS1). For the operating strategy 2 (OS2) shown for comparison, the rotor speed is also adjusted, as a result the step response is much slower, see (a)

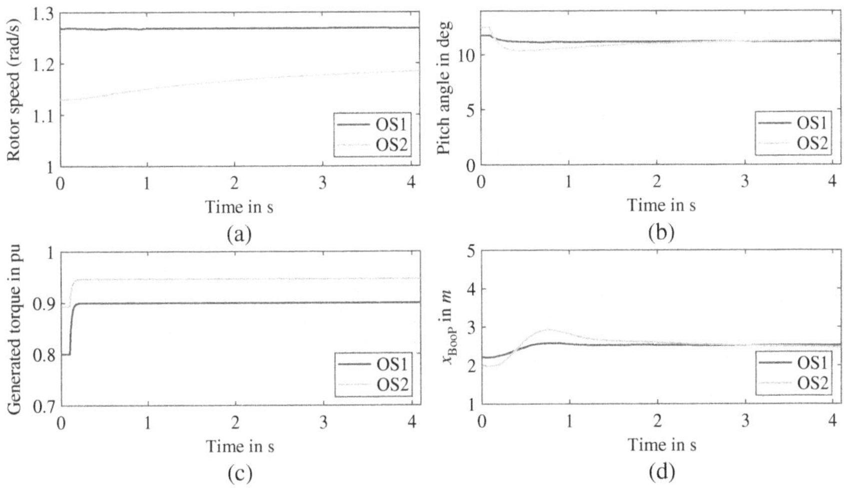

Figure 6.3 Test Case 1: States and system inputs (controller outputs) after step change of active power request P_{req} at $t_e = 0.1$ s: (a) Rotor speed of operating strategy 1 (OS1) proposed in Section 3.4 and operating strategy 2 (OS2) where the rotor and the torque is adjusted. (b) pitch angle of OS1 and OS2, (c) generator torque of OS1 and OS2, and (d) $x_{BooP} := Y_B$ denotes the out of plane motion of the blades

6.3.2 Decrease in Wind Inflow

A decrease of the wind inflow of $\Delta v = -0.3\, v_{rated}$ above the rated wind speed is considered. The results are shown in Figures 6.4 and 6.5. It can be seen that after a short drop in the rotor speed, the rated speed is reached after 8 sec by reducing the pitch angle through the controller. Since the rotor power changes are slow, this has no effect on the reactive power generated by the VSC (see Figure 6.4). Because of the slower time scale of the wind turbine, the primary power conversion in this case is decoupled from the secondary power conversion and fed into the grid. However, the change in active power due to the varying feed-in conditions leads to a minor change in the voltage at the PCC. The estimation of the network frequency by the PLL is not affected, it remains undisturbed.

6.3.3 Response to Reactive Power Request

In the Test Case 3, only the reactive power is adjusted. The active power demand is kept constant, which means that the power generation of the wind turbines remains constant. The active power injected by the converter shows an initial increase in power due to the fluctuations of PCC voltage caused by the reactive power injection. This transient active power quickly decays to the desired set point due to the controller actions in the converter, as shown in Figure 6.6. Depending on the GFL VSC mode (Section 5.7.2) and the used GFM type 2

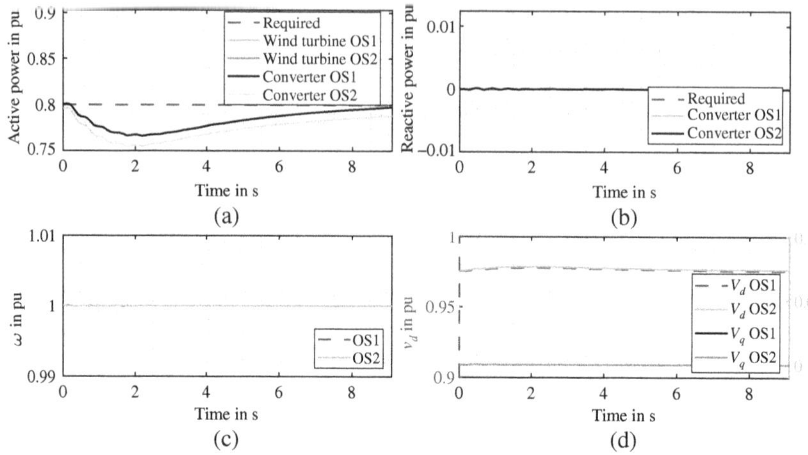

Figure 6.4 Test Case 2: System response after wind speed step at $t_e = 0.1$ s: (a) shows the active power of the generator P_g and converter P_{con} of the operating strategy 1 (OS1) proposed in Section 3.4 and operating strategy 2 (OS2), whereby the rotor speed and generator torque are adjusted simultaneously, (b) shows the reactive power feeded by the converter, (c) the estimated PLL frequency and (d) the voltage at PCC in dq-coordinates

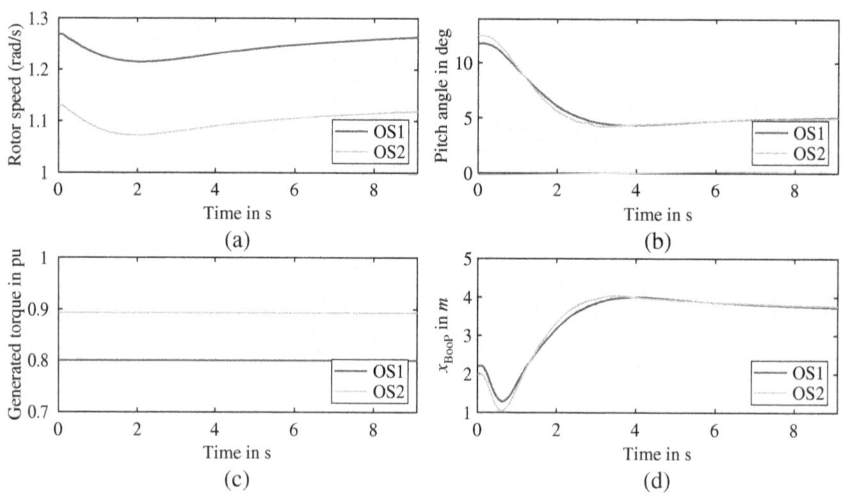

Figure 6.5 Test Case 2: System states and inputs (controller outputs) after wind speed step at $t_e = 0.1$ s: (a) Rotor speed of operating strategy 1 (OS1) proposed in Section 3.4 and operating strategy 2 (OS2), whereby the rotor speed and generator torque are adjusted simultaneously, (b) pitch angle of OS1 and OS2, (c) generator torque of OS1 and OS2, and (d) $x_{BooP} := Y_B$ denotes the out of plane motion of the blades

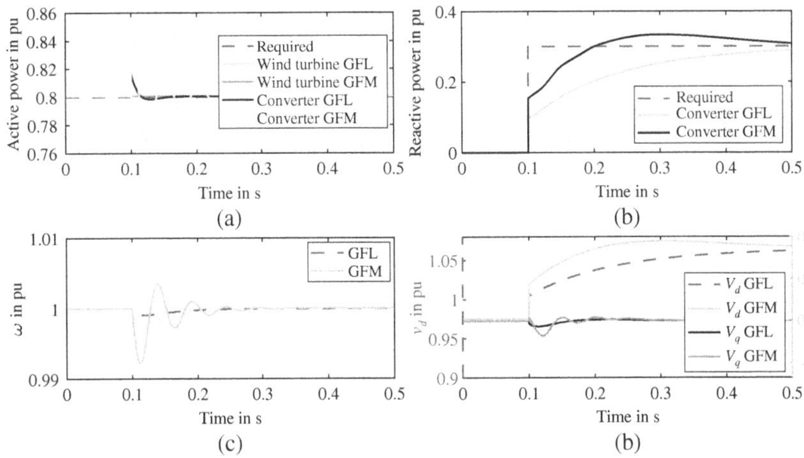

Figure 6.6 Test Case 3: System response after step change of reactive power request at $t_e = 0.1$ s: (a) shows the active power of the generator P_g and converter P_{con} of the operating strategy 1 (OS1) proposed in Section 3.4 and operating strategy 2 (OS2), whereby the rotor speed and generator torque are adjusted simultaneously. (b) shows the reactive power feeded by the converter, (c) the estimated PLL frequency and (d) the voltage at PCC in dq-coordinates

scheme from Section 5.7.4, the active power trajectory shows transient behavior with and without observable oscillation. The converter regulates the reactive power to the desired value at 0.3 pu and reaches the steady state about 500msec after the reactive power request. In both VSC control concepts, the course of the reactive power depends on the selected control parameters. As a result of the reactive power injection, the voltage at the PCC is raised from 0.97 to 1.07 pu, as shown in Figure 6.6. The transient in the PCC voltage results in a slight excitation of the PLL frequency, which decays without oscillation in the range of milliseconds. This has no influence on the estimation of the power frequency by the PLL, which remains undisturbed. The cascaded power and current controllers in the inverter adjust the inverter terminal voltage to force the desired reactive power injection. The resulting trajectories of the terminal voltage in the *dq*-reference frame rotating synchronously with the PLL frequency are shown in Figure 6.7. The reactive power change results in a change in voltage magnitude from 0.99 to 1.11 pu for the terminal voltage *e* in both converter modes. The reactive power injection occurs by supplying the *q*-component of the current, which is forced from an initial 0 to −0.28 pu, as shown in Figure 6.7. As the active power demand remains constant and the voltage magnitude at the PCC increases, the converter reduces the *d*-component of the current to supply the same amount of active power to the grid as before the event. For the GFM mode,

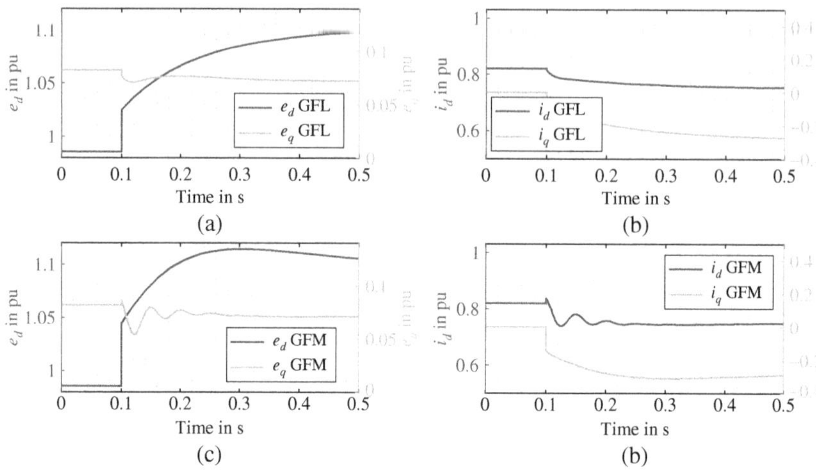

Figure 6.7 Test Case 3: System states and converter outputs after step change of reactive power request at $t_e = 0.1$ s: (a) Grid following (GFL) converter voltage $e_{d,q}$, (b) GFL converter current $i_{d,q}$, (c) Grid forming (GFM) converter voltage $e_{d,q}$, and (d) GFM converter current $i_{d,q}$

the oscillation in terms of power swings caused by mismatch in initial values can be seen in the q-component of the terminal voltage e and the d-component of the current i.

6.3.4 Grid Disturbance Rejection

The validation of grid disturbance rejection behavior of the DUT is based on two test cases. First, the effect of voltage drop (Test Case 4) at the grid point $\mathbf{v}_{g,abc}$ (e.g. illustrated in Figure 6.1) on the power injection of the VSC is investigated. The step-wise change of the grid voltage magnitude $\sqrt{2}V_g$ reduces the voltage at the PCC as shown in the diagram for $v_{d,q}$ of Figure 6.8. In GFL mode as a consequence of the feed-forward term (decoupling term) using the measurement of voltage at the PCC in the control law (5.84) the VSC quickly adjusts the terminal voltage. As a result, the converter injects reactive power (Figure 6.8) caused by the q-current component shown in Figure 6.9. Hence, the voltage level stabilizes for both controller scheme, as can be seen directly from the comparison of GFL and GFM in Figure 6.8. This also leads to a transient of the PLL-estimated ω in GFL mode, which quickly decays in the order of milliseconds shown in Figure 6.8.

Test Case 5: The second grid disturbance on the DUT is a step-wise change in the grid's voltage angle (phase-shift) given in (6.9). That event induces transients in all electrical quantities of the generation system clearly visible in Figures 6.10 and 6.11. Depending on applying the GFL or GFM mode, the transients show distinct dynamical properties.

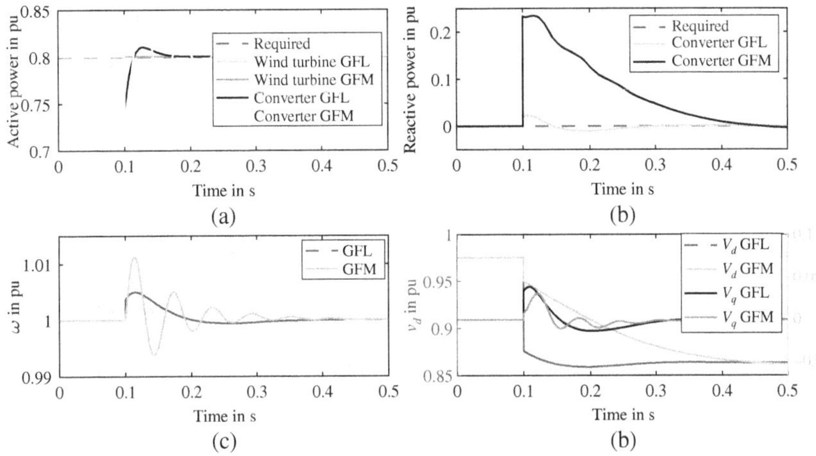

Figure 6.8 Test Case 4: System response after voltage magnitude drop at $t_e = 0.1$ s: (a) Active power of the generator P_g and converter P_{con} of the operating strategy 1 with grid following (GFL) and grid forming (GFM) converter, (b) Shows the reactive power feeded by the GFM and GFL converter, (c) the estimated PLL frequency and (d) the voltage at PCC in dq-coordinates for the GFM and GFL converter

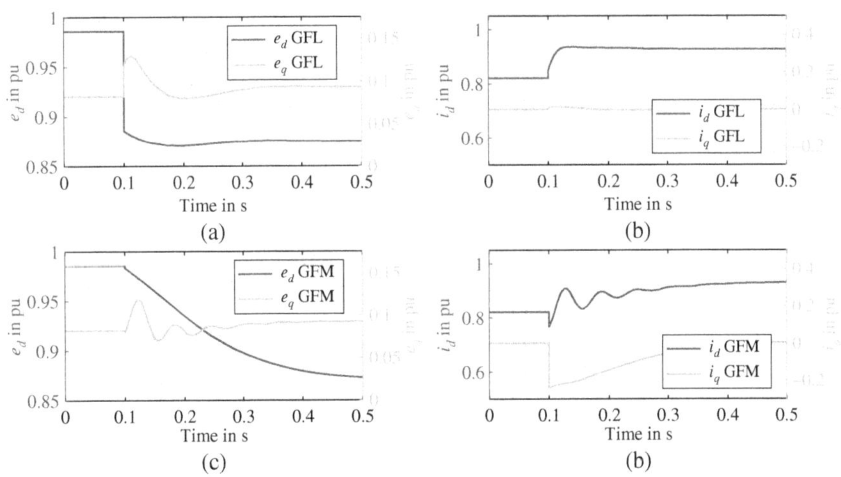

Figure 6.9 Test Case 4: System states and controller outputs after voltage magnitude drop at $t_e = 0.1$ s: (a) Grid following (GFL) converter voltage $e_{d,q}$, (b) GFL converter current $i_{d,q}$, (c) Grid forming (GFM) converter voltage $e_{d,q}$, and (d) GFM converter current $i_{d,q}$

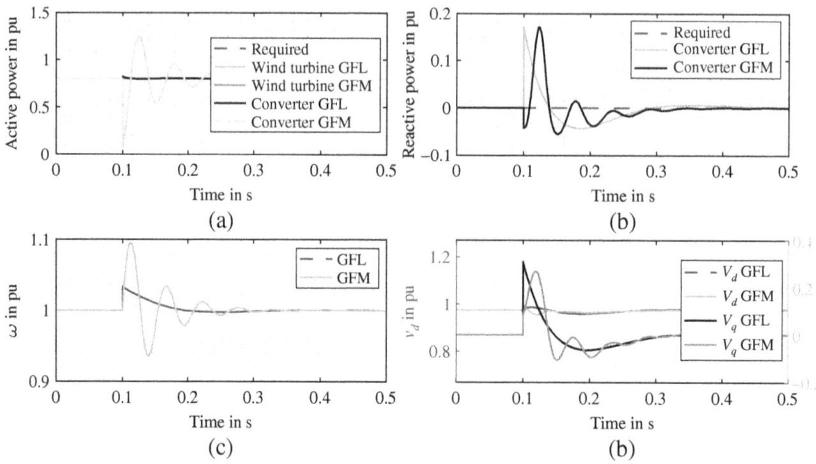

Figure 6.10 Test Case 5: Disturbance rejection after voltage phase shift at $t_e = 0.1$ s: (a) Active power of the wind turbine generator P_g and converter P_{con} with operating strategy 1 using grid following (GFL) and grid forming (GFM) converter, (b) Shows the reactive power feeded by the GFL and GFM converter, (c) the estimated PLL frequency related to GFL and GFM converter, and (d) the voltage at PCC in dq-coordinates for the GFL and GFM converter

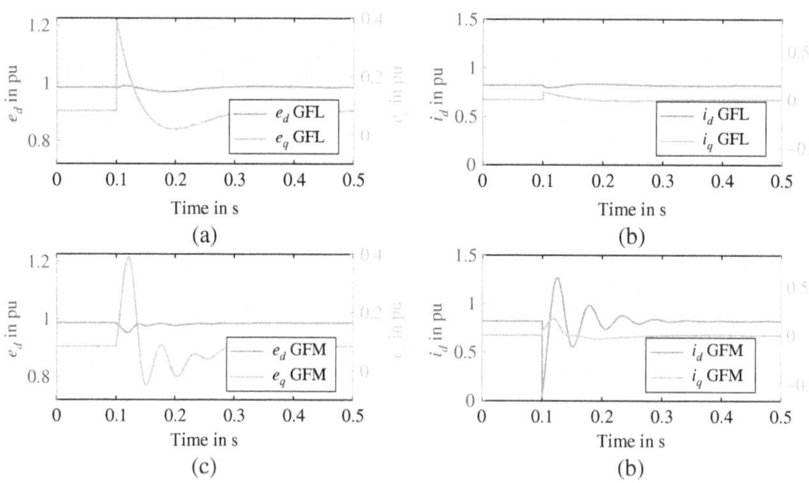

Figure 6.11 Test Case 5: Disturbance rejection after voltage phase shift at $t_e = 0.1$ s: (a) Grid following (GFL) converter voltage $e_{d,q}$, (b) GFL converter current $i_{d,q}$, (c) Grid forming (GFM) converter voltage $e_{d,q}$, and (d) GFM converter current $i_{d,q}$

Both VSC controller concepts operate in a similar time scale of tens of milliseconds, where the GFL trajectories decay without an oscillation. In contrast, the GFR mode shows inertial behavior. The transient of the voltage vector $e_{d,q}$ in the GFR concept is reflected by its frequency in Figure 6.11 and the oscillation visible in the q-component of the terminal voltage $v_{d,q}$ in Figure 6.10. Especially, the phase-locked loop experiences severe transients. Still, it remains in synchronicity with the electrical grid, as shown in Figure 6.10. In both modes, GFL and GFR, the active and reactive power show an initial increase in injected power that subsequently decays due to the control.

6.4 Integration of PV Power Plants

The analysis of the Test Cases 1–5 for the PV power plants follows the same structure as before for wind turbines. In the simulation studies the power generated by the PV-generator controlled by the DC–DC converter is the input of the DC link. The power injected into the secondary conversion process is based on the power fed into the DC link model, where the power balance of the generated PV solar power and the injected power of the converter determines the voltage state of the DC link. It is described by model Eqs. (5.51) to (5.54), which are used in the simulations.

The primary energy conversion of the PV system contains the PV cells, the DC–DC converter, and the power-tracking algorithm. The cells of PV modules consist of an electrical single-diode model. Inputs to the primary conversion are the irradiation, required active power, and the DC link voltage of the VSC. The DC link voltage is needed for the switched mode DC–DC converter, where the demanded power-tracking concept controls the primary conversion. The output of the primary sub-system is the power injected into the DC link. The VSC of the secondary power conversion is identical to the power converter model used in the wind power plant.

6.4.1 Response to Active Power Request

The PV power plant is de-rated in this test case before the event. Here, the power demand is set to 0.8 pu, where a simulated irradiation of $S = 1000\ \mathrm{W/m^2}$ is sufficient to provide the desired power level at standard test conditions, see Figure 6.12. Due to the increased power demand to 0.9 pu, the power-tracking algorithm controlling the primary conversion needs to transfer the system to a new operating point, respecting the non-linearity inherited in the power conversion characteristic of the PV cells. Based on the results with the model-free extremum-seeking controller in Section 4.5.3 (to slow) and the model-based

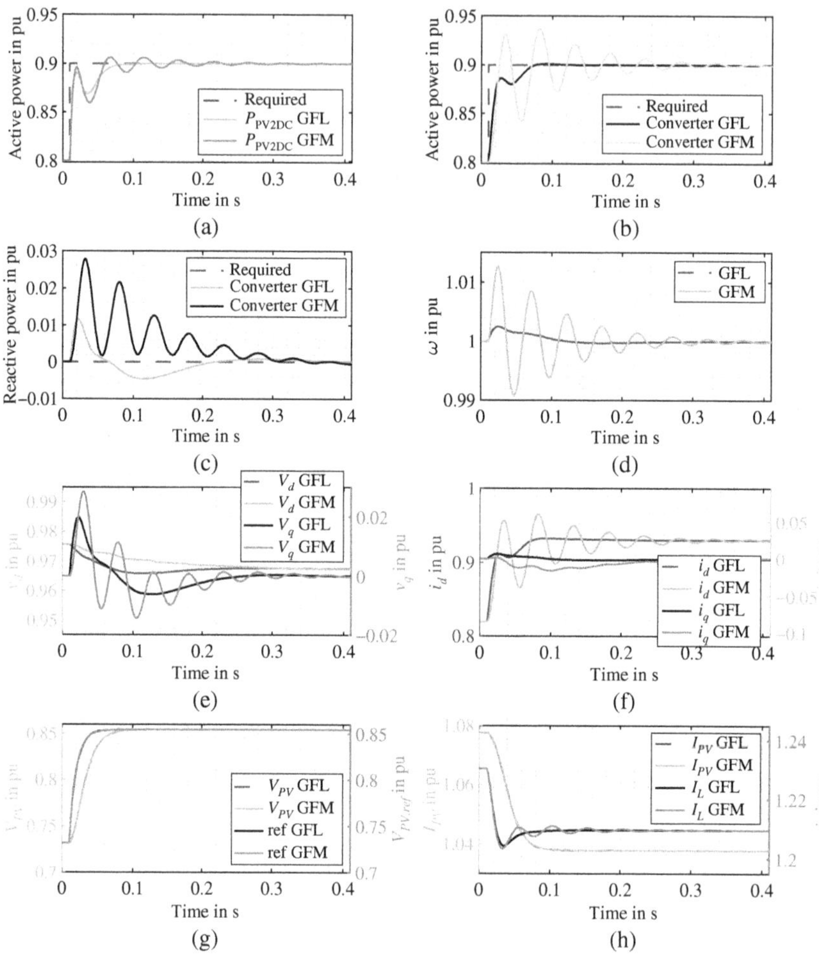

Figure 6.12 Test Case 1: System response after step change of active power request at $t_e = 0.01$ s related to generated PV power (primary power conversion) and converter states (secondary power conversion) with grid-following (GFL) and grid-forming (GFM) operation: (a) PV power, (b) active power at PCC, (c) reactive power at PCC, (d) estimated grid frequency at PCC, (e) voltage at PCC in dq-coordinates, (f) current in dq-coordinates, (g) PV voltage, (h) PV and load current in DC/DC converter

method 4.5.2 (sufficiently fast), we use a model-based approach equivalent to the latter presented in Kusche et al. [2022].

All results for Test Case 1 are shown in Figure 6.12. One can see well how the reference value for the PV voltage in the diagram v_{pv} over t is adjusted within 10 ms. The relatively fast change of the power fed into the DC link leads to some oscillations in the active power that decay after 300 ms. The fluctuations can also be seen on the secondary side. It can be improved by using a larger DC link capacitor to mitigate the oscillations. Or a filter state to reduce the dynamics of $v_{pv,ref}$ could help if necessary.

6.4.2 RES Fluctuation: Irradiation Step

In this scenario, Test Case 2, the power of the primary source decreases. In absolute terms, the initial setting starts with the nominal value at standard test conditions, which is then lowered by 20% for $t \geq t_e = 0.01$ s. After a negative jump in the primary source, the power supplied by the irradiance is no longer sufficient to provide the desired active power at the PCC. This leads in a steady-state offset where only approximately 73% power instead of 80% is generated. The simulations results are shown in Figure 6.13. Similar to the system behavior for "Test Case 1," the applied power-tracking algorithm has a significant influence of the overall power plant dynamics. Simulations with a model-based approach did not lead to valuable results. The fast model-based approach fulfills the requirements. Therefore, the results of this method are analyzed. Due to the shift of the maximum power point, the injected DC power decreases immediately before the applied power-tracking algorithm leads to a convergence toward the desired power set point. Furthermore, a distinct non-minimum phase behavior can be well recognized between the reference voltage $v_{PV,ref}$ and the step response of the actual voltage v_{PV}.

6.4.3 Response to Reactive Power Request

The step change in reactive power demand for Test Case 3 (6.7) leads to the same characteristic results discussed for the wind power plant. A small fluctuation of the DC link voltage based on the varying active power injection can be observed and is shown in Figure 6.14. This causes an oscillation of the power generated by the primary converter.

6.4.4 Grid Disturbance Rejection

The validation of grid disturbance rejection behavior is based on the same scenarios used for the wind power plant. First, *Test Case 4*, the effect of the voltage drop at grid point $\mathbf{v}_{g,abc}$ illustrated in Figure 6.1 on the VSC power injection is

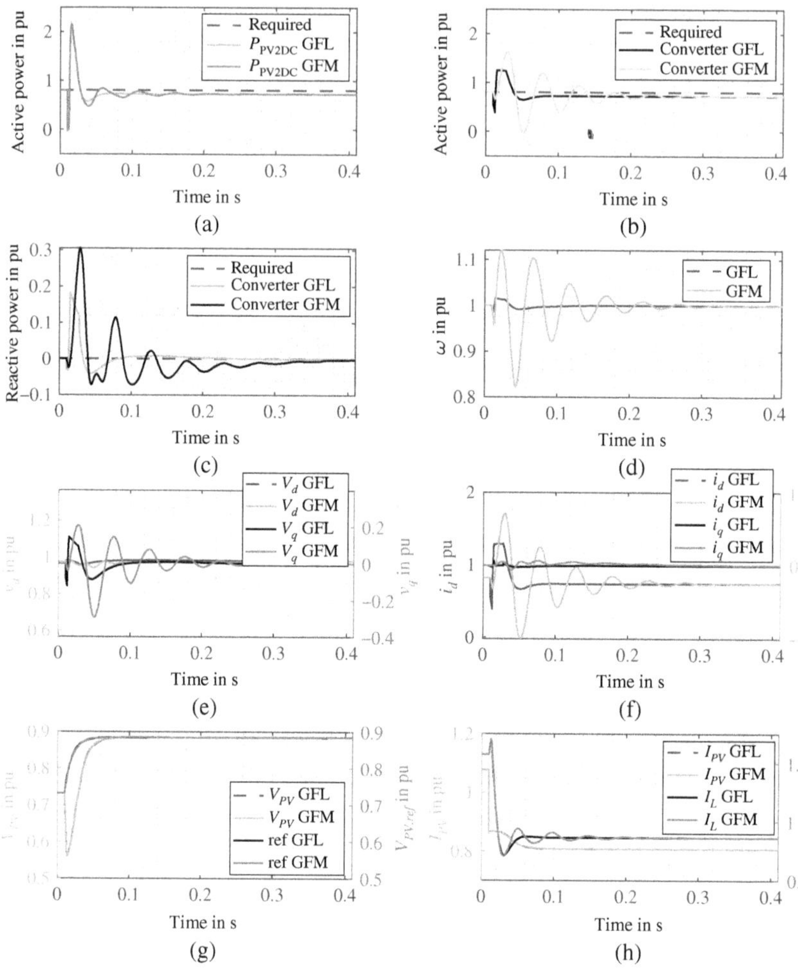

Figure 6.13 Test Case 2: System response after irradiation step at $t_e = 0.01$ s related to generated PV power (primary power conversion) and converter states (secondary power conversion) with grid-following (GFL) and grid-forming (GFM) operation: (a) PV power, (b) active power at PCC, (c) reactive power at PCC, (d) estimated grid frequency at PCC, (e) voltage at PCC in dq-coordinates, (f) current in dq-coordinates, (g) PV voltage, (h) PV and load current of the DC/DC converter

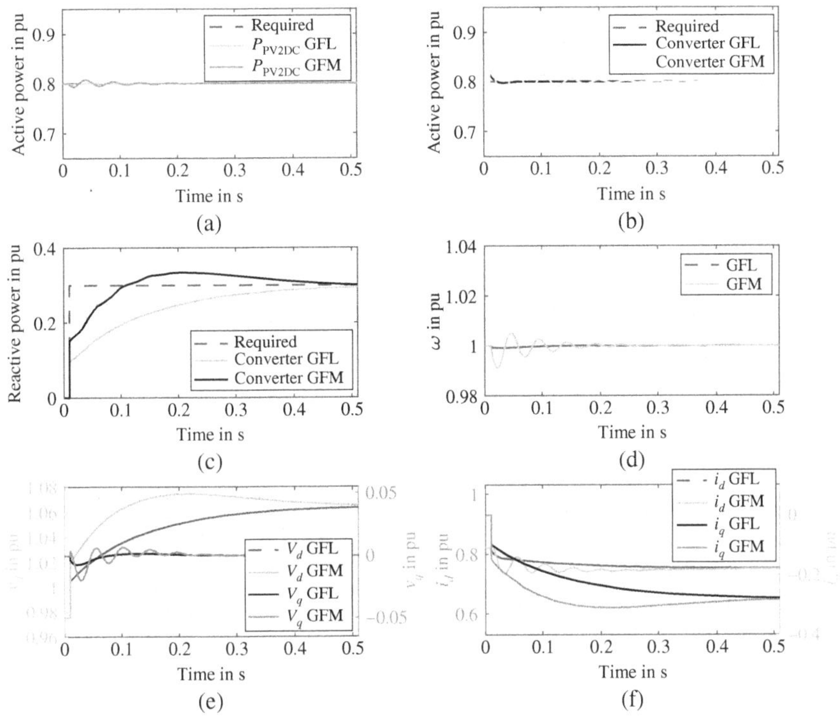

Figure 6.14 Test Case 3: System response after step change of reactive power request at $t_e = 0.1$ s related to generated PV power (primary power conversion) and converter states (secondary power conversion) with grid-following (GFL) and grid-forming (GFM) operation: (a) PV power, (b) active power at PCC, (c) reactive power at PCC, (d) estimated grid frequency at PCC, (e) voltage at PCC in dq-coordinates, (f) current in dq-coordinates

investigated. The step-wise change of the grid voltage magnitude $\sqrt{2}V_g$ reduces the voltage at the PCC as shown in the diagram for $v_{d,q}$ of Figure 6.8. The test scenario results shown in Figure 6.15 are almost identical to the curves of the signals observed for the wind power plant. The response is primarily determined by the power converter on the secondary side and the GFM and GFL control methods used. The slight difference to the wind turbine simulation result is caused by small fluctuations in the power generated by the primary converter due to the coupling with the DC link voltage.

Let us now examine *Test Case 5*, the PV power plant dynamics after a phase angle shift in the grid. As also shown for the wind turbines, the characteristic of secondary conversion, particularly with GFM operation, is important for the dynamics of the overall system. Due to the drastic change in active power supplied to the electrical grid by GFR operation the power variation on the primary side is

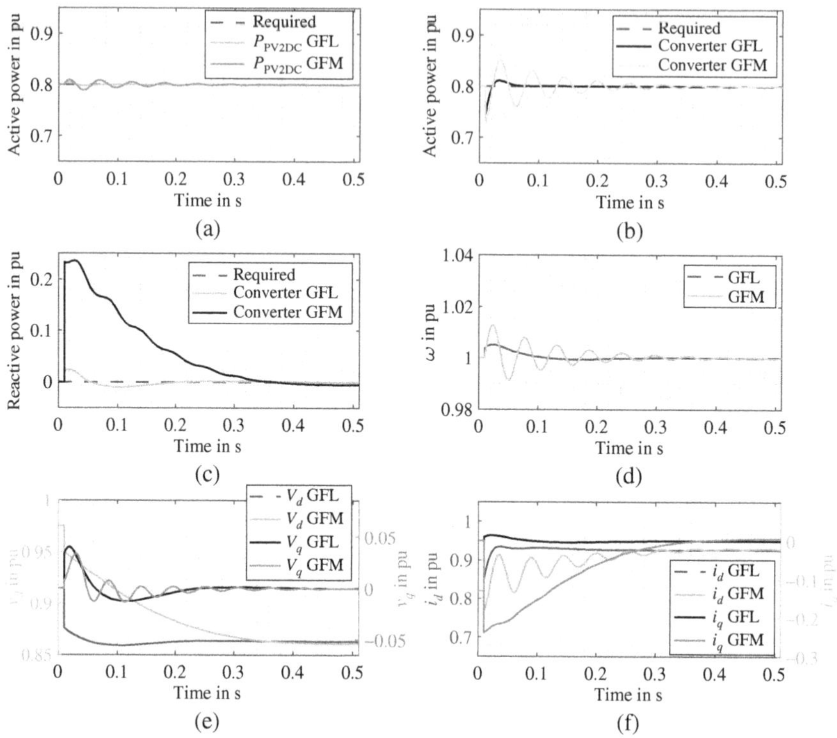

Figure 6.15 Test Case 4: System response after voltage magnitude drop at $t_e = 0.01$ s related to generated PV power (primary power conversion) and converter states (secondary power conversion) with grid-following (GFL) and grid-forming (GFM) operation: (a) PV power, (b) active power at PCC, (c) reactive power at PCC, (d) estimated grid frequency at PCC, (e) voltage at PCC in dq-coordinates, (f) current in dq-coordinates

affected and indicates an oscillation at the frequency of the active power injection by the voltage source converter, see the presented results in Figure 6.16. However, even in this drastic scenario, the power oscillations show a maximum magnitude of 0.1 pu due to the fast-acting DC-DC converter within the PV generator at the primary side decaying after 200 ms.

6.5 Summary

To investigate the dynamic properties and to allow a classification of the various dynamics based on different time scales, five test cases were selected to show the distinct responses of the renewable energy power plants (RPPs) considering relevant control and disturbance inputs. The actuating signals of the grid integrated power plant considered here are the desired active and reactive power

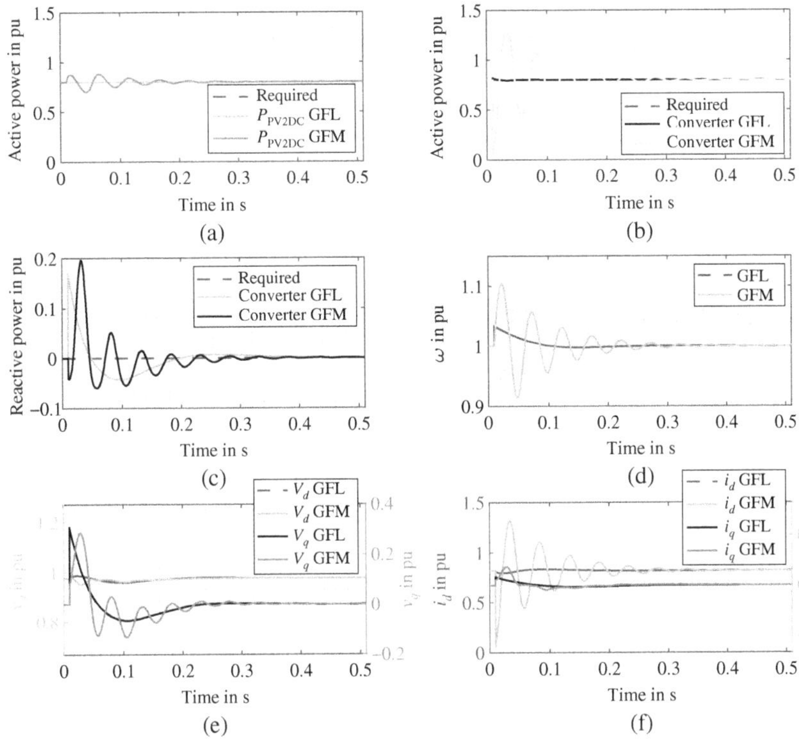

Figure 6.16 Test Case 5: System response after voltage phase shift at $t_e = 0.01$ s related to generated PV power (primary power conversion) and converter states (secondary power conversion) with grid-following (GFL) and grid-forming (GFM) operation: (a) PV power, (b) active power at PCC, (c) reactive power at PCC, (d) estimated grid frequency at PCC, (e) voltage at PCC in dq-coordinates, (f) current in dq-coordinates

injection. In contrast, the disturbance inputs are defined as the change in the primary source, and as grid disturbance a voltage magnitude drop, or voltage phase angle change. A comparison of the simulated internal states of the primary and secondary power conversion and systems outputs allows for a discussion of the working principles, grid interaction, and time scales of different power plant technology like PV systems and wind power.

6.6 Problems

6.1 Describe qualitatively the differences between PV and wind power plants in terms of the dynamics of primary generation and feed-in using VSC. What influence have the concepts of GFL and GFM on the overall dynamics?

6.2 Perform an eigenvalue analysis of the grid-integrated wind power plant model for a selection of operating points. Use the model parameter of the wind turbine listed in Tables D.2 and D.3, the voltage source converter parameters in Table D.5, and the grid model in Table D.6.

6.3 Perform an eigenvalue analysis of the grid-integrated PV power plant model for a selection of operating points. Use the model parameter of the PV system in Table D.4, the voltage source converter parameters in Table D.5, and the grid model in Table D.6.

6.4 Extend the single-machine test system in Figure 6.1 by one resp. three additional generating units and integrate them into the two-area model described in Kundur [1994]. Initially use one generating unit per area and then expand this to two units per area. Investigate different variants with GFL and GFM VSC systems and primary sources.

6.7 Bibliography

The next step, as already addressed in the problems in Section 6.6, is the integration of the presented single power plants into multi-machine systems. Further investigations on how to control inverter-based multi-machine systems in a decentralized manner with as less communication as possible is the subject of research. The concept of coordinated control of virtual power plants (VPP) provides one approach. A survey of current research activities on VPP is given in Juangarcia et al. [2018]. A special variant for improving short-term dynamics is the dynamic virtual power plant (DVPP), presented in Marinescu et al. [2022] and Björk et al. [2023]. A model-based decentralized approach to controller design for such distributed systems is proposed in Häberle et al. [2022] and investigated for GFM operation in Häberle et al. [2023] by simulations. The first experimental validation using GFL VSC for DVPPs with grid feed-in at one PCC is presented in Andrejewski et al. [2023]. The DVPP structure studied includes three power plants: a PV power plant, a wind power plant, and a Static Synchronous Compensator (STATCOM).

A

Modeling in the Takagi–Sugeno Framework

A.1 Introduction

In the literature, the terms Takagi–Sugeno (TS) model and TS system are not clearly distinguished from each other. Frequently both concepts are used as synonyms. For the sake of clarity, a clear distinction is made between both terms:

- **TS systems** stand for their own without a direct reference to an object in the real world. Their construction, mathematical properties, and use are described in this section.
- **TS models**: TS systems become TS models when referring to a real object in the world. The concrete use of TS models and the modeling approach to describe RESs and their components are described in the main part of the book.

Given is the nonlinear parameter variable dynamic system

$$\begin{aligned} \dot{\mathbf{x}} &= \mathbf{f}(\mathbf{x}, \mathbf{u}, \boldsymbol{\theta}), \qquad \mathbf{x}_0 = \mathbf{x}(t_0), \\ \mathbf{y} &= \mathbf{g}(\mathbf{x}), \end{aligned} \tag{A.1}$$

where $\mathbf{f}: \mathbb{R}^n \rightarrow \mathbb{R}^n$ and $\mathbf{g}: \mathbb{R}^n \rightarrow \mathbb{R}^p$ are smooth vector-valued functions, $\mathbf{x} \in \mathbb{R}^n$ denotes the state-space vector, $\mathbf{u} \in \mathbb{R}^m$ is the input vector with controllable and uncontrollable inputs, $\boldsymbol{\theta} \in \mathbb{R}^d$ denotes a time-variable parameter vector, and $\mathbf{y} \in \mathbb{R}^p$ denotes the output vector. In this study, an exact representation or approximations of (A.1) by TS systems [Takagi and Sugeno, 1985] are investigated. The class of TS systems provide a useful and uniform framework for nonlinear controller and observer design. Originally introduced in the context of fuzzy systems [Takagi and Sugeno, 1985], TS systems are weighted combinations of linear systems such as

$$\begin{aligned} \dot{\mathbf{x}} &= \sum_{i=1}^{N_r} h_i(\mathbf{z}) \left(\mathbf{A}_i \mathbf{x} + \mathbf{B}_i \mathbf{u} \right), \qquad \mathbf{x}_0 = \mathbf{x}(t_0), \\ \mathbf{y} &= \sum_{i=1}^{N_r} h_i(\mathbf{z}) \, \mathbf{C}_i \mathbf{x} \end{aligned} \tag{A.2}$$

Advanced Control of Grid-Integrated Renewable Energy Power Plants: LMI-based Design in the Takagi-Sugeno Framework, First Edition. Horst Schulte.

or affine systems

$$\begin{aligned}\dot{\mathbf{x}} &= \sum_{i=1}^{N_r} h_i(\mathbf{z})\left(\mathbf{A}_i\mathbf{x} + \mathbf{B}_i\mathbf{u} + \mathbf{a}_i\right), \qquad \mathbf{x}_0 = \mathbf{x}(t_0),\\ \mathbf{y} &= \sum_{i=1}^{N_r} h_i(\mathbf{z})\left(\mathbf{C}_i\mathbf{x} + \mathbf{c}_i\right)\end{aligned} \tag{A.3}$$

with $\mathbf{A}_i \in \mathbb{R}^{n\times n}$, $\mathbf{B}_i \in \mathbb{R}^{n\times m}$, $\mathbf{C}_i \in \mathbb{R}^{p\times n}$, $\mathbf{a}_i \in \mathbb{R}^n$, and $\mathbf{c}_i \in \mathbb{R}^p$. The functions $h_i : \mathbb{R}^l \to \mathbb{R}$ are the membership functions and fulfill the convex sum condition

$$0 \le h_i(\mathbf{z}) \le 1, \quad i = 1,2,\dots,N_r, \qquad \sum_{i=1}^{N_r} h_i(\mathbf{z}) = 1 \quad \forall \mathbf{z}. \tag{A.4}$$

The vector of the so-called premise variables may directly contain states x_k, inputs u_k, and time-variable parameters θ_k, or be a function $\mathbf{z} = \mathbf{z}(\mathbf{x}, \mathbf{u}, \boldsymbol{\theta})$.

A.2 Constructing TS Systems

There are two different methods to construct TS systems from a given nonlinear system (A.1): (i) The sector nonlinearity approach and (ii) a technique based on interpolation of N_r linear time invariant (LTI) systems obtained by Taylor linearization. With the latter method, the linearization can also be done numerically, where the original systems do not have to be available in a closed analytical form.

A.2.1 Sector-Nonlinearity Approach

The sector-nonlinearity (SL) approach provides an exact representation of a given nonlinear system (A.1) through a TS system (A.2) in a predefined compact set of the state-space, input, and output space. It was first introduced by Ohtake et al. [2001]. First (A.1) is represented in the form

$$\begin{aligned}\dot{\mathbf{x}} &= \mathbf{A}(\mathbf{x},\mathbf{u},\boldsymbol{\theta})\,\mathbf{x} + \mathbf{B}(\mathbf{x},\boldsymbol{\theta})\,\mathbf{u}, \qquad \mathbf{x}_0 = \mathbf{x}(t_0),\\ \mathbf{y} &= \mathbf{C}(\mathbf{x},\boldsymbol{\theta})\,\mathbf{x}.\end{aligned} \tag{A.5}$$

Thus the nonlinear system is set to an **A**, **B**, and **C** matrix structure as in LTI systems. In general, the matrices contain not only constant coefficients but also scalar-valued functions $f_j(\cdot)$, where these can depend on the components of **x**, **u**, and θ. Assuming that these functions have a minimum $\underline{f}_j = \min(f_j)$ and a maximum $\overline{f}_j = \max(f_j)$ in a predefined region $(\cdot)$ with $f_j(\cdot) \in [\underline{f}_j, \overline{f}_j]$ then these can be represented by sector functions

$$f_j(\cdot) = \frac{\bar{f}_j - f_j(\cdot)}{\bar{f}_j - \underline{f}_j}\underline{f}_j + \frac{f_j(\cdot) - \underline{f}_j}{\bar{f}_j - \underline{f}_j}\bar{f}_j. \tag{A.6}$$

Here the proportion of the minimum $\underline{f}_j$ and maximum $\bar{f}_j$ are weighted by the variable terms defined as

$$w_{1,j}(\cdot) = \frac{\bar{f}_j - f_j(\cdot)}{\bar{f}_j - \underline{f}_j}, \qquad w_{2,j}(\cdot) = \frac{f_j(\cdot) - \underline{f}_j}{\bar{f}_j - \underline{f}_j}. \tag{A.7}$$

These are denoted in the following as weighting functions with the convex sum condition

$$w_{1,j}(\cdot) \geq 0, \quad w_{2,j}(\cdot) \geq 0, \quad w_{1,j}(\cdot) + w_{2,j}(\cdot) = 1 \tag{A.8}$$

for any value in the predefined region of $(\cdot)$. By replacing all variables in $\mathbf{A}(\mathbf{x}, \mathbf{u}, \boldsymbol{\theta})$, $\mathbf{B}(\mathbf{x}, \boldsymbol{\theta})$, and $\mathbf{C}(\mathbf{x}, \boldsymbol{\theta})$ of (A.5) with $f_j(\cdot)$ and summarizing all weighting functions under use of the convex sum properties one obtains

$$\sum_{k=1}^{N_r} h_i(\mathbf{z}) = \prod_{j=1}^{N_l} \sum_{k=1}^{2} w_{k,j}(\cdot), \quad N_r = 2^{N_l}, \tag{A.9}$$

where $\mathbf{z} \in \mathbb{R}^l$ denotes the vector of premise variables, which summarize all variables of the $w_{k,j}$ functions (A.7). For illustration, we consider a system with $N_l = 2$ therefore is $N_r = 2^{N_l} = 4$. Using (A.9), we get the sum of all membership functions

$$\sum_{k=1}^{4} h_i(\mathbf{z}) = \left(w_{1,1}(\cdot) + w_{2,1}(\cdot)\right)\left(w_{1,2}(\cdot) + w_{2,2}(\cdot)\right)$$

containing the terms

$$h_1(\mathbf{z}) = w_{1,1}(\cdot)w_{1,2}(\cdot), \quad h_2(\mathbf{z}) = w_{1,1}(\cdot)w_{2,2}(\cdot),$$
$$h_3(\mathbf{z}) = w_{2,1}(\cdot)w_{1,2}(\cdot), \quad h_4(\mathbf{z}) = w_{2,1}(\cdot)w_{2,2}(\cdot).$$

A.2.2 Taylor Linearization and Interpolation

The second method is based on an approximation of a given nonlinear differentiable system

$$\begin{aligned} \dot{\mathbf{x}} &= \mathbf{f}(\mathbf{x}, \mathbf{u}), \qquad \mathbf{x}_0 = \mathbf{x}(t_0), \\ \mathbf{y} &= \mathbf{g}(\mathbf{x}), \end{aligned} \tag{A.10}$$

using an interpolation of N_r local LTI systems. In contrast to the SL method an approximation of the original system is used instead of an exact description.

The approximation error depends on the selected number and distribution of local systems in the input, state, and output space of (A.10). In Angelis [2001] it could be analytically shown that the right-hand side approximation error decreases with increasing number of local models. Whereby for a permitted error bound a necessary number is analytically calculable under the assumption of an equidistant model distribution [Schulte, 2005a].

The TS system construction by Taylor linearization and interpolation is based on three steps:

1. A set of N_r equilibrium points (EPs) $\{\mathbf{x}_i^e, \mathbf{u}_i^e\}_{i=1,2,\dots,N_r}$ is obtained from (A.10) with

$$\mathbf{f}(\mathbf{x}_i^e, \mathbf{u}_i^e) = \mathbf{0}. \tag{A.11}$$

These are parameterized with a minimum possible dimension of $\mathbf{z}$

$$\mathbf{z}_i = \{x_{i,k}^e, u_{i,k}^e\}, \quad k \in \mathbb{N}. \tag{A.12}$$

For example, to parameterize the equilibrium points of the wind turbine model in the full load region, one dimension $z \in \mathbb{R}^+$ is sufficient. Figure A.1 illustrates the variable coordinates of the EPs as a relation between the pitch angle β^e and effective wind speed v^e where the remaining coordinates are constant. In this case the rotor speed and generator torque are constant

$$\omega_r^e = \omega_{r,R} = const., \quad T_g^e = T_{g,R} = const.,$$

where $\omega_{r,R}$ denotes the rated rotor speed and $T_{g,R}$ the rated generator torque.

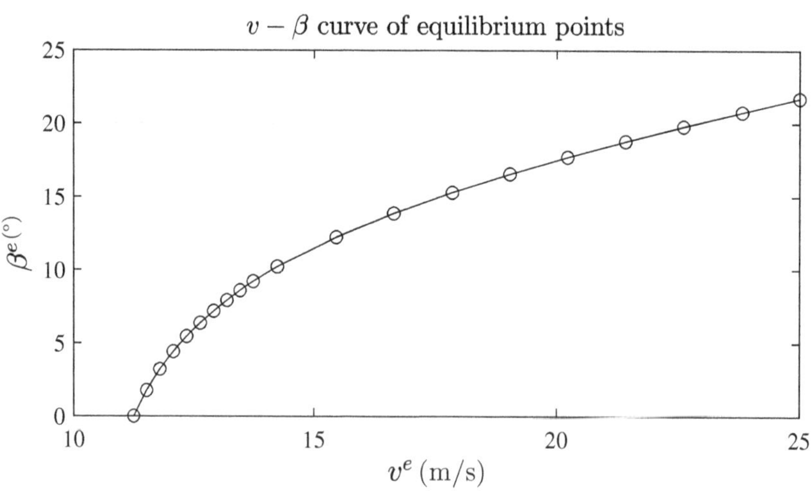

Figure A.1 Curve of equilibrium points $\{v_i^e, \beta_i^e\}$ related to the NREL 5-MW wind turbine. Source: Adapted from Jonkman et al. [2009]

2. The Taylor linearization is used to calculate a set of LTI state-space systems, which describe the dynamics of the original system (A.10) around $i = 1,2,\dots,N_r$ equilibrium points

$$\begin{aligned} \Delta\dot{\mathbf{x}}_i &= \mathbf{A}_i\Delta\mathbf{x}_i + \mathbf{B}_i\Delta\mathbf{u}_i \\ \Delta\mathbf{y}_i &= \mathbf{C}_i\Delta\mathbf{x}_i \end{aligned} \tag{A.13}$$

with

$$\begin{aligned} \Delta\mathbf{x}_i(t) &= \mathbf{x}(t) - \mathbf{x}_i^e, \qquad & \Delta\dot{\mathbf{x}}_i(t) &= \dot{\mathbf{x}}(t), \\ \Delta\mathbf{u}_i(t) &= \mathbf{u}(t) - \mathbf{u}_i^e, \qquad & \Delta\mathbf{y}_i(t) &= \mathbf{y}(t) - \mathbf{y}_i^e. \end{aligned}$$

The matrices $\mathbf{A}_i$ $\mathbf{B}_i$ and $\mathbf{C}_i$ for $i = 1,2,\dots,N_r$ in (A.13) are determined by the Jacobi matrices

$$\mathbf{A}_i = \left.\frac{\partial\mathbf{f}(\mathbf{x},\mathbf{u})}{\partial\mathbf{x}}\right|_{\mathbf{z}_i}, \qquad \mathbf{B}_i = \left.\frac{\partial\mathbf{f}(\mathbf{x},\mathbf{u})}{\partial\mathbf{u}}\right|_{\mathbf{z}_i}, \qquad \mathbf{C}_i = \left.\frac{\partial\mathbf{g}(\mathbf{x})}{\partial\mathbf{x}}\right|_{\mathbf{z}_i} \tag{A.14}$$

with $\mathbf{z}_i$ as the ith linearization point at the equilibrium (A.11) and (A.12).

3. The set of local linear systems are related to the common state and input vector

$$\begin{aligned} \dot{\mathbf{x}} &= \mathbf{A}_i\mathbf{x} + \mathbf{B}_i\mathbf{u} + \underbrace{\left(-\mathbf{A}_i\mathbf{x}_i^e - \mathbf{B}_i\mathbf{u}_i^e\right)}_{\mathbf{a}_i}, \\ \mathbf{y} &= \mathbf{C}_i\mathbf{x} + \underbrace{\left(-\mathbf{C}_i\mathbf{x}_i^e\right)}_{\mathbf{c}_i}, \end{aligned} \tag{A.15}$$

where $\mathbf{a}_i$ and $\mathbf{c}_i$ denote the biases of state and output equation. Finally all linear systems are interpolated in the multidimensional $\mathbf{z}$ space, see (A.3),

$$\begin{aligned} \dot{\mathbf{x}} &= \sum_{i=1}^{N_r} h_i(\mathbf{z})\left(\mathbf{A}_i\mathbf{x} + \mathbf{B}_i\mathbf{u} + \mathbf{a}_i\right), \\ \mathbf{y} &= \sum_{i=1}^{N_r} h_i(\mathbf{z})\left(\mathbf{C}_i\mathbf{x} + \mathbf{c}_i\right) \end{aligned}$$

by local membership functions $h_i(\mathbf{z})$ based on products of sums of local functions $w_{k,j}(z_j)$

$$\sum_{i=1}^{N_r} h_i(\mathbf{z}) = \prod_{j=1}^{N_l}\sum_{k=1}^{N_j} w_{k,j}(z_j) \quad \text{with} \quad N_r = \prod_{j=1}^{N_l} N_j, \tag{A.16}$$

where $\mathbf{z} = [z_1,\dots,z_j,\dots,z_{N_l}]^T$.

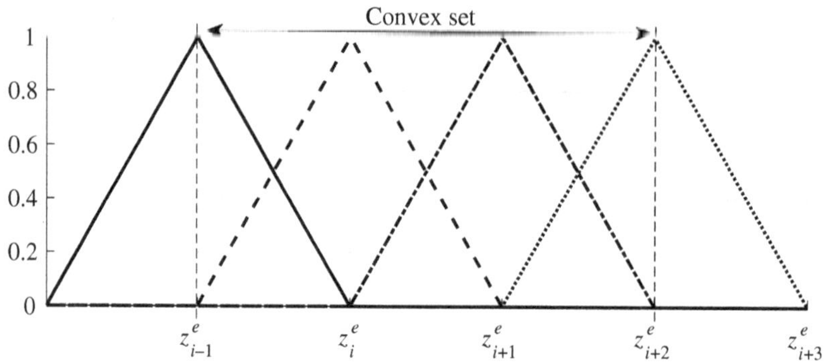

Figure A.2 Test-functions (A.17) of four linearization points $z^e_{i-1}, \ldots, z^e_{i+2}$

In cases where the Taylor linearization and interpolation is being used for the TS construction, the triangular function is utilized

$$w(z) = \begin{cases} \frac{z - z_{\min}}{z^e - z_{\min}}, & \text{if} \quad z_{\min} \leq z < z^e, \\ 1 - \frac{z - z^e}{z_{\max} - z^e}, & \text{if} \quad z^e \leq z \leq z_{\max}, \\ 0, & \text{otherwise,} \end{cases} \tag{A.17}$$

where z^e denotes the linearization point of the associated LTI system for which $w(z^e) = 1$. Furthermore the parameters $z_{\min}$ and $z_{\max}$ in (A.17) define the range of z in which $w(z)$ is not equal to zero. For the case of a scalar-valued $z \in \mathbb{R}$ triangular functions (A.17) of four linearization points are illustrated in Figure A.2. In this case $h_i(z) = w_k(z)$, $N_r = N_1$ which follows directly from (A.16) with $l = 1$. Note that the convex sum condition (A.4) with $\sum_{i=1}^{4} h_i(z) = 1$ holds in the range $z = [z^e_{i-1}, z^e_{i+2}]$.

B

LMI Conditions for Stability Analysis and Controller Design

The following conditions for stability analysis and control design are standard in the literature on system and control theory using convex optimization. In the treatment of Boyd et al. [1994] it was shown "that a wide variety of problems arising in system and control theory can be reduced to a handful of standard convex and quasiconvex optimization problems that involve matrix inequalities." The application of convex optimization to the class of Takagi–Sugeno fuzzy systems then took place somewhat later, i.e. Tanaka et al. [1996, 1998], and was first summarized in the monograph by Tanaka and Wang [2001]. Since then, considerable efforts have been made to transfer results from robust control theory using linear matrix inequality (LMI) regions [Chilali and Gahinet, 1996a] to the Takagi–Sugeno approach.

B.1 Introduction

As shown in Appendix A, TS systems (A.2), (A.3) are used to obtain a convex representation of any nonlinear system of the form (A.1). It is now shown how LMIs can be used to analyze the stability of TS systems and to design model-based controllers and observers that guarantee not only stability but also guaranteed performance. The LMI conditions are in general derived from analysis using the direct method of Lyapunov among and extended by pole region and H_∞ constraints. There are efficient numerical methods to determine whether an LMI is feasible, i.e. interior-point methods [Nesterov and Nemirovskii, 1994].

Advanced Control of Grid-Integrated Renewable Energy Power Plants: LMI-based Design in the Takagi-Sugeno Framework, First Edition. Horst Schulte.

B.2 Linear Matrix Inequalities

B.2.1 Basics

A LMI with given constant symmetric matrices $\mathbf{F}_i = \mathbf{F}_i^T \in \mathbb{R}^{n \times n}$, $i = 0, 1, \dots, m$ has the form

$$\mathbf{F}(\mathbf{x}) \triangleq \mathbf{F}_0 + \sum_{i=1}^{m} x_i \mathbf{F}_i \succ 0, \tag{B.1}$$

where $\mathbf{x} = [x_1, x_2, \dots, x_i, \dots, x_m]^T$ is the vector variable [Boyd et al., 1994] and $\mathbf{F}(\mathbf{x}) \succ 0$ denotes that $\mathbf{F}(\mathbf{x})$ is positive definite. The inequality (B.1) is fulfilled if $\exists \mathbf{x}$ so that $\mathbf{F}(\mathbf{x})$ is positive definite, i.e. $\mathbf{y}^T \mathbf{F}(\mathbf{x}) \mathbf{y} > 0$ for all nonzero $\mathbf{y} \in \mathbb{R}^n$. The LMI (B.1) is equivalent to a set of n polynomial inequalities in $\mathbf{x}$, where the constituent principal minors of $\mathbf{F}(\mathbf{x})$ must be positive. In the analogous way, an LMI for the negative definiteness of $\mathbf{H}(\mathbf{x})$ with $\mathbf{H}(\mathbf{x}) \prec 0$ can also be specified

$$\mathbf{H}(\mathbf{x}) \triangleq \mathbf{H}_0 + \sum_{i=1}^{m} x_i \mathbf{H}_i \prec 0, \tag{B.2}$$

where (B.2) is equivalent to a set of n polynomial inequalities in $\mathbf{x}$, where the signs of the principal minors of $\mathbf{H}(\mathbf{x})$ must be alternate. An example will illustrate this circumstance.

Example B.1 *Given the matrix*

$$\mathbf{H}(\mathbf{x}) = \begin{pmatrix} x_1 - x_2 & 0 & 0 \\ 0 & x_2 - 1 & x_1 \\ 0 & x_1 & -1 \end{pmatrix},$$

with $m = 2$ variables, i.e. $\mathbf{x} = [x_1, x_2]^T$. We are looking for the solution set of $\mathbf{H}(\mathbf{x}) \prec 0$, which can be written in an LMI form (B.2)

$$\mathbf{H}(\mathbf{x}) = \mathbf{H}_0 + x_1 \mathbf{H}_1 + x_2 \mathbf{H}_2 \prec 0, \tag{B.3}$$

where

$$\mathbf{H}_0 = \begin{pmatrix} 0 & 0 & 0 \\ 0 & -1 & 0 \\ 0 & 0 & -1 \end{pmatrix}, \quad \mathbf{H}_1 = \begin{pmatrix} 1 & 0 & 0 \\ 0 & 0 & 1 \\ 0 & 1 & 0 \end{pmatrix}, \quad \mathbf{H}_2 = \begin{pmatrix} -1 & 0 & 0 \\ 0 & 1 & 0 \\ 0 & 0 & 0 \end{pmatrix}.$$

To determine the solution $\{\mathbf{x} \,|\, \mathbf{H}(\mathbf{x}) \prec 0\}$ of (B.3) *by $n = 3$ polynomial inequalities in $\mathbf{x}$, we use the minor criteria of negative definite matrices*

$$\Delta_1 = \det\left(x_1 - x_2\right) = x_1 - x_2 < 0, \qquad \Delta_2 = \det\begin{pmatrix} x_1 - x_2 & 0 \\ 0 & x_2 - 1 \end{pmatrix} > 0,$$

$$\Delta_3 = \det\left(\mathbf{H}(\mathbf{x})\right) < 0.$$

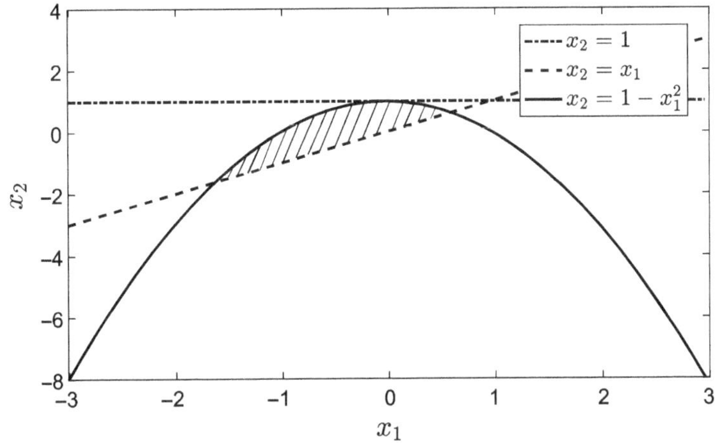

Figure B.1 Feasible set of (B.3) (hatched area)

This results in

$$\underbrace{(x_1 - x_2)}_{<0}\underbrace{(x_2 - 1)}_{<0} > 0\,, \qquad \underbrace{(x_1 - x_2)}_{<0}\underbrace{\left((x_2 - 1)(-1) - x_1^2\right)}_{>0} < 0,$$

which determine the feasible set illustrated in Figure B.1.

Based on the previous example, for illustration, a graphical solution was demonstrated. However, LMIs that occur in control theoretical and application problems can often not be solved analytically (probably a solution cannot be found). This was outlined in the work of Pyatnitskii and Skorodinskii [1982]. Therefore LMIs are usually solved numerically. An efficient numerical approach the interior-point algorithms for LMI problems (B.1), (B.2) was introduced by Nesterov and Nemirovskii [1994]. Numerous solvers based on this method are available to the control engineer such as LMILAB, SeDuMi, SDPT3, VSDP, or LMIRank. In this book, the SeDuMi and SDPT3 solvers within the Yalmip toolbox [Löfberg, 2004] are used.

In system and control theory we will often encounter LMI problems in which the variables are matrices, e.g. the Lyapunov inequality

$$\mathbf{A}^T\mathbf{P} + \mathbf{P}\mathbf{A} \prec 0\,, \qquad \mathbf{P} \succ 0, \tag{B.4}$$

where $\mathbf{A} \in \mathbb{R}^{n \times n}$ is given and $\mathbf{P} = \mathbf{P}^T$ is the variable. Usually, the LMIs are not written in the explicit form (B.1), (B.2), but can be directly converted into it. For example, the Lyapunov inequality for $n = 2$ can be converted to

$$\mathbf{P} = \begin{pmatrix} p_{11} & p_{12} \\ p_{12} & p_{22} \end{pmatrix} = \mathbf{F}(\mathbf{x}) = p_{11}\underbrace{\begin{pmatrix} 1 & 0 \\ 0 & 0 \end{pmatrix}}_{\mathbf{F}_1} + p_{12}\underbrace{\begin{pmatrix} 0 & 1 \\ 1 & 0 \end{pmatrix}}_{\mathbf{F}_2} + p_{22}\underbrace{\begin{pmatrix} 0 & 0 \\ 0 & 1 \end{pmatrix}}_{\mathbf{F}_3} \succ 0,$$

where $\mathbf{F}_0 = \mathbf{0}$ with $\mathbf{x} = [x_1, x_2, x_3]^T = [p_{11}, p_{12}, p_{22}]^T$. Furthermore

$$\begin{pmatrix} a_{11} & a_{12} \\ a_{21} & a_{22} \end{pmatrix} \mathbf{P} + \mathbf{P} \begin{pmatrix} a_{11} & a_{12} \\ a_{21} & a_{22} \end{pmatrix} = \mathbf{H}(\mathbf{x})$$

$$= p_{11}\underbrace{\begin{pmatrix} 2a_{11} & a_{12} \\ a_{12} & 0 \end{pmatrix}}_{\mathbf{H}_1} + p_{12}\underbrace{\begin{pmatrix} 2a_{21} & a_{11} + a_{22} \\ a_{11} + a_{22} & 2a_{12} \end{pmatrix}}_{\mathbf{H}_2} + p_{22}\underbrace{\begin{pmatrix} 0 & a_{21} \\ a_{21} & 2a_{22} \end{pmatrix}}_{\mathbf{H}_3} \prec 0,$$

where $\mathbf{H}_0 = \mathbf{0}$.

In sum, LMIs can be characterized in terms of

1. (B.1), (B.2) are convex constraints on $\mathbf{x}$, i.e. the sets $\{\mathbf{x} \mid \mathbf{F}(\mathbf{x}) \succ 0\}$, $\{\mathbf{x} \mid \mathbf{H}(\mathbf{x}) \prec 0\}$ are convex, see Example B.1. It follows that LMIs can be formulated as convex optimization problems that are amenable to computer solution [Boyd et al., 1994].
2. (B.1), (B.2) can represent a wide variety of convex constraints on $\mathbf{x}$. In particular, linear inequalities, convex quadratic inequalities, and matrix norm inequalities can all be cast in the form of an LMI, see Section B.2.2.
3. Multiple LMIs $\mathbf{F}^{(1)}(\mathbf{x}) \succ 0, \ldots, \mathbf{F}^{(p)}(\mathbf{x}) \succ 0$ can be expressed as the single LMI $\mathbf{diag}\left(\mathbf{F}^{(1)}(\mathbf{x}) \succ 0, \ldots, \mathbf{F}^{(p)}(\mathbf{x})\right) \succ 0$.

B.2.2 Suitable Auxiliary Lemmas

1. **Congruence equivalence relation**: The definiteness of a matrix is invariant with respect to pre- and post-multiplication by a full-rank real matrix and its transpose

$$\mathbf{Q} \succ 0,\ \mathbf{Q} \in \mathbb{R}^{n \times n} \ \Leftrightarrow\ \mathbf{W}\mathbf{Q}\mathbf{W}^T \succ 0,\ \mathbf{W} \in \mathbb{R}^{n \times n} \quad \text{if} \quad \operatorname{rank}(\mathbf{W}) = n. \tag{B.5}$$

The main use is to rearrange matrix inequalities in such a way to get an LMI formulation with a change of variables.

2. **Schur complement**: The Schur complement states that the following three statements are equivalent

$$\begin{pmatrix} \mathbf{M}_{11} \mathbf{M}_{12} \\ \mathbf{M}_{21} \mathbf{M}_{22} \end{pmatrix} \prec 0 \quad \Leftrightarrow \quad \mathbf{M}_{11} \prec 0,\ \mathbf{M}_{22} - \mathbf{M}_{12}\mathbf{M}_{11}^{-1}\mathbf{M}_{21} \prec 0 \tag{B.6}$$
$$\Leftrightarrow \quad \mathbf{M}_{22} \prec 0,\ \mathbf{M}_{11} - \mathbf{M}_{12}\mathbf{M}_{22}^{-1}\mathbf{M}_{21} \prec 0,$$

where $\mathbf{M}_{11}$ and $\mathbf{M}_{22}$ are square matrices. The main use is to convert quadratic matrix inequalities (QMIs) into LMIs. A well-known example is the conversion of the Riccati inequality into an LMI using the equivalence in (B.6), see Boyd et al. [1994].

B.3 Stability Analysis of TS Systems

The asymptotic stability of TS systems is investigated using the direct method of Lyapunov [1992]. Based on this, additional constraints are introduced for a more specific control-oriented system analysis that also establish the foundation of controller design in Section B.5. The direct method makes use of a Lyapunov function $V : \mathbb{R}^n \rightarrow \mathbb{R}$ with the properties

1. $V(\mathbf{x}) = 0$ if and only if (iff) $\mathbf{x} = \mathbf{0}$
2. $V(\mathbf{x}) > 0$ iff $\mathbf{x} \neq \mathbf{0}$
3. $\dot{V}(\mathbf{x}) = \frac{dV}{dt} = \sum_{i=1}^{n} \frac{\partial V}{\partial x_i}\dot{\mathbf{x}} < 0$ for $\mathbf{x} \neq \mathbf{0}$

The stability analysis of a dynamic system $\dot{\mathbf{x}} = \mathbf{f}(\mathbf{x})$ is reduced to finding a Lyapunov function $V(\mathbf{x})$. Lyapunov realized that stability of the dynamic system described by a differential equation can be proven without requiring knowledge of its solutions $\mathbf{x}(t)$ for $t \in [t_0, \infty]$. In this investigation, we restrict the analysis to dynamic systems in Takagi–Sugeno form (A.2) and (A.3).

B.3.1 Stability of Unforced TS Systems

B.3.1.1 Asymptotic Stability

A common Lyapunov function candidate is of the quadratic form

$$V(\mathbf{x}) = \mathbf{x}^T \mathbf{P}\mathbf{x}, \tag{B.7}$$

with a positive definite matrix $\mathbf{P} = \mathbf{P}^T \succ 0$. The time derivative of $V(\mathbf{x})$ gives

$$\dot{V}(\mathbf{x}) = \dot{\mathbf{x}}^T \mathbf{P}\mathbf{x} + \mathbf{x}^T \mathbf{P}\dot{\mathbf{x}}. \tag{B.8}$$

The unforced TS system (Eq. (A.2) with $\mathbf{u} = 0$)

$$\dot{\mathbf{x}} = \sum_{i=1}^{N_r} h_i(\mathbf{z})\mathbf{A}_i\mathbf{x}, \tag{B.9}$$

is asymptotically stable if (B.7) fulfills the above-mentioned properties (1) to (3). To verify this, the derivative $\dot{\mathbf{x}}$ is replaced by the right side of (B.9).

$$\begin{aligned}\dot{V}(\mathbf{x}) &= \left(\sum_{i=1}^{N_r} h_i(\mathbf{z})\,\mathbf{A}_i\,\mathbf{x}\right)^T \mathbf{P}\mathbf{x} + \mathbf{x}^T\mathbf{P}\left(\sum_{i=1}^{N_r} h_i(\mathbf{z})\,\mathbf{A}_i\,\mathbf{x}\right)\\ &= \sum_{i=1}^{N_r} h_i(\mathbf{z})\mathbf{x}^T\left(\mathbf{A}_i^T\mathbf{P} + \mathbf{P}\mathbf{A}_i\right)\mathbf{x}.\end{aligned} \tag{B.10}$$

Thus one obtains the following theorem that contains a sufficient stability criterion based on an LMI condition.

Theorem B.1 ***(Wang et al. [1996])*** *The unforced Takagi–Sugeno system $\dot{\mathbf{x}} = \sum_{i=1}^{N_r} h_i(\mathbf{z})\mathbf{A}_i\mathbf{x}$ is global asymptotically stable (GAS) if there exists a common matrix $\mathbf{P} = \mathbf{P}^T \succ 0$ such that the LMI problem*

$$\mathbf{A}_i^T\mathbf{P} + \mathbf{P}\mathbf{A}_i \prec 0, \qquad i = 1, 2, \ldots, N_r, \tag{B.11}$$

is feasible.

Note that a system is GAS if for every trajectory $\mathbf{x}(t) \rightarrow \mathbf{x}_e$ for $t \rightarrow \infty$. The vector $\mathbf{x}_e$ denotes an equilibrium point of the system (B.9), $\mathbf{f}(\mathbf{x}_e) = 0$.

Proof: $V(\mathbf{x})$ (B.7) fulfills the three properties of a Lyapunov function: (1) $V(\mathbf{x}) = 0$ iff $\mathbf{x} = 0$, (2) $V(\mathbf{x}) > 0$ iff $\mathbf{x} \neq 0$ with $\mathbf{P} = \mathbf{P}^T \succ 0$ and using the previous derivation (B.9) is (3) $\dot{V}(\mathbf{x}) < 0$ for $x \neq \mathbf{0}$ due to

$$\dot{V}(\mathbf{x}) = \sum_{i=1}^{N_r} h_i(\mathbf{z})\mathbf{x}^T \underbrace{\left(\mathbf{A}_i^T\mathbf{P} + \mathbf{P}\mathbf{A}_i\right)}_{\mathbf{\Gamma}_i}\mathbf{x} < 0, \tag{B.12}$$

if $\mathbf{\Gamma}_i \prec 0$ considering the property $h_i(\mathbf{z}) \geq 0$, see (A.4). Note that the Theorem B.1 is based on the requirement that all summand $\mathbf{\Gamma}_i$, $i = 1, 2, \ldots, N_r$ must be negatively definite.

B.3.1.2 $\mathcal{D}$-Stability

Let $\mathcal{D}$ be a subregion of the complex left-half plane which is described by an LMI region.

Definition B.1 ***(LMI region [Chilali and Gahinet, 1996a].)*** A subset $\mathcal{D}$ of the complex plane is called an LMI region if there exists a symmetric quadratic matrix $\boldsymbol{\alpha} \in \mathbb{R}^{\gamma\times\gamma}$ and $\boldsymbol{\beta} \in \mathbb{R}^{\gamma\times\gamma}$ such that

$$\mathcal{D} = \{z \in \mathbb{C} : \quad \mathbf{F}_\mathcal{D}(z) \prec 0\}, \tag{B.13}$$

with the characteristic matrix function $\mathbf{F}_\mathcal{D}(z) = \boldsymbol{\alpha} + z\boldsymbol{\beta} + \bar{z}\boldsymbol{\beta}^T$ where $\bar{z} \in \mathbb{C}$ denotes the conjugated complex value to z. Thus an LMI region is a subset of the complex plane that is specified by the LMI $\mathbf{F}_\mathcal{D}(z) \prec 0$ in z and $\bar{z}$. As a consequence LMI regions are convex and symmetric with respect to the real axis.

According to Gutman's theorem for LMI regions in Gutman and Jury [1981], the locations of the eigenvalues of a matrix $\mathbf{A}$ in $\mathcal{D}$ (B.13) can be also characterized as follows

Theorem B.2 ***(Chilali and Gahinet [1996a])*** *The eigenvalues of the matrix* $\mathbf{A}$ *are in the* $\mathcal{D}$*-subregion* (B.13) *iff there exists a symmetric matrix* $\mathbf{X}$ *such that*

$$\mathcal{M}_{\mathcal{D}}(\mathbf{A}, \mathbf{X}) \prec 0, \quad \mathbf{X} \succ 0, \tag{B.14}$$

with the block matrix $\mathcal{M}_{\mathcal{D}}(\mathbf{A}, \mathbf{X}) \in \mathbb{R}^{\gamma \times \gamma}$

$$\mathcal{M}_{\mathcal{D}}(\mathbf{A}, \mathbf{X}) = \boldsymbol{\alpha} \otimes \mathbf{X} + \boldsymbol{\beta} \otimes (\mathbf{AX}) + \boldsymbol{\beta}^T \otimes (\mathbf{AX})^T. \tag{B.15}$$

Proof: See the Appendix in Chilali and Gahinet [1996a] where $\otimes$ denotes the Kronecker product of matrices.

Note that $\mathbf{F}_{\mathcal{D}}(z)$ in (B.13) and $\mathcal{M}_{\mathcal{D}}(\mathbf{A}, \mathbf{X})$ in (B.15) are related by the list of correspondences

$$\left(\mathbf{X},\, (\mathbf{AX}),\, (\mathbf{AX})^T\right) \longleftrightarrow \left(1,\, z,\, \bar{z}\right). \tag{B.16}$$

Applying Theorem B.2, we can define the $\mathcal{D}$-Stability of an unforced TS system.

Definition B.2 ***($\mathcal{D}$-Stability)*** A TS system $\dot{\mathbf{x}} = \sum_{i=1}^{N_r} h_i(\mathbf{z})\mathbf{A}_i\mathbf{x}$ is called $\mathcal{D}$-stable if there exists a common matrix $\mathbf{X} = \mathbf{X}^T \succ 0$ such that the LMI problem (B.15)

$$\mathcal{M}_{\mathcal{D}}(\mathbf{A}_i, \mathbf{X}) \prec 0, \quad \mathbf{X} \succ 0, \tag{B.17}$$

is feasible for all $i = 1, 2, \dots, N_r$.

If the $\mathcal{D}$-region consists of the entire left-half plane where $\mathbf{F}_{\mathcal{D}}(z) = z + \bar{z} \prec 0$ the notion of (B.15) with $\boldsymbol{\alpha} = 0$ and $\boldsymbol{\beta} = 1$ reduces to the LMI condition of the asymptotic stability (B.11) with

$$\mathcal{M}_{\mathcal{D}}(\mathbf{A}_i, \mathbf{X}) = 1 \otimes (\mathbf{A}_i\mathbf{X}) + 1 \otimes (\mathbf{X}\mathbf{A}_i^T) \prec 0, \quad \mathbf{X} \succ 0,$$

where $\mathbf{P}^{-1} = \mathbf{X}$. In the following, $\mathcal{D}$-stability criteria for various regions that can be parameterized with a maximum of three values will be specified. In particular, two LMI criteria for striped and conic-shaped regions with upper and lower bounds are presented:

(1) Pole region $\mathcal{S}(\alpha_{\text{min}}, \alpha_{\text{max}})$: The striped pole region $\mathcal{S}(\alpha_{\text{min}}, \alpha_{\text{max}})$ illustrated in Figure B.2 is defined by two parameters where $\alpha_{\text{max}} > \alpha_{\text{min}} > 0$.

Theorem B.3 ***(Pole region*** $\mathcal{S}(\alpha_{min}, \alpha_{max})$***)*** *The Takagi–Sugeno system* $\dot{\mathbf{x}} - \sum_{i=1}^{N_r} h_l(\mathbf{z})\mathbf{A}_i\mathbf{x}$ *has all its poles in* $\mathcal{S}(\alpha_{min}, \alpha_{max})$ *where* $0 < \alpha_{min} < \alpha_{max}$ *if there exists* $\mathbf{X} \succ 0$ *such that the LMI problem*

$$\begin{aligned} \mathbf{A}_i\mathbf{X} + \mathbf{X}\mathbf{A}_i^T &\prec -2\alpha_{min}\mathbf{X} \\ \mathbf{A}_i\mathbf{X} + \mathbf{X}\mathbf{A}_i^T &\succ -2\alpha_{max}\mathbf{X}, \end{aligned} \tag{B.18}$$

is feasible for all $i = 1, 2, \dots, N_r$.

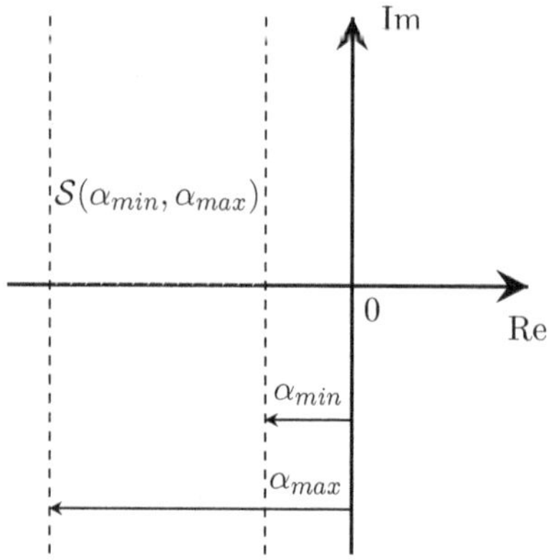

Figure B.2 Parameterized stripe-shaped pole region $S(\alpha_{min}, \alpha_{max})$

Proof: The pole region $S(\alpha_{min}, \alpha_{max})$ is the intersection $S_I(\alpha_{\min}) \cap S_{II}(\alpha_{max})$ of the sectors

$$S_I(\alpha_{min}) = \{(x, y) \mid z < -\alpha_{min} + iy,\ y \in \{-\infty, \infty\}\},$$
$$S_{II}(\alpha_{max}) = \{(x, y) \mid z > -\alpha_{max} + iy,\ y \in \{-\infty, \infty\}\}.$$

With $z = -x + iy$ and $-x = \frac{z+\bar{z}}{2}$ we get

$$\begin{aligned} S(\alpha_{min}, \alpha_{max}) &= S_I(\alpha_{min}) \cap S_{II}(\alpha_{max}) \\ &= \{(x, y) \mid (z + \bar{z} < -2\alpha_{min}) ... (z + \bar{z} > -2\alpha_{max})\}. \end{aligned}$$

Using the equivalent relation according to Gutman [1979]

$$\left(1,\ z,\ \bar{z}\right) \leftrightarrow \left(\mathbf{X},\ \mathbf{A}_i\mathbf{X},\ \mathbf{X}\mathbf{A}_i^T\right), \tag{B.19}$$

we obtain (B.18).

Note that Theorem B.3 is related to the direct method of Lyapunov [1992] if the function $V(\mathbf{x})$ possessing the following properties

$$\dot{V}(\mathbf{x}) + 2\alpha_{\min} V(\mathbf{x}) < 0, \tag{B.20}$$

$$\dot{V}(\mathbf{x}) + 2\alpha_{\max} V(\mathbf{x}) > 0 \quad \text{for} \quad \mathbf{x} \neq \mathbf{0}, \tag{B.21}$$

in addition to the properties mentioned above. With the substitution of $V(\mathbf{x})$ by (B.7) and $\dot{V}(\mathbf{x})$ by (B.10) in (B.20) and the convex sum condition (A.4) one obtains

$$\sum_{i=1}^{N_r} h_i(\mathbf{z}) \mathbf{x}^T \left(\mathbf{A}_i^T \mathbf{P} + \mathbf{P}\mathbf{A}_i\right) \mathbf{x} + \underbrace{\sum_{i=1}^{N_r} h_i(\mathbf{z})}_{=1} 2\alpha_{\min}\, \mathbf{x}^T \mathbf{P} \mathbf{x} < 0$$

$$\mathbf{x}^T \sum_{i=1}^{N_r} h_i(\mathbf{z}) \left[\underbrace{\left(\mathbf{A}_i^T \mathbf{P} + \mathbf{P}\mathbf{A}_i\right) + 2\alpha_{\max} \mathbf{P}}_{\mathbf{\Gamma}_i} \right] \mathbf{x} < 0,$$

with the sufficient condition $\mathbf{\Gamma}_i \prec 0,\ i = 1, 2, \dots, N_r$. Using the congruence relation (B.5) $\mathbf{\Gamma}_i \prec 0 \ \Leftrightarrow\ \mathbf{X}\mathbf{\Gamma}_i\mathbf{X} \prec 0$ we get with $\mathbf{X} = \mathbf{P}^{-1}$

$$\mathbf{X}\mathbf{A}_i^T + \mathbf{A}_i\mathbf{X} + 2\alpha_{\min}\mathbf{X} \prec 0.$$

In the same way, we obtain (B.21).

Remark B.1 The parameter α_{min} is the exponential decay rate of $\mathbf{x}(t)$ for $t \geq 0$. Here the function $V(\mathbf{x}) \leq V(\mathbf{x}_0)e^{-2\alpha_{min}t}$ fulfills $\dot{V}(\mathbf{x}) + 2\alpha_{min}V(\mathbf{x}) \leq 0,\ \mathbf{x}_0 = \mathbf{x}(0)$. Based on the quadratic Lyapunov function $V(\mathbf{x})$ (B.7) results in

$$\begin{aligned}
&V(\mathbf{x}) \leq V(\mathbf{x}_0)e^{-2\alpha_{min}t}\\
&\mathbf{x}^T\mathbf{P}\mathbf{x} \leq \mathbf{x}_0^T\mathbf{P}\mathbf{x}_0\, e^{-2\alpha_{min}t} = \mathbf{x}_0^T e^{-\alpha_{min}t}\,\mathbf{P}\,\mathbf{x}_0\, e^{-\alpha_{min}t}\\
&\quad \Rightarrow \mathbf{x}(t) \leq \mathbf{x}_0\, e^{-\alpha_{min}t}.
\end{aligned}$$

(2) Pole region $S(\alpha, r, \theta)$: Another given polar region where the damping of the system is also considered is treated in the following theorem.

Theorem B.4 *(Pole region* $S(\alpha, r, \theta)$*) The TS system* $\dot{\mathbf{x}} = \sum_{i=1}^{N_r} h_i(\mathbf{z})\mathbf{A}_i\mathbf{x}$ *has all its poles in* $S(\alpha, r, \theta)$ *illustrated in Figure* B.3 *where* $0 < \alpha < r$ *and* $0 < \theta < \pi/2$ *if there exists* $\mathbf{X} \succ 0$ *such that the LMI problem*

$$\begin{aligned}
&\begin{pmatrix} (\mathbf{A}_i\mathbf{X} + \mathbf{X}\mathbf{A}_i^T)\sin\theta & (\mathbf{A}_i\mathbf{X} - \mathbf{X}\mathbf{A}_i^T)\cos\theta \\ -(\mathbf{A}_i\mathbf{X} - \mathbf{X}\mathbf{A}_i^T)\cos\theta & (\mathbf{A}_i\mathbf{X} + \mathbf{X}\mathbf{A}_i^T)\sin\theta \end{pmatrix} \prec 0,\\
&\begin{pmatrix} -r\mathbf{X} & \mathbf{A}_i\mathbf{X} \\ \mathbf{X}\mathbf{A}_i^T & -r\mathbf{X} \end{pmatrix} \prec 0, \qquad \mathbf{A}_i\mathbf{X} + \mathbf{X}\mathbf{A}_i^T \prec -2\alpha\mathbf{X},
\end{aligned} \tag{B.22}$$

is feasible for all $i = 1, 2, \dots, N_r$.

Proof: Part I: Let us first consider the conical sector (without the α stability region and disc sector with radius r). The conical sector is a subset of the complex number plane $z \in \mathbb{C}$ with $z = -x + iy = |z|e^{i\theta}$, $\theta = \arctan(-y/x)$ and defined as

$$S_I(\theta) = \{(x, y) \mid |y| < -\tan\theta\, x\}.$$

This inequality can be rearranged to

$$|y|^2 < (-\tan\theta\, x)^2 = \frac{(\sin\theta\, x)^2}{\cos^2\theta} \quad \Leftrightarrow \quad \frac{(y\cos\theta)^2}{x\sin\theta} < \sin\theta\, x,$$

$$\Leftrightarrow -x\ \sin\theta - i\,(y\cos\theta)\,\frac{1}{-x\sin\theta}\,(-i\,y\cos\theta) < 0. \tag{B.23}$$

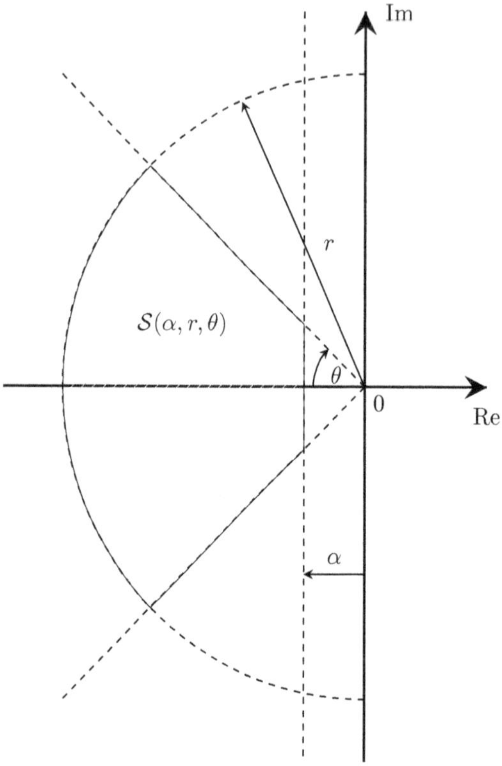

Figure B.3 Parameterized pole region $S(\alpha, r, \theta)$

To decompose the quadratic form (B.23) the Schur complement (B.6) can now be applied and by the definitions $M_{11} = -x\ \sin\theta$, $M_{12} = -i\,(y\cos\theta)$, $M_{21} = iy\cos\theta$, and $M_{22} = -x\ \sin\theta$ we get

$$\begin{pmatrix} -x\sin\theta & iy\cos\theta \\ -iy\cos\theta & -x\sin\theta \end{pmatrix} \prec 0.$$

With the conjugated complex $\bar{z} = -x - iy$ one obtains for

$$-x = \frac{z+\bar{z}}{2}, \qquad iy = \frac{z-\bar{z}}{2},$$

and it follows that

$$\begin{pmatrix} \frac{z+\bar{z}}{2}\sin\theta & \frac{z-\bar{z}}{2}\cos\theta \\ -\frac{z-\bar{z}}{2}\cos\theta & \frac{z+\bar{z}}{2}\sin\theta \end{pmatrix} \prec 0.$$

Using the equivalent relationship (B.19) we obtain equation (B.22) of Theorem B.4.

Part II: The α-stability constraint $\mathbf{A}_i\mathbf{X} + \mathbf{X}\mathbf{A}_i^T \prec -2\alpha\mathbf{X}$ results from the following reasoning: Given is the sector in the complex plane

$$S_{II}(\alpha) = \{(x, y) \mid z < -\alpha + iy,\ y \in \{-\infty, \infty\}\}.$$

with $z = -x + iy$ and $-x = \frac{z+\bar{z}}{2}$ results in

$$z + \bar{z} + 2\alpha < 0.$$

Using (B.19) we get $\mathbf{A}_i\mathbf{X} + \mathbf{X}\mathbf{A}_i^T \prec -2\alpha\mathbf{X}$.

Part III: The LMI disk region

$$\begin{pmatrix} -r\mathbf{X} & \mathbf{A}_i\mathbf{X} \\ \mathbf{X}\mathbf{A}_i^T & -r\mathbf{X} \end{pmatrix} \prec 0, \qquad \mathbf{X} \succ 0, \tag{B.24}$$

can be derived as follows: Given is the disk sector

$$S_{III}(r) = \{(x, y) \mid |z| = \sqrt{x^2 + y^2} < r\},$$

with the equivalent relations

$$x^2 + y^2 < r^2 \quad \Leftrightarrow \quad (-x + iy)(-x - iy) < r^2 \quad \Leftrightarrow \quad z\bar{z} < r^2$$
$$\Leftrightarrow -r - \bar{z}(-r)^{-1}z < 0.$$

The application of (B.19) results in

$$\underbrace{-r\mathbf{X}}_{\mathbf{M}_{11}} - \underbrace{\mathbf{A}_i\mathbf{X}}_{\mathbf{M}_{12}} \underbrace{(-r\mathbf{X})^{-1}}_{\mathbf{M}_{22}^{-1}} \underbrace{\mathbf{X}\mathbf{A}_i^T}_{\mathbf{M}_{12}^T} \prec 0.$$

Finally using the Schur complement (B.6) with $\mathbf{M}_{11} \prec 0$ gives (B.24).

Summary: The pole region $S(\alpha, r, \theta)$ results from the intersection of $S_I(\theta)$, $S_{II}(\alpha)$, and $S_{III}(r)$, which corresponds to a combination of the LMI constraints in (B.22).

B.4 Relaxations

In the TS framework, many control and observer design tasks result in a negative definite double sum problem

$$\mathbf{\Gamma}(\mathbf{z}, \chi) = \sum_{i=1}^{N_r}\sum_{j=1}^{N_r} h_i(\mathbf{z})h_j(\mathbf{z})\mathbf{\Gamma}_{ij}(\chi) \prec 0, \tag{B.25}$$

with $\chi \in \{\mathbf{M}_j, \mathbf{N}_j, \mathbf{X}\}$ where each term $\mathbf{\Gamma}_{ij}(\chi)$ depends linearly or affinely on the matrix variables $\mathbf{X}$,$\mathbf{M}_j$, or $\mathbf{N}_j$. A trivial sufficient condition of $\mathbf{\Gamma}(\mathbf{z}, \chi) \prec 0$ is that all sum terms with $h_i(\mathbf{z})h_j(\mathbf{z}) \neq 0,\ \exists\mathbf{z}$ are negative definite

$$\begin{aligned} \mathbf{\Gamma}_{ij}(\chi) \prec 0 \quad & \text{for all} \quad i, j = 1, 2, \dots N_r \\ & s.t. \quad h_i(\mathbf{z})h_j(\mathbf{z}) \neq 0,\ \exists\mathbf{z}, \end{aligned} \tag{B.26}$$

where $h_j(\mathbf{z})$ fulfills the convex sum property (A.4). Here only the coupled expressions with $h_i(\mathbf{z})h_j(\mathbf{z}) \neq 0$ have to be considered. This means that the condition should hold for all $i, j = 1, 2, \ldots, N_r$ expecting $h_i(\mathbf{z})h_j(\mathbf{z}) = 0\,\forall\mathbf{z}$ [Tanaka and Wang, 2001]. With many local valid h-functions, the consideration of this fact can lead to significantly less number of LMI constraints. The above condition (B.26) can be relaxed as follows by considering that with $h_i(\mathbf{z}) \geq 0$ and $h_i(\mathbf{z})h_j(\mathbf{z}) = h_j(\mathbf{z})h_i(\mathbf{z})$ [Wang et al., 1996]

$$\begin{aligned} &\boldsymbol{\Gamma}_{ij}(\chi) + \boldsymbol{\Gamma}_{ji}(\chi) \prec 0\,, \\ &\boldsymbol{\Gamma}_{ii}(\chi) \prec 0 \quad \text{for all} \quad i = 1, 2, \ldots, N_r\,, \quad j = i+1, i+2, \ldots, N_r \\ &\qquad\qquad s.t. \quad h_i(\mathbf{z})h_j(\mathbf{z}) \neq 0,\ \exists\mathbf{z}. \end{aligned} \tag{B.27}$$

Remark B.2 The maximum number of LMI terms N_{LMI} in (B.26) is

$$N_{\mathrm{LMI}} = N_r^2, \tag{B.28}$$

which holds for the worst case where all h_i-functions are globally valid, i.e. $h_i(\mathbf{z})h_j(\mathbf{z}) \neq 0,\ \forall\mathbf{z}$. By the relaxation (B.27) the number of LMIs is reduced to

$$N_{\mathrm{LMI}} = \frac{N_r\,(N_r + 1)}{2}. \tag{B.29}$$

B.5 State Feedback Design for TS Systems

The states feedback design for TS systems in this study is based on the parallel distributes compensation (PDC) scheme

$$\mathbf{u} = -\sum_{i=1}^{N_r} h_i(\mathbf{z})\,\mathbf{K}_i\,\mathbf{x}, \tag{B.30}$$

with the state feedback gains $\mathbf{K}_i \in \mathbb{R}^{m \times b}$ where $h_i(\mathbf{z})$ is equivalent to the h_i-functions of the TS system (A.2), (A.3). As for LTI systems, this control law can be extended by an external reference signal $\mathbf{w} \in \mathbb{R}^p$ without an integral term such as

$$\mathbf{u} = \sum_{i=1}^{N_r} h_i(\mathbf{z})\,\left(-\mathbf{K}_i\mathbf{x} + \mathbf{F}_i\mathbf{w}\right), \tag{B.31}$$

or with an integral term

$$\mathbf{u} = \sum_{i=1}^{N_r} h_i(\mathbf{z})\,\left(-\mathbf{K}_{x,i}\mathbf{x} + \mathbf{K}_{I,i}\mathbf{x}_I\right), \quad \mathbf{x}_I = \int_0^t (\mathbf{w}(\tau) - \mathbf{y}(\tau))\,d\tau. \tag{B.32}$$

The design with the latter control law is based on an augmented TS system with $\tilde{\mathbf{x}} = \left[\mathbf{x}^T\, \mathbf{x}_I^T\right]^T$. The following method of pole region specification with LMI-based criteria provides a design method for calculating the desired $\mathbf{K}_i$ values in (B.30), (B.31) and $\tilde{\mathbf{K}}_i = \left[\mathbf{K}_{x,i} - \mathbf{K}_{I,i}\right]$ in (B.32).

The starting point of the design is the closed-loop system that consists of the TS system (A.2) as the plant and the control law (B.30). Substitution of the plant input in TS form by the control law results in

$$\dot{\mathbf{x}} = \sum_{i=1}^{N_r} h_i(\mathbf{z}) \left(\mathbf{A}_i \mathbf{x} - \mathbf{B}_i \sum_{j=1}^{N_r} h_j(\mathbf{z}) \mathbf{K}_j \mathbf{x} \right)$$

which implies that utilizing the sum condition of h_i

$$\dot{\mathbf{x}} = \sum_{i=1}^{N_r} \sum_{j=1}^{N_r} h_i(\mathbf{z}) h_j(\mathbf{z}) \left(\mathbf{A}_i - \mathbf{B}_i \mathbf{K}_j\right) \mathbf{x}. \tag{B.33}$$

Therefore, the closed-loop system is a weighted combination of N_r^2 linear models. In the specific case that the TS system consists of common $\mathbf{B}$ matrices there are only N_r linear models, due to

$$\dot{\mathbf{x}} = \sum_{i=1}^{N_r} h_i(\mathbf{z}) \left(\mathbf{A}_i - \mathbf{B}\mathbf{K}_i\right) \mathbf{x}. \tag{B.34}$$

B.5.1 Pole Region Specification

Like the techniques for LTI systems, the design objective for TS systems is to find a control law (B.30) that satisfies the closed-loop design specification. The desired closed-loop dynamic is specified by the location of all eigenvalues $\lambda_{i,j} = \mathrm{eig}\left(\mathbf{A}_i - \mathbf{B}_i \mathbf{K}_j\right)$, $\lambda_{i,j} \in \mathbb{C}^n$. To ensure a satisfactory transient response of the closed-loop system (B.33), all eigenvalues $\lambda_{i,j}$ must lie in a prescribed pole region like the stripe- or conic-shaped area illustrated in Figures B.2 and B.3.

Theorem B.5 ***(Pole region specification*** $S(\alpha_{min}, \alpha_{max})$***)*** *The closed-loop system* (B.33) *with the PDC control law* $\mathbf{u} = -\sum_{i=j}^{N_r} h_j(\mathbf{z})\,\mathbf{K}_j\,\mathbf{x}$ *is global asymptotically stable where all eigenvalues* $\lambda_{i,j} = \mathrm{eig}\left(\mathbf{A}_i - \mathbf{B}_i \mathbf{K}_j\right)$ *are located in the given pole region* $S(\alpha_{min}, \alpha_{max})$ *illustrated in Figure* B.2 *if there exists a common* $\mathbf{X} > 0$ *and* $\mathbf{M}_j$, $j = 1, 2, \dots, N_r$ *such that the LMI problem*

$$\begin{aligned} \mathbf{\Gamma}_{ij}^1 &= \mathbf{A}_i \mathbf{X} + \mathbf{X}\mathbf{A}_i^T - \mathbf{B}_i \mathbf{M}_j - \mathbf{M}_j^T \mathbf{B}_i^T + 2\alpha_{min}\mathbf{X} < 0, \\ \mathbf{\Gamma}_{ij}^2 &= \mathbf{A}_i \mathbf{X} + \mathbf{X}\mathbf{A}_i^T - \mathbf{B}_i \mathbf{M}_j - \mathbf{M}_j^T \mathbf{B}_i^T + 2\alpha_{max}\mathbf{X} > 0, \end{aligned} \tag{B.35}$$

with the relaxed conditions (B.27)

$$\begin{aligned}
&\boldsymbol{\Gamma}_{ij}^{1}(\mathbf{X},\mathbf{M}_j)+\boldsymbol{\Gamma}_{ji}^{1}(\mathbf{X},\mathbf{M}_i)\prec 0,\\
&\boldsymbol{\Gamma}_{ii}^{1}(\mathbf{X},\mathbf{M}_i)\prec 0 \quad \text{for all} \quad i=1,2,\dots,N_r\,, \quad j=i+1,i+2,\dots,N_r\\
&\qquad\qquad s.t. \quad h_i(\mathbf{z})h_j(\mathbf{z})\neq 0,\ \exists \mathbf{z},\\
&\boldsymbol{\Gamma}_{ij}^{2}(\mathbf{X},\mathbf{M}_j)+\boldsymbol{\Gamma}_{ji}^{2}(\mathbf{X},\mathbf{M}_i)\prec 0,\\
&\boldsymbol{\Gamma}_{ii}^{2}(\mathbf{X},\mathbf{M}_i)\prec 0 \quad \text{for all} \quad i=1,2,\dots,N_r\,, \quad j=i+1,i+2,\dots,N_r,\\
&\qquad\qquad s.t. \quad h_i(\mathbf{z})h_j(\mathbf{z})\neq 0,\ \exists \mathbf{z},
\end{aligned}$$

is feasible for all $j=1,2,\dots,N_r$. *So that the feedback gains can be obtained as*

$$\mathbf{K}_j=\mathbf{M}_j\mathbf{X}^{-1}\,, \quad j=1,2,\dots,N_r. \tag{B.36}$$

Proof: The proof results directly by the application of Theorem B.3, where the system matrix $\mathbf{A}_i$ is substituted by the matrix of the closed-loop system $\left(\mathbf{A}_i-\mathbf{B}\,\mathbf{K}_i\right)$. Here the resulting bilinear term $\mathbf{K}_j\mathbf{X}$ is replaced by the variable $\mathbf{M}_j$. Afterward (B.27) is used to relax the negative definite sum problem.

Theorem B.6 ***(Pole region specification*** $S(\alpha,r,\theta)$) *The closed-loop system* (B.33) *with the PDC control law* $\mathbf{u}=-\sum_{i=j}^{N_r}h_j(\mathbf{z})\,\mathbf{K}_j\,\mathbf{x}$ *is global asymptotically stable where all eigenvalues* $\lambda_{i,j}=\text{eig}\left(\mathbf{A}_i-\mathbf{B}_i\,\mathbf{K}_j\right)$ *are located in the given pole region* $S(\alpha,r,\theta)$ *illustrated in Figure* B.3 *if there exists a common* $\mathbf{X}\succ 0$ *and* $\mathbf{M}_j$, $j=1,2,\dots,N_r$ *such that the LMI problem*

$$\begin{aligned}
\boldsymbol{\Gamma}_{ij}^{1}&=\mathbf{A}_i\mathbf{X}+\mathbf{X}\mathbf{A}_i^T-\mathbf{B}_i\,\mathbf{M}_j-\mathbf{M}_j^T\mathbf{B}_i^T+2\alpha\mathbf{X}\,,\\
\boldsymbol{\Gamma}_{ij}^{2}&=\begin{pmatrix}(\mathbf{A}_i\mathbf{X}+\mathbf{X}\mathbf{A}_i^T-\mathbf{B}_i\mathbf{M}_j-\mathbf{M}_j^T\mathbf{B}_i^T)\sin\theta & (\mathbf{A}_i\mathbf{X}-\mathbf{X}\mathbf{A}_i^T-\mathbf{B}_i\mathbf{M}_j+\mathbf{M}_j^T\mathbf{B}_i^T)\cos\theta\\ (\mathbf{X}\mathbf{A}_i^T-\mathbf{A}_i\mathbf{X}-\mathbf{M}_j^T\mathbf{B}_i^T+\mathbf{B}_i\mathbf{M}_j)\cos\theta & (\mathbf{A}_i\mathbf{X}+\mathbf{X}\mathbf{A}_i^T-\mathbf{B}_i\mathbf{M}_j-\mathbf{M}_j^T\mathbf{B}_i^T)\sin\theta\end{pmatrix},\\
\boldsymbol{\Gamma}_{ij}^{3}&=\begin{pmatrix}-r\mathbf{X} & \mathbf{A}_i\mathbf{X}-\mathbf{B}_i\mathbf{M}_j\\ \mathbf{X}\mathbf{A}_i^T-\mathbf{M}_j^T\mathbf{B}_i^T & -r\mathbf{X}\end{pmatrix},
\end{aligned} \tag{B.37}$$

with the relaxed conditions (B.27)

$$\begin{aligned}
&\boldsymbol{\Gamma}_{ij}^{k}(\mathbf{X},\mathbf{M}_j)+\boldsymbol{\Gamma}_{ji}^{k}(\mathbf{X},\mathbf{M}_i)\prec 0\,,\\
&\boldsymbol{\Gamma}_{ii}^{k}(\mathbf{X},\mathbf{M}_i)\prec 0\,, \quad k=1,2,3 \quad \text{for all} \quad i=1,2,\dots,N_r\,, \quad j=i+1,i+2,\dots,N_r\\
&\qquad\qquad s.t. \quad h_i(\mathbf{z})h_j(\mathbf{z})\neq 0,\ \exists \mathbf{z},
\end{aligned}$$

is feasible. So that, the feedback gains can be obtained as

$$\mathbf{K}_j=\mathbf{M}_j\mathbf{X}^{-1}\,, \quad j=1,2,\dots,N_r. \tag{B.38}$$

Proof: The proof results directly by the application of Theorem B.4. Further steps are analogous to the proof of Theorem B.5.

Note that the existence of a common positive-definite matrix $\mathbf{X}$ in (B.35) and (B.37) is crucial for the stability of the whole closed-loop TS system. Otherwise, you would have only stable single LTI systems with eigenvalues in the desired pole regions, which is not a sufficient criterion for overall stability.

B.5.2 Minimization of $\mathbf{M}_j$

To reduce the manipulated variable effort within the predefined pole region of the closed-loop system, a minimization of the design variable $\mathbf{M}_j$ is conducted in terms of its Euclidean norm given as

$$\mathbf{M}_j^T \mathbf{M}_j \prec \gamma\, \mathbf{I}, \tag{B.39}$$

where the aim of optimization is to minimize the scalar variable $\gamma > 0$. This is formulated as an additional constraint based on an LMI expression as follows. First thanks to the Schur complement (B.6) we obtain from (B.39)

$$-\gamma \mathbf{I} - \mathbf{M}_j^T\,(-\mathbf{I})^{-1}\,\mathbf{M}_j \prec 0. \tag{B.40}$$

This results in

$$\begin{aligned} &\text{minimize} \quad \gamma \\ &\text{subject to} \quad \begin{pmatrix} -\gamma \mathbf{I} & \mathbf{M}_j^T \\ \mathbf{M}_j & -\mathbf{I} \end{pmatrix} \prec 0\,, \quad j = 1, 2, \dots, N_r, \end{aligned} \tag{B.41}$$

where γ and $\mathbf{M}_j$ are the optimization variables. It should be noted, that via the minimization of $\mathbf{M}_j$ the matrices of control coefficients $\mathbf{K}_j$ can be influenced only indirectly by (B.38). However, this often leads to better results in reducing the control effort.

B.5.3 Stability of Forced Systems

B.5.3.1 Disturbance Rejection

An important characteristic of closed-loop systems is the disturbance rejection capability. For a systematic model-based design a TS system with disturbance $\mathbf{d} \in \mathbb{R}^{m_d}$ and controllable input $\mathbf{u} \in \mathbb{R}^m$ is given

$$\begin{aligned} \dot{\mathbf{x}} &= \sum_{i=1}^{N_r} h_i(\mathbf{z})\,\left(\mathbf{A}_i\,\mathbf{x} + \mathbf{B}_i\,\mathbf{u} + \mathbf{B}_{d,i}\,\mathbf{d}\right), \qquad \mathbf{x}_0 = \mathbf{x}(t_0), \\ \mathbf{y} &= \sum_{i=1}^{N_r} h_i(\mathbf{z})\,\mathbf{C}_i\,\mathbf{x}. \end{aligned} \tag{B.42}$$

The objective of disturbance rejection by state feedback controller (B.30)

$$\mathbf{u} = -\sum_{i=1}^{N_r} h_i(\mathbf{z})\, \mathbf{K}_i\, \mathbf{x},$$

can be formalized by minimizing the disturbance-to-output L_2 gain in the time domain

$$\begin{aligned} &\text{minimize} \quad \gamma \\ &\text{subject to} \quad \sup_{\|\mathbf{d}\|_2 \neq 0} \frac{\|\mathbf{y}\|_2}{\|\mathbf{d}\|_2} \leq \gamma, \end{aligned} \tag{B.43}$$

where $\|\mathbf{d}\|_2$ and $\|\mathbf{y}\|_2$ are defined as the L_2 norm of the vector-valued function $\mathbf{z}(t)$ denoted as disturbance input and $\mathbf{y}(t)$ as system output:

$$\|\mathbf{d}\|_2 = \sqrt{\int_0^\infty \mathbf{d}^T(t)\, \mathbf{d}(t)\, dt}\,, \qquad \|\mathbf{y}\|_2 = \sqrt{\int_0^\infty \mathbf{y}^T(t)\, \mathbf{y}(t)\, dt}. \tag{B.44}$$

Theorem B.7 ***(Controller design for disturbance rejection)*** *The control gains* $\mathbf{K}_j$, $j = 1, 2, \ldots, N_r$ *with* $\mathbf{M}_j = \mathbf{K}_j\mathbf{X}$ *of the state feedback controller* (B.30) *stabilize the TS system* (B.42) *and reject the disturbances* $\mathbf{d}$ *realized by minimizing of* γ *if* (B.27) *is fulfilled for* $\mathbf{X} \succ 0$*, where*

$$\boldsymbol{\Gamma}_{ij} = \begin{pmatrix} \mathbf{A}_i\mathbf{X} + \mathbf{X}\mathbf{A}_i^T - \mathbf{B}_i\mathbf{M}_j - \mathbf{M}_j^T\mathbf{B}_i^T & \mathbf{B}_{d,i} & \mathbf{X}\,\mathbf{C}_i^T \\ \mathbf{B}_{d,i}^T & -\gamma^2\mathbf{I} & \mathbf{0} \\ \mathbf{C}_i\,\mathbf{X} & \mathbf{0} & -\mathbf{I} \end{pmatrix}. \tag{B.45}$$

Proof: The proof is given in Lendek et al. [2011], Theorem 3.4 or Tanaka and Wang [2001], Theorem 14.

B.5.3.2 Input-to-State Stability

In contrast to Section B.5.3.1, not the input and output characteristics but the state stability concerning external disturbances as non-adjustable inputs are studied. Given is a state feedback TS system with an external input $\mathbf{d}$

$$\dot{\mathbf{x}} = \sum_{i=1}^{N_r} h_i(\mathbf{z}) \left(\mathbf{A}_i\,\mathbf{x} + \mathbf{B}_i\,\mathbf{u} + \mathbf{B}_{d,i}\,\mathbf{d}\right), \qquad \mathbf{x}_0 = \mathbf{x}(t_0),$$

$$\mathbf{u} = -\sum_{j=1}^{N_r} h_j(\mathbf{z})\, \mathbf{K}_i\, \mathbf{x}.$$

From that follows the closed-loop system

$$\dot{\mathbf{x}} = \sum_{i=1}^{N_r}\sum_{j=1}^{N_r} h_i(\mathbf{z})h_j(\mathbf{z}) \left[\left(\mathbf{A}_i - \mathbf{B}_i\mathbf{K}_j\right)\mathbf{x} + \mathbf{B}_{d,i}\,\mathbf{d}\right]. \tag{B.46}$$

This leads to the question whether the closed-loop TS system (B.46) can become unstable by an external excitation $\mathbf{d}(t)$. What must be fulfilled so that (B.46) remains input-to-state stable (ISS) in the terms of Lyapunov?

First, we assume that for the TS system (B.46) without an external input, i.e. for $\mathbf{d} = \mathbf{0}$ exists a quadratic Lyapunov function

$$V(\mathbf{x}) = \mathbf{x}^T\mathbf{P}\mathbf{x}, \quad \mathbf{P} \succ 0, \tag{B.47}$$

with

$$V(\mathbf{x}) > 0 \quad \forall \mathbf{x} \neq \mathbf{0}, \qquad \dot{V}(\mathbf{x}) < -2\alpha V(\mathbf{x}) = -2\alpha\,\mathbf{x}^T\mathbf{P}\mathbf{x}. \tag{B.48}$$

Here α is identical to the decay of the LMI region (B.18). Since for $\mathbf{P} \succ 0$ it holds that

$$\lambda_{min}(\mathbf{P})||\mathbf{x}||^2 \leq \mathbf{x}^T\mathbf{P}\mathbf{x} \leq \lambda_{max}(\mathbf{P})||\mathbf{x}||^2, \tag{B.49}$$

where $\lambda_{\min}(\mathbf{P})$ and $\lambda_{max}(\mathbf{P})$ denote the smallest and largest eigenvalue of $\mathbf{P}$ we obtain

$$\dot{V}(\mathbf{x}) < -2\alpha\,\mathbf{x}^T\mathbf{P}\mathbf{x} \leq -2\alpha\,\lambda_{min}(\mathbf{P})\,||\mathbf{x}||^2 = -\lambda\,||\mathbf{x}||^2.$$

Note that $\lambda = 2\alpha\,\lambda_{min}(\mathbf{P})$ has been defined for the sake of simplicity. After this preliminary work, the stability of the closed-loop system disturbed by $\mathbf{d} \neq \mathbf{0}$ will be investigated by

$$\begin{aligned}
\dot{V}(\mathbf{x}) &= \dot{\mathbf{x}}^T\mathbf{P}\mathbf{x} + \mathbf{x}^T\mathbf{P}\dot{\mathbf{x}} \\
&= \mathbf{x}^T\sum_{i=1}^{N_r}\sum_{j=1}^{N_r} h_i(\mathbf{z})h_j(\mathbf{z})\left[\left(\mathbf{A}_i - \mathbf{B}_i\,\mathbf{K}_j\right)^T\mathbf{P} + \mathbf{P}\left(\mathbf{A}_i - \mathbf{B}_i\,\mathbf{K}_j\right)\right]\mathbf{x} \\
&\quad + 2\sum_{i=1}^{N_r} h_i(\mathbf{z})\,\mathbf{d}^T\mathbf{B}_{d,i}^T\mathbf{P}\mathbf{x} \\
&\leq -\lambda\,||\mathbf{x}||^2 + \sum_{i=1}^{N_r} h_i(\mathbf{z})\,\mathbf{d}^T\mathbf{B}_{d,i}^T\mathbf{P}\mathbf{x} \\
&\leq -\lambda\,||\mathbf{x}||^2 + 2\delta||\mathbf{d}||\,||\mathbf{x}|| \\
&= -\left(\frac{\lambda}{2}||\mathbf{x}||^2 - 2\delta||\mathbf{d}||\,||\mathbf{x}|| + \frac{2\delta^2}{\lambda}||\mathbf{d}||^2\right) + \frac{2\delta^2}{\lambda}||\mathbf{d}|| - \frac{\lambda}{2}||\mathbf{x}||^2 \\
&= -\left(\sqrt{\frac{\lambda}{2}}||\mathbf{x}|| + \delta\sqrt{\frac{2}{\lambda}}||\mathbf{d}||\right)^2 + \frac{2\delta^2}{\lambda}||\mathbf{d}|| - \frac{\lambda}{2}||\mathbf{x}||^2 \\
&\leq -\frac{\lambda}{2}\,||\mathbf{x}||^2 + \frac{2\delta^2}{\lambda}||\mathbf{d}||^2,
\end{aligned} \tag{B.50}$$

where $\delta = \max_i \|\mathbf{B}_{d,i}^T \mathbf{P}\|$. The closed-loop TS system (B.46) has the ISS property because $\alpha(\|\mathbf{x}\|) = \frac{\lambda}{2}\|\mathbf{x}\|^2$ and $\theta(\|\mathbf{d}\|) = \frac{2\delta^2}{\lambda}\|\mathbf{d}\|^2$ are $\mathcal{K}^\infty$ functions $f : \mathbb{R}^+ \to \mathbb{R}^+$ introduced by Sontag and Wang [1995]. That means both are strictly increasing, fulfill $f(0) = 0$ and $f(x) \to \infty$ when $x \to \infty$.

In Schulte and Gauterin [2014, 2015] this ISS property (B.50) of nonlinear systems in TS form is used for the dynamic analysis of small wind turbines with mechanical pitch systems.

B.6 Observer Design for TS Systems

Estimation of non-measurable system states and disturbances is based in this work on the observer law for TS systems (A.2)

$$\begin{aligned} \dot{\hat{\mathbf{x}}} &= \sum_{i=1}^{N_r} h_i(\mathbf{z}) \left(\mathbf{A}_i \hat{\mathbf{x}} + \mathbf{B}_i \mathbf{u} + \mathbf{L}_i (\mathbf{y} - \hat{\mathbf{y}})\right), \\ \hat{\mathbf{y}} &= \sum_{i=1}^{N_r} h_i(\mathbf{z}) \mathbf{C}_i \hat{\mathbf{x}}, \qquad \hat{\mathbf{x}}_0 = \hat{\mathbf{x}}(t_0), \end{aligned} \tag{B.51}$$

and affine TS systems (A.3)

$$\begin{aligned} \dot{\hat{\mathbf{x}}} &= \sum_{i=1}^{N_r} h_i(\mathbf{z}) \left(\mathbf{A}_i \hat{\mathbf{x}} + \mathbf{B}_i \mathbf{u} + \mathbf{a}_i + \mathbf{L}_i (\mathbf{y} - \hat{\mathbf{y}})\right), \\ \hat{\mathbf{y}} &= \sum_{i=1}^{N_r} h_i(\mathbf{z}) \left(\mathbf{C}_i \hat{\mathbf{x}} + \mathbf{c}_i\right), \qquad \hat{\mathbf{x}}_0 = \hat{\mathbf{x}}(t_0), \end{aligned} \tag{B.52}$$

with the observer gains $\mathbf{L}_i \in \mathbb{R}^{n \times m}$ and the membership function $h_i(\mathbf{z})$ of the related TS systems. The estimation error is

$$\mathbf{e} = \mathbf{x} - \hat{\mathbf{x}} \quad \text{with} \quad \dot{\mathbf{e}} = \dot{\mathbf{x}} - \dot{\hat{\mathbf{x}}}. \tag{B.53}$$

By replacing $\dot{\mathbf{x}}$ with the left hand side (LHS) of (A.2) and $\dot{\hat{\mathbf{x}}}$ with (B.51) we obtain the error differential equation

$$\dot{\mathbf{e}} = \sum_{i=1}^{N_r} \sum_{j=1}^{N_r} h_i(\mathbf{z}) h_j(\mathbf{z}) \left(\mathbf{A}_i - \mathbf{L}_i \mathbf{C}_j\right) \mathbf{e}, \qquad \hat{\mathbf{e}}_0 = \hat{\mathbf{e}}(t_0). \tag{B.54}$$

The same result is obtained if instead of (A.2) and (B.51) the affine TS system (A.3) and the corresponding observer law (B.52) are used.

B.6.1 Pole Region Specification for Observer Design

The objective of the observer design is to reconstruct the system states as quickly as possible with simultaneously low measurement noise sensitivity. This can be achieved by a suitable choice of the location of all eigenvalues

$$\lambda_{i,j} = \text{eig}\left(\mathbf{A}_i - \mathbf{L}_i\,\mathbf{C}_j\right), \quad \lambda_{i,j} \in \mathbb{C}^n, \tag{B.55}$$

of the error state equation (B.54). Here we use the same parameterized regions as for the controller design in Section (B.5.1).

Theorem B.8 ***(Pole region specification*** $S(\alpha_{\min}, \alpha_{\max})$ ***for obs. design)*** *The error system* (B.54) *related to the TS system* (A.2) *and TS observer* (B.51) *or TS affine system* (A.3) *and* (B.52) *is global asymptotically stable where all eigenvalues* $\lambda_{i,j} = \text{eig}\left(\mathbf{A}_i - \mathbf{L}_i\,\mathbf{C}_j\right)$ *are located in the given pole region* $S(\alpha_{\min}, \alpha_{\max})$ *illustrated in Figure* B.2 *if there exists* $\mathbf{P} \succ 0$ *and* $\mathbf{N}_i, j = 1, 2, \ldots, N_r$ *such that the LMI problem*

$$\begin{aligned}
\boldsymbol{\Gamma}^1_{ij} &= \mathbf{A}_i^T\mathbf{P} + \mathbf{P}\mathbf{A}_i - \mathbf{C}_j^T\,\mathbf{N}_i^T - \mathbf{N}_i\mathbf{C}_j + 2\alpha_{\min}\,\mathbf{P} \prec 0\,,\\
\boldsymbol{\Gamma}^2_{ij} &= \mathbf{A}_i^T\mathbf{P} + \mathbf{P}\mathbf{A}_i - \mathbf{C}_j^T\,\mathbf{N}_i^T - \mathbf{N}_i\mathbf{C}_j + 2\alpha_{\max}\,\mathbf{P} \succ 0,
\end{aligned} \tag{B.56}$$

with the relaxed conditions (B.27)

$$\begin{aligned}
&\boldsymbol{\Gamma}^1_{ij}(\mathbf{P},\mathbf{N}_i) + \boldsymbol{\Gamma}^1_{ji}(\mathbf{P},\mathbf{N}_j) \prec 0\,,\\
&\boldsymbol{\Gamma}^1_{ii}(\mathbf{P},\mathbf{N}_i) \prec 0 \quad \text{for all} \quad i = 1,2,\ldots,N_r\,, \quad j = i+1, i+2, \ldots, N_r\\
&\qquad\qquad s.t. \quad h_i(\mathbf{z})h_j(\mathbf{z}) \neq 0, \ \exists \mathbf{z}\,,\\
&\boldsymbol{\Gamma}^2_{ij}(\mathbf{P},\mathbf{N}_i) + \boldsymbol{\Gamma}^2_{ji}(\mathbf{P},\mathbf{N}_j) \prec 0\,,\\
&\boldsymbol{\Gamma}^2_{ii}(\mathbf{P},\mathbf{N}_i) \prec 0 \quad \text{for all} \quad i = 1,2,\ldots,N_r\,, \quad j = i+1, i+2, \ldots, N_r\\
&\qquad\qquad s.t. \quad h_i(\mathbf{z})h_j(\mathbf{z}) \neq 0, \ \exists \mathbf{z},
\end{aligned}$$

is feasible for all $j = 1, 2, \ldots, N_r$*. So that the observer gains can be obtained as*

$$\mathbf{L}_i = \mathbf{P}^{-1}\,\mathbf{N}_i\,, \quad i = 1, 2, \ldots, N_r. \tag{B.57}$$

Proof: Initial step is the Theorem B.3 for unforced TS systems. Here we substitute the system matrices $\mathbf{A}_i$ by the error matrices $\mathbf{A}^e_{ij} = \mathbf{A}_i - \mathbf{L}_i\,\mathbf{C}_j$ from (B.54) and get

$$\begin{aligned}
\mathbf{Q}^1_{ij} &= \mathbf{A}_i\mathbf{X} - \mathbf{L}_i\mathbf{C}_j\mathbf{X} + \mathbf{X}\mathbf{A}_i^T - \mathbf{X}\mathbf{C}_j^T\mathbf{L}_i^T + 2\alpha_{\min}\,\mathbf{P} \prec 0,\\
\mathbf{Q}^2_{ij} &- \mathbf{A}_i\mathbf{X} - \mathbf{L}_i\mathbf{C}_j\mathbf{X} + \mathbf{X}\mathbf{A}_i^T - \mathbf{X}\mathbf{C}_j^T\mathbf{L}_i^T + 2\alpha_{\max}\,\mathbf{P} \succ 0.
\end{aligned}$$

Using the congruence equivalence relation (B.5)

$$\mathbf{P}\,\mathbf{Q}^1_{ij}\,\mathbf{P}^T \prec 0, \quad \mathbf{P}\,\mathbf{Q}^2_{ij}\,\mathbf{P}^T \succ 0 \quad \text{where} \quad \mathbf{P} = \mathbf{P}^T, \quad \mathbf{P} = \mathbf{X}^{-1},$$

and with the new variable $\mathbf{N}_i = \mathbf{P}\mathbf{L}_i$ (B.57) follows (B.56).

Theorem B.9 ***(Pole region specification*** $S(\alpha, r, \theta)$ ***for observer design)*** *The error system* (B.54) *related to the TS system* (A.2) *and TS observer* (B.51) *or TS affine system* (A.3) *and* (B.52) *is global asymptotically stable where all eigenvalues* $\lambda_{i,j} = \text{eig}\left(\mathbf{A}_i - \mathbf{L}_i\,\mathbf{C}_j\right)$ *are located in the given pole region* $S(\alpha, r, \theta)$ *illustrated in Figure* B.3 *if there exists* $\mathbf{P} \succ 0$ *and* $\mathbf{N}_i, j = 1, 2, \ldots, N_r$ *such that the LMI problem with the relaxed conditions* (B.27)

$$\begin{aligned}
&\boldsymbol{\Gamma}_{ij}^{k}(\mathbf{X}, \mathbf{N}_i) + \boldsymbol{\Gamma}_{ji}^{k}(\mathbf{X}, \mathbf{M}_j) \prec 0, \\
&\boldsymbol{\Gamma}_{ii}^{k}(\mathbf{X}, \mathbf{N}_i) \prec 0, \quad k = 1, 2, 3 \quad \text{for all} \quad i = 1, 2, \ldots, N_r\,, \quad j = i+1, i+2, \ldots, N_r \\
&\qquad\qquad s.t. \quad h_i(\mathbf{z})h_j(\mathbf{z}) \neq 0, \ \exists \mathbf{z},
\end{aligned} \tag{B.58}$$

where

$$\boldsymbol{\Gamma}_{ij}^{1} = \mathbf{A}_i^T\mathbf{P} + \mathbf{P}\mathbf{A}_i - \mathbf{C}_j^T\,\mathbf{N}_i^T - \mathbf{N}_i\mathbf{C}_j + 2\alpha\mathbf{P}\,,$$

$$\boldsymbol{\Gamma}_{ij}^{2} = \begin{pmatrix} \left(\mathbf{P}\mathbf{A}_i + \mathbf{A}_i^T\mathbf{P} - \mathbf{N}_i\mathbf{C}_j - \mathbf{C}_j^T\mathbf{N}_i^T\right)\sin\theta & \left(\mathbf{P}\mathbf{A}_i - \mathbf{A}_i^T\mathbf{P} - \mathbf{N}_i\mathbf{C}_j + \mathbf{C}_j^T\mathbf{N}_i^T\right)\cos\theta \\ \left(-\mathbf{P}\mathbf{A}_i + \mathbf{A}_i^T\mathbf{P} + \mathbf{N}_i\mathbf{C}_j - \mathbf{C}_j^T\mathbf{N}_i^T\right)\cos\theta & \left(\mathbf{P}\mathbf{A}_i + \mathbf{A}_i^T\mathbf{P} - \mathbf{N}_i\mathbf{C}_j - \mathbf{C}_j^T\mathbf{N}_i^T\right)\sin\theta \end{pmatrix}$$

$$\boldsymbol{\Gamma}_{ij}^{3} = \begin{pmatrix} -r\mathbf{P} & \mathbf{P}\mathbf{A}_i - \mathbf{N}_i\mathbf{C}_j \\ \mathbf{A}_i^T\mathbf{P} - \mathbf{C}_j^T\mathbf{N}_i^T & -r\mathbf{P} \end{pmatrix},$$

is feasible. So that the observer gains can be obtained by $\mathbf{L}_i = \mathbf{P}^{-1}\,\mathbf{N}_i$.

Proof:

- Proof for $k = 1$ in (B.58) follows directly the proof of Theorem B.8.
- Proof for $k = 2, 3$ in (B.58): Initial step is the Theorem B.4 for unforced TS systems. Here we substitute in $\boldsymbol{\Gamma}_{ij}^2$ and $\boldsymbol{\Gamma}_{ij}^3$ the system matrices $\mathbf{A}_i$ by the error matrices $\mathbf{A}_{ij}^e = \mathbf{A}_i - \mathbf{L}_i\,\mathbf{C}_j$ from (B.54) and get

$$\mathbf{Q}_{ij}^2 = \begin{pmatrix} \left(\mathbf{A}_i\mathbf{X} - \mathbf{L}_i\mathbf{C}_j\mathbf{X} + \mathbf{X}\mathbf{A}_i^T - \mathbf{X}\mathbf{C}_j^T\mathbf{L}_i^T\right)\sin\theta & \left(\mathbf{A}_i\mathbf{X} - \mathbf{L}_i\mathbf{C}_j\mathbf{X} - \mathbf{X}\mathbf{A}_i^T + \mathbf{X}\mathbf{C}_j^T\mathbf{L}_i^T\right)\cos\theta \\ \left(-\mathbf{A}_i\mathbf{X} + \mathbf{L}_i\mathbf{C}_j\mathbf{X} + \mathbf{X}\mathbf{A}_i^T - \mathbf{X}\mathbf{C}_j^T\mathbf{L}_i^T\right)\cos\theta & \left(\mathbf{A}_i\mathbf{X} - \mathbf{L}_i\mathbf{C}_j\mathbf{X} + \mathbf{X}\mathbf{A}_i^T - \mathbf{X}\mathbf{C}_j^T\mathbf{L}_i^T\right)\sin\theta \end{pmatrix} \prec 0,$$

$$\mathbf{Q}_{ij}^3 = \begin{pmatrix} -r\mathbf{X} & \mathbf{A}_i\mathbf{X} - \mathbf{L}_i\mathbf{C}_j\mathbf{X} \\ \mathbf{X}\mathbf{A}_i^T - \mathbf{X}\mathbf{C}_j^T\mathbf{L}_i^T & -r\mathbf{X} \end{pmatrix} \prec 0.$$

  Using the congruence equivalence relation (B.5)

$$\tilde{\mathbf{P}}\,\mathbf{Q}_{ij}^2\,\tilde{\mathbf{P}}^T \prec 0, \quad \tilde{\mathbf{P}}\,\mathbf{Q}_{ij}^3\,\tilde{\mathbf{P}}^T \succ 0 \ \text{with} \ \tilde{\mathbf{P}} = \begin{pmatrix} \mathbf{P} & \mathbf{0} \\ \mathbf{0} & \mathbf{P} \end{pmatrix} \quad \text{where} \ \mathbf{P} = \mathbf{X}^{-1}, \ \tilde{\mathbf{P}} = \tilde{\mathbf{P}}^T,$$

and with the new variable $\mathbf{N}_i = \mathbf{P}\mathbf{L}_i$ (B.57) we get (B.58).

C

Renewable Energy Sources

C.1 Wind Energy Systems

C.1.1 Froude's Actuator Disc Theory (Froude–Rankine Theorem)

Froude's actuator disc theory also known as Froude–Rankine Theorem is used in Section 2.2 with the result (2.8) that the velocity through the disc is the average of the velocities far upstream denoted by v_∞ and downstream denoted by v_W

$$v_D = \frac{v_\infty + v_W}{2}. \tag{C.1}$$

The question becomes why is this the case? Let's study the appropriate explanation of Froude: the thrust force F_T on the actuator disc is given using the simple momentum theory and balance of forces

$$F_T = \frac{d}{dt}(m\,v_\infty) - \frac{d}{dt}(m\,v_W) = \dot{m}(v_\infty - v_W) = (v_\infty - v_W)\rho A_D\,v_D, \tag{C.2}$$

where $\dot{m} = \rho A_D\,v_D$ and with the assumption $\dot{v}_\infty = \dot{v}_W = 0$. Otherwise, this force can be expressed by the difference of static pressure directly before p_D^+ and after p_D^- the rotor plane

$$F_T = (p_D^+ - p_D^-)A_D. \tag{C.3}$$

To get the pressure difference $(p_D^+ - p_D^-)$ Bernoulli equation (energy balance) is applied separately to the upstream and downstream sections. Assuming the flow is incompressible $\rho_\infty = \rho_W = \rho$ and the static pressure energy is equal in the upstream and downstream sections we obtain

$$\frac{1}{2}\rho v_\infty^2 + p_\infty = \frac{1}{2}\rho v_D^2 + p_D^+ \qquad \text{upstream section,} \tag{C.4}$$

$$\frac{1}{2}\rho v_W^2 + p_W = \frac{1}{2}\rho v_D^2 + p_D^- \qquad \text{downstream section.} \tag{C.5}$$

Advanced Control of Grid-Integrated Renewable Energy Power Plants: LMI-based Design in the Takagi-Sugeno Framework, First Edition. Horst Schulte.

Since the pressures far upstream p_W and far downstream p_∞ are equal, the difference between (C.4) and gives (C.5)

$$p_D^+ - p_D^- = \frac{1}{2}\rho(v_\infty^2 - v_W^2). \tag{C.6}$$

Finally with (C.2) and (C.3) as well as substituting the pressure difference by (C.6) follows the expression (C.1)

$$v_D = \frac{1}{2}\frac{v_\infty^2 - v_W^2}{v_\infty - v_W} = \frac{v_\infty + v_W}{2}.$$

C.1.2 Rotor-Equivalent Wind Speed

Wind turbines (WT) face stochastic, intermittent wind excitations that cover multiple spatial and temporal domains [Burton et al., 2021; Wächter et al., 2012]. The stochastic characteristics of the wind field vary with, for example, topographic conditions or thermal effects [Burton et al., 2021]. For wind turbine certification by simulation studies in the industry, various wind profiles and models are used to test turbine dynamics with different controllers.

Extreme operating gusts in the wind profiles are usually used to assess the mechanical design of the turbine with relevant loading scenarios [GL Wind, 2010], which feature a drastic variation of the wind speed within a time range of seconds. In this work, we also use the gust wind profiles to validate the reduced wind turbine models compared to the high-fidelity Fatigue, Aerodynamics, Structures, and Turbulence (FAST) model (as e.g. shown in Figure 2.18) and to validate the wind speed observer (see Figure 3.11).

The certification process [GL Wind, 2010] of wind turbine units, particularly the fatigue analysis, turbulent spatially distributed wind fields along the rotor which interacting with the blade segments are used in long-term simulation. These profiles may be produced by tools like TurbSim [Jonkman and Buhl, 2006]. That allows the generation of input data for simulation studies at different wind conditions, resulting in either a single wind velocity in extreme operating gusts or a fully distributed wind field in turbulent long-term operation.

For the objective of system investigation specifically for the controller and observer design in Chapter 3, it is thus a useful approach to model the primary source as a rotor-equivalent wind speed, which aggregates all aerodynamic effects and considers an equivalent rotor excitation compared to a spatially distributed wind field excitation [Andrejewski et al., 2021]. The rotor is effectively a filter that concentrates the spatially distributed power into an aggregate rotor power. To run the simulations in Chapters 2, 3, and 6, the time-varying and spatially distributed wind field data from TurbSim [Jonkman and Buhl, 2006] are averaged over the rotor plane. This results in a rotor-equivalent wind speed [Ostergaard et al., 2007] as a temporally (no longer spatially) variable wind speed far in front of the rotor v_∞, which is used as an input for simulation runs based on analytical wind turbine description.

C.1.3 Membership Functions and Matrices of TS Multi-region Model

The membership functions h_m of the multi-region WT model proposed in Section 2.3.5, Eq. (2.157) are obtained from the product of the combination of

$$h_m \in \{h_{I,1}, h_{I,2}\} \times \{h_{II,1}, h_{II,2}, h_{II,3}, h_{II,4}\} \times \{\tilde{h}_1, \tilde{h}_2\}.$$

These are with the related $\mathbf{A}_m$ and $\mathbf{B}_m$ matrices as follows

$$\begin{aligned}
h_1 &= h_{I,1}\, h_{II,1}\, \tilde{h}_1, & \mathbf{A}_1 &= \mathbf{A}_{I,1}, & \mathbf{B}_1 &= \mathbf{B}_I,\\
h_2 &= h_{I,2}\, h_{II,1}\, \tilde{h}_1, & \mathbf{A}_2 &= \mathbf{A}_{I,2}, & \mathbf{B}_2 &= \mathbf{B}_I,\\
h_3 &= h_{I,1}\, h_{II,2}\, \tilde{h}_1, & \mathbf{A}_3 &= \mathbf{A}_{I,1}, & \mathbf{B}_3 &= \mathbf{B}_I,\\
h_4 &= h_{I,2}\, h_{II,2}\, \tilde{h}_1, & \mathbf{A}_4 &= \mathbf{A}_{I,2}, & \mathbf{B}_4 &= \mathbf{B}_I,\\
h_5 &= h_{I,1}\, h_{II,3}\, \tilde{h}_1, & \mathbf{A}_5 &= \mathbf{A}_{I,1}, & \mathbf{B}_5 &= \mathbf{B}_I,\\
h_6 &= h_{I,2}\, h_{II,3}\, \tilde{h}_1, & \mathbf{A}_6 &= \mathbf{A}_{I,2}, & \mathbf{B}_6 &= \mathbf{B}_I,\\
h_7 &= h_{I,1}\, h_{II,4}\, \tilde{h}_1, & \mathbf{A}_7 &= \mathbf{A}_{I,1}, & \mathbf{B}_7 &= \mathbf{B}_I,\\
h_8 &= h_{I,2}\, h_{II,4}\, \tilde{h}_1, & \mathbf{A}_8 &= \mathbf{A}_{I,2}, & \mathbf{B}_8 &= \mathbf{B}_I,\\
h_9 &= h_{I,1}\, h_{II,1}\, \tilde{h}_1, & \mathbf{A}_9 &= \mathbf{A}_{II,1}, & \mathbf{B}_9 &= \mathbf{B}_{II},\\
h_{10} &= h_{I,1}\, h_{II,2}\, \tilde{h}_2, & \mathbf{A}_{10} &= \mathbf{A}_{II,2}, & \mathbf{B}_{10} &= \mathbf{B}_{II},\\
h_{11} &= h_{I,1}\, h_{II,3}\, \tilde{h}_2, & \mathbf{A}_{11} &= \mathbf{A}_{II,3}, & \mathbf{B}_{11} &= \mathbf{B}_{II},\\
h_{12} &= h_{I,1}\, h_{II,4}\, \tilde{h}_2, & \mathbf{A}_{12} &= \mathbf{A}_{II,4}, & \mathbf{B}_{12} &= \mathbf{B}_{II},\\
h_{13} &= h_{I,2}\, h_{II,1}\, \tilde{h}_2, & \mathbf{A}_{13} &= \mathbf{A}_{II,1}, & \mathbf{B}_{13} &= \mathbf{B}_{II},\\
h_{14} &= h_{I,2}\, h_{II,2}\, \tilde{h}_2, & \mathbf{A}_{14} &= \mathbf{A}_{II,2}, & \mathbf{B}_{14} &= \mathbf{B}_{II},\\
h_{15} &= h_{I,2}\, h_{II,3}\, \tilde{h}_2, & \mathbf{A}_{15} &= \mathbf{A}_{II,3}, & \mathbf{B}_{15} &= \mathbf{B}_{II},\\
h_{16} &= h_{I,2}\, h_{II,4}\, \tilde{h}_2, & \mathbf{A}_{16} &= \mathbf{A}_{II,4}, & \mathbf{B}_{16} &= \mathbf{B}_{II}.
\end{aligned}$$

D

Parameters of Renewable Energy Power Plants

D.1 Physical Constants

Physical constants used in the renewable energy power plant models are given in Table D.1.

Table D.1 Physical constants.

Symbol	Value	Unit	Description
ρ	1.225	$\frac{kg}{m^3}$	Air density at 15 °C
k	1.381×10^{-23}	J/K	Boltzmann constant
q	1.6×10^{-19}	$\frac{kg}{m^3}$	Unit charge

D.2 Wind Power Plant Parameters

The model parameters for the NREL 5-MW turbine and the related structural parameters of reduced-order control-oriented wind turbine models are listed in Tables D.2 and D.3.

D.3 PV Power Plant Parameters

Table D.4 contains the model parameters of the generic 3-MW photovoltaic system illustrated in Figure 4.12 (primary power conversion) consisting of a photo voltaic (PV) generator with monocrystalline silicon solar cells manufactured by JA-Solar, JACM6SR-3, and a generic DC–DC converter.

That system with a 3-MW PV-generator in combination with a DC–DC converter and so-called central inverter can be considered as a sub-unit of a power plant with

Advanced Control of Grid-Integrated Renewable Energy Power Plants: LMI-based Design in the Takagi-Sugeno Framework, First Edition. Horst Schulte.

Table D.2 Parameters of NREL 5-MW reduced wind turbine model (part I).

Symbol	Value	Unit	Description
$\dot{\beta}_{max}$	8.0	$\frac{deg}{s}$	Maximum pitch rate
λ_d	7.0895		Design tip-speed or rated-tip speed ratio λ_{rated}
$\omega_{r,rated}$	1.2671	$\frac{rad}{s}$	Rated rotor speed (angular velocity)
$\omega_{g,rated}$	122.9087	$\frac{rad}{s}$	Rated generator speed (angular velocity)
ρ	1.225	$\frac{kg}{m^3}$	Air density at 15 °C
τ_β	0.1	s	Time constant of first order pitch system
τ_v	4	s	Time constant of first-order Effective wind speed disturbance model
d_B	20000	$\frac{Ns}{m}$	Flap-wise blade damping constant
d_S	660.5378	Nms	Drive train damping constant
d_T	70000	$\frac{Ns}{m}$	Fore–aft tower damping constant
J_g	534.1	kg m^2	Generator inertia
J_r	38759.227×10^3	kg m^2	Rotor inertia
v_d	11.26	$\frac{m}{s}$	Design wind speed
k_B	40000×1.4	$\frac{N}{m}$	Effective blade flap-wise stiffness parameter
k_S	92214	Nm	Drive train stiffness parameter
k_T	1981900	$\frac{N}{m}$	Effective tower fore–aft stiffness parameter
m_B	4435	kg	Effective mass of blade-tip motion
m_{blade}	17740	kg	Blade mass
m_{hub}	56780	kg	Hub mass
$m_{nacelle}$	240000	kg	Nacelle mass
m_{rotor}	110000	kg	Rotor mass
m_T	436865	kg	Effective mass of nacelle-tower motion
m_{tower}	347640	kg	Tower mass

Table D.3 Parameters of NREL 5-MW reduced wind turbine model (part II).

Symbol	Value	Unit	Description
N	3		Number of blades
n_g	97		Gearbox ratio
$P_{r,rated}$	5296610	W	Rated wind turbine rotor power
R	63	m	Rotor radius
$T_{r,rated}$	4.1801×10^6	Nm	Rated aerodynamic rotor torque
$T_{g,rated}$	4.3094×10^4	Nm	Rated generator torque

Table D.4 Parameters of a generic 3 MW photovoltaic power plant.

Symbol	Value	Unit	Description
α_T	0.06×10^{-2}		Temperature coefficient on PV current
β_T	-0.36×10^{-2}		Temperature coefficient on PV voltage
A_n	1.374		Diode ideality factor in the single diode model
C	17700	μF	Capacitor of the buck converter, see Figure 4.12
$i_{ph,sc,STC}$	3.1154×10^3	A	Short-circuit current of PV generator at standard test condition (STC)
$i_{s,STC}$	1.1004×10^{-7}	A	Diode's saturation current at STC
L	1.8329	mH	Induction of the buck converter, see Figure 4.12
$N_{cell,p}$	1		PV cells per module connected in parallel
$N_{cell,s}$	72		PV cells per module connected in series
$N_{mod,p}$	336		Number of PV modules connected in parallel
$N_{mod,s}$	27		Number of PV modules connected in series
N_p	336		Total number of PV cells connected in parallel
N_s	1944		Total number of PV cells connected in series
P_{MPP}	3.0450	MW	Peak power of PV generator at STC
$P_{MPP,c}$	4.72	W	Peak power of one cell at STC
$V_{pv,MPP}$	1062.2	V	Terminal voltage of PV generator at STC
$I_{pv,MPP}$	2902.1	A	Output current of PV generator at STC
R_h	10.196	V/A	Shunt resistor of the single diode model
$v_{oc,STC}$	0.644	V	PV open-circuit voltage of one cell at STC
$v_{T,STC}$	25.7	mV	Thermal voltage of p–n junction at STC
$T_{c,STC}$	298	°K	Cell temperature at STC
S_{STC}	1000	W/m^2	Irradiation at STC

several 10 MW such as the PV power plant “Solarpark Waldpolenz” near Polenz, Germany, with $P = 52$ MW peak power.

D.4 VSC Parameters

Table D.5 contains the model parameters of the voltage source converter (see Figure 5.1) proposed in Chapter 5 used for the grid integration of the PV system and wind turbine in Chapter 6.

Table D.5 Parameters of the voltage source converter.

Symbol	Value	Unit	Description
$V_{pp,RMS}$	3306.8	V	Rated voltage phase to phase root mean square (RMS) value
S_b	5	MW	Rated three-phase voltamperes (base power)
i_b	1234.6	A	Rated current
v_b	2700	V	Rated voltage
z_b	2.187	V/A	Reference impedance (base value)
i_{max}	1.3	pu	Maximum current value
f_{rated}	50	Hz	Nominal fundamental frequency of the AC side
R_{ft}	0.0273	V/A	Resistance (LCL filter and transformer)
L_{ft}	0.6914	mH	Longitudinal induction (LCL filter and transformer)
$k_{p,PLL}$	0.1094		Proportional gain of phase-locked loop (PLL) controller
$k_{i,PLL}$	1.8786		Integral gain of PLL controller
τ_{pf}	0.01	s	Time constants for active power filtering
$k_{p,p}$	0.5		Proportional gain of active power controller
$k_{i,p}$	10		Integral gain of active power controller
τ_{qf}	0.01	s	Time constants for reactive power filtering
$k_{p,q}$	0.5		Proportional gain of reactive power controller
$k_{i,q}$	10		Integral gain of reactive power controller

D.5 Equivalent Grid Model Parameters

Table D.6 contains the model parameters of the equivalent grid model illustrated in Figure 5.3 used for the grid integration of the PV system and wind turbine in Chapter 6.

Table D.6 Parameters of the equivalent grid model.

Symbol	Value	Unit	Description
$V_{pp,RMS}$	3306.8	V	Rated voltage phase to phase RMS value
v_b	2700	V	Rated voltage
f_{rated}	50	Hz	Nominal fundamental frequency of the grid
R_g	0.0164	V/A	Ohmic resistance of equivalent grid model
L_g	2.1	mH	Longitudinal induction equivalent grid model

E

Control Concept for Bumpless Transition Between Operating Regions

For a bumpless control between partialload and fullload, the control laws of these regions are merged using the Takagi–Sugeno (TS) approach. To this purpose, a coupling filter for bumpless transition is introduced

$$\begin{pmatrix} \tilde{\beta}_{ref} \\ \tilde{T}_g \end{pmatrix} = \begin{pmatrix} f_1(\omega_r) & 0 \\ 0 & f_2(T_g) \end{pmatrix} \begin{pmatrix} \beta_{ref} \\ T_g \end{pmatrix} = \mathbf{F}(\omega_r, T_g) \begin{pmatrix} \beta_{ref} \\ T_g \end{pmatrix}, \tag{E.1}$$

where

$$f_1(\omega_r) = w_{11}(\omega_r)\underline{f}_1 + w_{21}(\omega_r)\bar{f}_1,$$
$$f_2(T_g) = T_{g,max}\frac{1}{T_g} = w_{12}(T_g)\underline{f}_2 + w_{22}(T_g)\bar{f}_2$$

according to the index notation of w_{ij} in (A.7) with the lower and upper sector bounds

$$\underline{f}_1 = 0, \quad \bar{f}_1 = 1, \quad \underline{f}_2 = \frac{T_{g,max}}{T_{g,max}} = 1, \quad \bar{f}_2 = \frac{T_{g,max}}{T_{g,min}}.$$

Due to the convex sum condition $w_{1j}(\cdot) + w_{2j}(\cdot) = 1$ determined by (A.7), illustrated in Figures E.1 and E.2, the filter matrix (E.1) can be written as

$$\mathbf{F}(\omega_r, T_g) = \begin{pmatrix} \left(w_{12}(T_g) + w_{22}(T_g)\right)\left(w_{11}(\omega_r)\underline{f}_1 + w_{21}(\omega_r)\bar{f}_1\right) & 0 \\ 0 & \left(w_{11}(\omega_r) + w_{21}(\omega_r)\right)\left(w_{12}(T_g)\underline{f}_2 + w_{22}(T_g)\bar{f}_2\right) \end{pmatrix}$$

Advanced Control of Grid-Integrated Renewable Energy Power Plants: LMI-based Design in the Takagi-Sugeno Framework, First Edition. Horst Schulte.

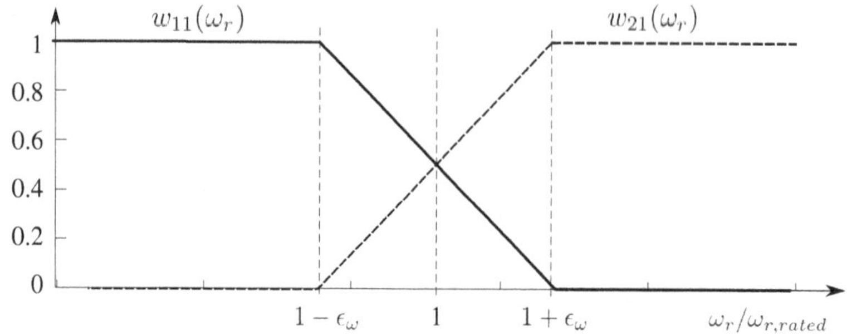

Figure E.1 Weighting functions $w_{i1}(\omega_r)$ for $i = 1{,}2$ with the overlapping factor ϵ_ω

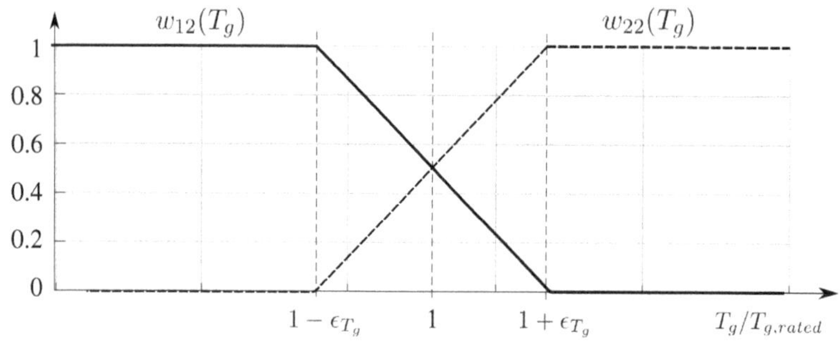

Figure E.2 Weighting functions $w_{i2}(T_g)$ for $i = 1{,}2$ with the overlapping factor ϵ_{T_g}

and rearranged in

$$\begin{aligned}\mathbf{F}(\omega_r, T_g) = {} & \underbrace{w_{11}(\omega_r)w_{12}(T_g)}_{h_1^a(\omega_r,T_g)}\underbrace{\begin{pmatrix}\underline{f}_1 & 0\\ 0 & \underline{f}_2\end{pmatrix}}_{\mathbf{F}_1} + \underbrace{w_{11}(\omega_r)w_{22}(T_g)}_{h_2^a(\omega_r,T_g)}\underbrace{\begin{pmatrix}\underline{f}_1 & 0\\ 0 & \bar{f}_2\end{pmatrix}}_{\mathbf{F}_2}\\ & + \underbrace{w_{21}(\omega_r)w_{12}(T_g)}_{h_3^a(\omega_r,T_g)}\underbrace{\begin{pmatrix}\bar{f}_1 & 0\\ 0 & \underline{f}_2\end{pmatrix}}_{\mathbf{F}_3} + \underbrace{w_{21}(\omega_r)w_{22}(T_g)}_{h_4^a(\omega_r,T_g)}\underbrace{\begin{pmatrix}\bar{f}_1 & 0\\ 0 & \bar{f}_2\end{pmatrix}}_{\mathbf{F}_4},\end{aligned}$$

which gives the compact sum form

$$\mathbf{F}(\omega_r, T_g) = \sum_{i=1}^{N_r^a=4} h_i^a(\mathbf{z}^a)\,\mathbf{F}_i \tag{E.2}$$

with $\mathbf{z}^a = (\omega_r, T_g)^T$. Note that the index a is introduced for the functions h_i^x in order to be able to differentiate between function type $x \in \{a, b, c\}$ in the following.

Control law of the partial-load region from (3.45) with that nomenclature is

$$u_1 = -\sum_{i=1}^{N_r^b} h_i^b(\mathbf{z}^b)\,(k_{x,i}^b,\ 0)\,\Delta\mathbf{x} + \sum_{i=1}^{N_r^b} h_i^b(\mathbf{z}^b)\,k_{I,i}^b\,x_I, \tag{E.3}$$

where $u_1 = T_g$ with $\mathbf{z}^b = (\omega_r,\ \hat{v})^T$. And the control law of the full-load region from (3.35) with $u_2 = \beta$ as controller output is

$$u_2 = -\sum_{i=1}^{N_r^c} h_i^c(\mathbf{z}^c)\,(\mathbf{k}_{x,i}^c)^T\,\Delta\mathbf{x} + \sum_{i=1}^{N_r^c} h_i^c(\mathbf{z}^c)\,k_{I,i}^c\,x_I \tag{E.4}$$

with $\mathbf{z}^c = (\omega_r,\ \beta,\ \hat{v})^T$. The membership functions $h_i^b(\mathbf{z}^b)$ and $h_i^c(\mathbf{z}^c)$ correspond to the previously introduced functions contained in (3.45) and (3.35). Both control laws are related to the common vector $\Delta\mathbf{x}$ (3.29) and integral state (3.28)

$$\Delta\mathbf{x} = \begin{pmatrix} \omega_r - \omega_{r,ref} \\ \beta \end{pmatrix}, \qquad x_I = \int_0^t \left(\omega_{r,ref} - \omega_r\right) d\tau.$$

These are merged into a common control law

$$\begin{pmatrix} u_1 \\ u_2 \end{pmatrix} = -\sum_{i=1}^{N_r^b}\sum_{j=1}^{N_r^c} h_i^b(\mathbf{z}^b) h_j^c(\mathbf{z}^c) \left[\begin{pmatrix} (k_{x,i}^b,\ 0) \\ (\mathbf{k}_{x,j}^c)^T \end{pmatrix} \Delta\mathbf{x} + \begin{pmatrix} k_{I,i}^b \\ k_{I,j}^c \end{pmatrix} x_I \right] \tag{E.5}$$

and finally combined with the bumpless coupling filter (E.2)

$$\begin{pmatrix} \tilde{u}_1 \\ \tilde{u}_2 \end{pmatrix} = -\sum_{m=1}^{N_r^a}\sum_{i=1}^{N_r^b}\sum_{j=1}^{N_r^c} h_m^a(\mathbf{z}^a) h_i^b(\mathbf{z}^b) h_j^c(\mathbf{z}^c) \mathbf{F}_m \left[\begin{pmatrix} (k_{x,i}^b,\ 0) \\ (\mathbf{k}_{x,j}^c)^T \end{pmatrix} \Delta\mathbf{x} + \begin{pmatrix} k_{I,i}^b \\ k_{I,j}^c \end{pmatrix} x_I \right]. \tag{E.6}$$

References

Jacob Aho, Paul Fleming, and Lucy Y. Pao. Active power control of wind turbines for ancillary services: A comparison of pitch and torque control methodologies. In *2016 American Control Conference (ACC)*, Boston Marriott Copley Place, Boston, MA, USA, July 2016.

Airfoil-Tools. Airfoil tools - search 1638 airfoils. http://airfoiltools.com, 08.11.2020.

Hirofumi Akagi, Edson Hirokazu Watanabe, and Mauricio Aredes. *Instantaneous Power Theory and Applications to Power Conditioning*. Number 13 in IEEE Press Series on Power and Energy Systems. John Wiley & Sons, 2007. ISBN 047011892X 9780470118924

B. Amlang, D. Arsudis, W. Leonhard, W. Vollstedt, and K. Wefelmeier. Elektrische Energieversorgung mit Windkraftanlagen. Technical report, BMFT 032-8265-B, Technische Universität Braunschweig, 1992.

Moritz Andrejewski, Stephan Kusche, Florian Pöschke, and Horst Schulte. POSYTYF, Deliverable 2.1 - Models for dispatchable and non-dispatchable RPPs. Technical report, HTW Berlin, 2021.

Moritz Andrejewski, Verena Häberle, Nico Goldschmidt, Florian Dörfler, and Horst Schulte. Experimental Validation of a Dynamic Virtual Power Plant Concept Based on Multiple-Converter Power Hardware-In-the-Loop Test Bench. Technical report, Ithaca, New York, 2023. URL https://doi.org/10.48550/arXiv.2309.00882.

G. Z. Angelis. *System Analysis, Modelling and Control with Polytopic Linear Models.* PhD thesis, Eindhoven University of Technology, Netherlands, 2001.

Harald Aschemann and Julia Kersten. Active damping and drive train control for a wind turbine with a hydrostatic transmission and a PMSG. *IFAC-PapersOnLine*, 50(1):9920–9925, 2017. ISSN 2405-8963. doi: 10.1016/j.ifacol.2017.08.1634 URL https://www.sciencedirect.com/science/article/pii/S2405896317322358. 20th IFAC World Congress.

Hans-Peter Beck and Ralf Hesse. Virtual synchronous machine. In *2007 9th International Conference on Electrical Power Quality and Utilisation*, pages 1–6, 2007. doi: 10.1109/EPQU.2007.4424220

Advanced Control of Grid-Integrated Renewable Energy Power Plants: LMI-based Design in the Takagi-Sugeno Framework, First Edition. Horst Schulte.

Mouhacine Benosman. *Learning-Based Adaptive Control, An Extremum Seeking Approach -Theory and Applications.* Elsevier, 2016.

Pontus Bergsten, R. Palm, and D. Driankov. Fuzzy observers. In *IEEE International Conference on Fuzzy Systems*, pages 700–703, Melbourne, Australia, 2001.

Albert Betz. *Wind-Energie und ihre Ausnutzung durch Windmühlen.* Nachdr. [d. Ausg.] Vandenhoeck u. Ruprecht, Göttingen, Germany, 1926.

Fernando D. Bianchi, Hernán De Battista, and Ricardo J. Mantz. *Wind Turbine Control Systems - Principles, Modelling and Gain Scheduling Design.* Springer-Verlag, London Limited, 2007.

Joakim Björk, Karl Henrik Johansson, and Florian Dörfler. Dynamic virtual power plant design for fast frequency reserves: Coordinating hydro and wind. *IEEE Transactions on Control of Network Systems*, 10(3):1266–1278, 2023. doi: 10.1109/TCNS.2022.3181553

E. A. Bossanyi. Wind turbine control for load reduction. *Wind Energy*, 6(3):229–244, 2003. doi: 10.1002/we.95

Stephen Boyd, Laurent El Ghaoui, Eric Feron, and Venkataramanan Balakrishnan. *Linear Matrix Inequalities in System and Control Theory.* SIAM, Philadelphia, PA, 1994.

Steven L. Brunton, Clarence W. Rowley, Sanjeev R. Kulkarni, and Charles Clarkson. Maximum power point tracking for photovoltaic optimization using ripple-based extremum seeking control. *IEEE Transactions on Power Electronics*, 25(10):2531–2540, Oct 2010. ISSN 1941-0107. doi: 10.1109/TPEL.2010.2049747

Eduard Bullich-Massagué, Ricard Ferrer-San-José, Mònica Aragüés-Peñalba, Luis Serrano-Salamanca, Carlos Pacheco-Navas, and Oriol Gomis-Bellmunt. Power plant control in large-scale photovoltaic plants: Design, implementation and validation in a 9.4 MW photovoltaic plant. *IET Renewable Power Generation*, 10(1):50–62, 2016. doi: 10.1049/iet-rpg.2015.0113 URL https://ietresearch.onlinelibrary.wiley.com/doi/abs/10.1049/iet-rpg.2015.0113.

Tony Burton, Nick Jenkins, David Sharpe, and Ervin A. Bossanyi. *Wind Energy Handbook.* John Wiley & Sons, Ltd, 3rd edition, April 2021. ISBN 978-1-119-45109-9

Ana Cabrera-Tobar, Eduard Bullich-Massagué, Mònica Aragüés-Peñalba, and Oriol Gomis-Bellmunt. Topologies for large scale photovoltaic power plants. *Renewable and Sustainable Energy Reviews*, 59:309–319, 2016. ISSN 1364-0321. doi: 10.1016/j.rser.2015.12.362 URL https://www.sciencedirect.com/science/article/pii/S1364032116000289.

Carlo Enrico Carcangiu, Iciar Font Balaguer, Stoyan Kanev, and Michele Rossetti. Closed-loop system identification of Alstom 3MW wind turbine. In Tom Proulx, editor, *Civil Engineering Topics, Volume 4*, pages 121–128, New York, NY, 2011. Springer New York. ISBN 978-1-4419-9316-8

Mahmoud Chilali and Pascal Gahinet. $H\infty$ design with pole placement constraints: An LMI approach. *IEEE Transactions on Automatic Control*, 41(3):358–367, 1996a. ISSN 0018-9286 doi: 10.1109/9.486637

Christian Clemens, Eckhard Gauterin, Florian Pöschke, and Horst Schulte. Assessment criteria for the mechanical loads of wind turbines applied to the example of active power control. In IFAC (International Federation of Automatic Control), editor, *Proceedings of IFAC World Congress 2020, IFAC PapersOnLine*, volume 53, page 341–347, Berlin, Germany, 2020. URL https://doi.org/10.1016/j.ifacol.2020.12.183. Paper.

Slobodan Cuk and R. D. Middlebrook. A general unified approach to modelling switching DC-to-DC converters in discontinuous conduction mode. In *1977 IEEE Power Electronics Specialists Conference*, pages 36–57, 1977. doi: 10.1109/PESC.1977.7070802

Kaushik Das, Müfit Altin, Anca Daniela Hansen, Poul Ejnar Sørensen, Damian Flynn, and Hans Abildgaard. Wind power support during overfrequency emergency events. *CIGRE Science & Engineering*, 9:73–83, 2018.

T. J. Dekker. Finding a zero by means of successive linear interpolation. In *Constructive Aspects of the Fundamental Theorem of Algebra*, London, 1969. Wiley-Interscience. ISBN 978-0-471-20300-1

DENA. Ancillary Services Study 2030 - Security and reliability of a power supply with a high percentage of renewable energy. Technical report, German Energy Agency (Dena), Berlin, 2014.

A. Díaz de Corcuera, A. Pujana-Arrese, J. M. Ezquerra, A. Milo, and J. Landaluze. Linear models-based LPV modelling and control for wind turbines. *Wind Energy*, 18(7):1151–1168, 2015. doi: 10.1002/we.1751 URL https://onlinelibrary.wiley.com/doi/abs/10.1002/we.1751.

Felix Nathan Dietrich, Steffen Borchers-Tigasson, Till Naumann, and Horst Schulte. Adaptive extremum seeking control of urban area wind turbines. *Energies 2021: Special Issue Modeling, Simulation and Control of Electric Drive Systems*, 14(5):1–12 (1356), 2021. ISSN 1996-1073. URL https://doi.org/10.3390/en14051356

C. S. Draper and Y. T. Li. Principles of optimalizing control systems and an application to the internal combustion engine. In *American Society of Mechanical Publication*, 1951.

W. C. Duesterhoeft, Max W. Schulz, and Edith Clarke. Determination of instantaneous currents and voltages by means of alpha, beta, and zero components. *Transactions of the American Institute of Electrical Engineers*, 70(2):1248–1255, 1951. doi: 10.1109/T-AIEE.1951.5060554

Thommy Ekelund. *Modeling and Linear Quadratic Optimal Control of Wind Turbines*. PhD thesis, Chalmers University of Technology, 1997.

ENTSO-E. Rate of Change of Frequency (RoCoF) withstand capability. Technical report, ENTSO-E guidance document for national implementation for network codes on grid connection, 2018.

Robert W. Erickson. DC–DC power converters. In *Wiley Encyclopedia of Electrical and Electronics Engineering*. John Wiley & Sons, Ltd, 2007. ISBN 9780471346081 doi: 10.1002/047134608X.W5808.pub2

I. Erlich, A. Korai, and F. Shewarega. Study on the minimum share of conventional generation units required for stable operation of future converter-dominated grids. In *2018 IEEE Power & Energy Society General Meeting (PESGM)*, pages 1–5, 2018. doi: 10.1109/PESGM.2018.8585943

Trishan Esram and Patrick L. Chapman. Comparison of photovoltaic array maximum power point tracking techniques. *IEEE Transactions on Energy Conversion*, 22(2):439–449, 2007. doi: 10.1109/TEC.2006.874230

Gene F. Franklin, J. David Powell, and Abbas Emami-Naeini. *Feedback Control of Dynamic Systems*. Addison-Wesley Publishing Company, Inc., 1994.

Christos Galinos, Albert M. Urban, and Wai Hou Lio. Optimised de-rated wind turbine response and loading through extended controller gain-scheduling. *Journal of Physics: Conference Series*, 1222:012020 1–8, 2019. doi: 10.1088/1742-6596/1222/1/012020

Robert Gasch and Jochen Twele, editors. *Wind Power Plants*. Springer-Verlag, New York, 2012. doi: 10.1007/978-3-642-22938-1

Eckhard Gauterin, P. Kammerer, Martin Kühn, and Horst Schulte. Effective wind speed estimation: Comparison between Kalman filter and Takagi–Sugeno observer techniques. *ISA Transactions*, 62:60–72, 2015.

Eckhard Gauterin, Florian Pöschke, and Horst Schulte. Global versus local Lyapunov approach used in disturbance observer-based wind turbine control. *Frontiers in Control Engineering (Section Adaptive, Robust and Fault Tolerant Control)*, 3:1–23, February 2023. doi: 10.3389/fcteg.2022.787530

Sören Georg. *Fault Diagnosis and Fault-Tolerant Control of Wind Turbines - Nonlinear Takagi–Sugeno and Sliding Mode Techniques*. PhD thesis, HTW Berlin, Control Engineering Group / University Rostock, Fakultät für Maschinenbau und Schiffstechnik, 2015.

Sören Georg, Matthias Müller, and Horst Schulte. Wind turbine model and observer in Takagi–Sugeno model structure. In *EAWE Conference 'The Science of Making Torque from Wind'*, volume IOP Journal of Physics: Conference Series (JPCS), Oldenburg, October 2012a.

Sören Georg, Horst Schulte, and Harald Aschemann. Control-oriented modelling of wind turbines using a Takagi–Sugeno model structure. In *IEEE International Conference on Fuzzy Systems*, pages 1737–1744, Brisbane, Australia, 2012b.

Martin Geyler and Peter Caselitz. Lastreduzierende Pitchregelung für Windenergieanlagen (Load Reducing Pitch Control for Wind Turbines). *at - Automatisierungstechnik*, 56(12):627–635, 2008. doi: 10.1524/auto.2008.0741

GL Wind. Guideline for the Certification of Wind Turbines. Technical report, Standard, Germanischer Lloyd, 2010.

Saeed Golestan, Josep M. Guerrero, and Juan C. Vasquez. Three-phase PLLs: A review of recent advances. *IEEE Transactions on Power Electronics*, 32(3):1894–1907, 2017. doi: 10.1109/TPEL.2016.2565642

Ruben Gomez-Merchan, Sergio Vazquez, Abraham Marquez Alcaide, Hossein Dehghani Tafti, Jose I. Leon, Josep Pou, Christian A. Rojas, Samir Kouro, and Leopoldo G. Franquelo. Binary search based flexible power point tracking algorithm for photovoltaic systems. *IEEE Transactions on Industrial Electronics*, 68(7):5909–5920, 2021. doi: 10.1109/TIE.2020.2998743

Shaul Gutman. Root clustering of a complex matrix in an algebraic region. *IEEE Transactions on Automatic Control*, 24(4):647–650, August 1979.

Shaul Gutman and E. Jury. A general theory for matrix root-clustering in subregions of the complex plane. *IEEE Transactions on Automatic Control*, 26(4):853–863, 1981. doi: 10.1109/TAC.1981.1102764

Verena Häberle, Michael W. Fisher, Eduardo Prieto-Araujo, and Florian Dörfler. Control design of dynamic virtual power plants: An adaptive divide-and-conquer approach. *IEEE Transactions on Power Systems*, 37(5):4040–4053, 2022. doi: 10.1109/TPWRS.2021.3139775

Verena Häberle, Ali Tayyebi, Xiuqiang He, Eduardo Prieto-Araujo, and Florian Dörfler. Grid-forming and spatially distributed control design of dynamic virtual power plants. *IEEE Transactions on Smart Grid*, 2023. doi: 10.1109/TSG.2023.3311481

Martin O. L. Hansen. *Aerodynamics of Wind Turbines*. Earthscan, 2nd edition, 2008.

Siegfried Heier. *Grid Integration of Wind Energy Conversion Systems*. John Wiley & Sons, Ltd, 2006.

E. L. van der Hooft and T. G. van Engelen. Estimated wind speed feed forward control for wind turbine operation optimisation. In *European Wind Energy Conference*, 2004.

Bernabé Ibáñez, F. A. Inthamoussou, and Hernán De Battista. Wind turbine load analysis of a full range LPV controller. *Renewable Energy*, 145:2741–2753, 2020. ISSN 0960-1481. doi: 10.1016/j.renene.2019.08.016 URL https://www.sciencedirect.com/science/article/pii/S0960148119312030.

D. Ichalal, B. Marx, D. Maquin, and J. Rago. On observer design for nonlinear Takagi-Sugeno systems with unmeasurable premise variables. In *International Symposium on Advanced Control of Industrial Processes (ADCONIP 2011)*, Hangzhou, China, May 2011.

Fernando A. Inthamoussou, Hernan De Battista, and Ricardo J. Mantz. LPV-based active power control of wind turbines covering the complete wind speed range. *Renewable Energy*, 99:996–1007, December 2016. doi: 10.1016/j.renene.2016.07.064 https://doi.org/j.renene.2016.07.064.

IRENA. Renewable Energy Statistics 2020. Technical report, The International Renewable Energy Agency, Abu Dhabi, 2020.

Joi. *Modelling of inverter-based generation for power system dynamic studies, Report 727*. Joint working group, C4/C6.35/CIRED, May 2018.

Jason Jonkman. FAST, https://www.nrel.gov/wind/nwtc/fastv8.html, accessed 2th August 2020. Technical report, NREL, 2020.

Jason M. Jonkman and M. L. Buhl. TurbSim User's Guide. Technical report, National Renewable Energy Laboratory, 2006.

Jason M. Jonkman, J. R. Buhl, and L. Marshall. FAST User's Guide, 2005.

J. Jonkman, S. Butterfield, W. Musial, and G. Scott. Definition of a 5-MW Reference Wind Turbine for Offshore System Development. Technical report, NREL/TP-500-38060, National Renewable Energy Laboratory, Golden, Colorado, 2009.

D. Astrain Juangarcia, I. Eguinoa, and T. Knudsen. Derating a single wind farm turbine for reducing its wake and fatigue. *Journal of Physics: Conference Series*, 1037(3):032039 Jun 2018. doi: 10.1088/1742-6596/1037/3/032039

Hassan K. Khalil. *Nonlinear Systems*. Prentice Hall, 2002.

Kwansu Kim, Hyun-Gyu Kim, Cheol Jin Kim, Insu Paek, Carlo L. Bottasso, and Filippo Campagnolo. Design and validation of demanded power point tracking control algorithm of wind turbine. *International Journal of Precision Engineering and Manufacturing-Green Technology*, 5:387–400, July 2018.

Arne Kisser, Marcel Engel, Laila Rezai, Moritz Andrejewski, Jens Fortmann, and Horst Schulte. A Test-bed System for Validation of Ancillary Services of Wind Farms under Realistic Conditions. In *Proceedings of 16th Wind Integration Workshop*, Berlin, 2017. ISBN 978-3-9816549-6-7

Norbert Klaes, Nico Goldschmidt, and Jens Fortmann. Voltage fed control of distributed power generation inverters with inherent service to grid stability. *Energies*, 13(10), 2020. ISSN 1996-1073. doi: 10.3390/en13102579 URL https://www.mdpi.com/1996-1073/13/10/2579.

Przemyslaw Komarnicki, Michael Kranhold, and Zbigniew A. Styczynski. *Sector Coupling - Energy-Sustainable Economy of the Future*. Springer Nature, 2023. ISBN 978-3-658-38111-0 doi: 10.1007/978-3-658-38111-0

Miroslav Krstić and Hsin-Hsiung Wang. Stability of extremum seeking feedback for general nonlinear dynamic systems. *Automatica*, 36(4):595–601, 2000. ISSN 0005-1098. doi: 10.1016/S0005-1098(99)00183-1 URL https://www.sciencedirect.com/science/article/pii/S0005109899001831.

Thomas S. Kuhn. *The Structure of Scientific Revolutions*. University of Chicago Press, 2nd edition, 1962. ISBN 9780226458113

G. A. M. van Kuik. The Lanchester–Betz–Joukowsky limit. *Wind Energy*, 10:289–291, 2007.

G. A. M. van Kuik, J. Peinke, R. Nijssen, D. Lekou, J. Mann, J. N. Sørensen, C. Ferreira, J. W. van Wingerden, D. Schlipf, P. Gebraad, H. Polinder, A. Abrahamsen, G. J. W. van Bussel, J. D. Sørensen, P. Tavner, C. L. Bottasso, M. Muskulus, D. Matha, H. J. Lindeboom, S. Degraer, O. Kramer, S. Lehnhoff, M. Sonnenschein, P. E. Sørensen, R. W. Künneke, P. E. Morthorst, and K. Skytte. Long-term research challenges in wind energy–a research agenda by the European Academy of Wind Energy. *Wind Energy Science*, 1(1):1–39, 2016. doi: 10.5194/wes-1-1-2016 URL https://wes.copernicus.org/articles/1/1/2016/.

Ritesh Kumar, Balakrushna Sahu, Chandan Kumar Shiva, and B. Rajender. A control topology for frequency regulation capability in a grid integrated PV system. *Archives of Electrical Engineering*, 69(2):389–401, 2020. doi: 10.24425/aee.2020.133033 URL http://journals.pan.pl/Content/116233/PDF/art11a.pdf.

Prabha Kundur. *Power System Stability And Control*. EPRI Power System Engineering Series. McGraw-Hill, 1994. ISBN 9780070635159

Stephan Kusche, Florian Pöschke, and Horst Schulte. POSYTYF, Deliverable 2.2 - Objectives and Assessment Criteria for Controller Design. Technical report, HTW Berlin, February 2022.

F. W. Lanchester. A contribution to the theory of propulsion and the screw propeller. *Transactions of the Institution of Naval Architects*, 57:98–116, 1915.

L. D. Landau and E. M. Lifschitz. *Lehrbuch der Theoretischen Physik, Band 1: Mechanik*. Akademie Verlag Berlin, 1990.

R. H. Lasseter. Microgrids. In *2002 IEEE Power Engineering Society Winter Meeting. Conference Proceedings*, volume 1, pages 305–308, 2002. doi: 10.1109/PESW.2002.985003

M. Leblanc. Sur lélectrification des chemins de fer au moyen de courants alternatifs de fréquence élevée. *Revue Générale de lElectricité*, 12(8):275–277, 1922.

Zsófia Lendek, Thierry Marie Guerra, Robert Babuška, and Bart de Schutter. *Stability Analysis and Nonlinear Observer Design Using Takagi-Sugeno Fuzzy Models*, volume 262 of *Studies in Fuzziness and Soft Computing*. Springer-Verlag Berlin, Heidelberg, 2011. ISBN 978-3-642-16775-1

Lennart Ljung. *System Identification*. Prentice Hall PTR, 2nd edition, 1999.

J. Löfberg. YALMIP: A toolbox for modeling and optimization in MATLAB®. In *Proceedings of the IEEE CACSD*, pages 284–289, Taipei, Taiwan, 2004.

D. G. Luenberger. Observing the state of a linear system. *IEEE Transactions on Military Electronics*, 8(2):74–80, 1964.

A. M. Lyapunov. The general problem of the stability of motion. *International Journal of Control*, 55(3):531–773, 1992.

J. Machowski, J. Bialek, and J. Bumby. *Power System Dynamics: Stability and Control.* John Wiley & Sons, Ltd, 2nd edition, 2008.

Marc Cheah Mañé, Jaume Girona Badia, Oriol Gomis Bellmunt, Eduardo Prieto Araujo, Vinícius Albernaz Lacerda, Florian Pöschke, Stephan Kusche, and Horst Schulte. POSYTYF, Deliverable D1.4 –Power System Modelling, Simulation and Co-simulation. Technical report, CITCEA-UPC, HTW Berlin, 2023.

B. Marinescu, O. Gomis-Bellmunt, F. Dörfler, H. Schulte, and L. Sigrist. Dynamic virtual power plant: A new concept for grid integration of renewable energy sources. *IEEE Access*, 2022:1–12, 2022. ISSN 2169-3536. doi: 10.1109/ACCESS.2022.3205731

David Marten, Joseph Saverin, Robert Behrens de Luna, and Sebastian Perez-Becker. QBlade User's Guide. Technical report, TU Berlin, 2021. URL https://docs.qblade.org/src/user/indexue.html.

Mathworks. *LMI Solvers.* URL https://www.mathworks.com/help/robust/lmis.html, 2021.

C. Messasma, S. E. Chouaba, B. Sari, and A. Barakat. An approach for Power Reserve Control (PRC) Strategy Based on a Novel ANN Model. In Mustapha Hatti, editor, *Advanced Computational Techniques for Renewable Energy Systems*, pages 594–601, Cham, 2023a. Springer International Publishing. ISBN 978-3-031-21216-1.

Chaouki Messasma, Abdallah Barakat, Seif eddine Chouaba, and Bilal Sari. PV system frequency regulation employing a new power reserve control approach and a hybrid inertial response. *Electric Power Systems Research*, 223:109556 2023b. ISSN 0378-7796. doi: 10.1016/j.epsr.2023.109556 URL https://www.sciencedirect.com/science/article/pii/S0378779623004455.

R. D. Middlebrook and Slobodan Cuk. A general unified approach to modelling switching-converter power stages. In *1976 IEEE Power Electronics Specialists Conference*, pages 18–34, 1976. doi: 10.1109/PESC.1976.7072895

Federico Milano and Alvaro Ortega Manjavacas. *Frequency Variations in Power Systems: Modeling, State Estimation and Control.* Wiley-IEEE Press, 2020. ISBN 978-1119551843.

Oliver Nelles. *Nonlinear System Identification with Local Linear Neuro-Fuzzy Models.* PhD thesis, TU Darmstadt, Shaker Verlag, Aachen, Germany, 1999.

Yurii Nesterov and Arkadii Nemirovskii. Interior-point polynomial methods in convex programming. *Studies in Applied Mathematics SIAM*, 13, 1994. URL https://epubs.siam.org/doi/book/10.1137/1.9781611970791.

David Niemann, Jing Li, Hua O. Wang, and Kazuo Tanaka. Parallel distributed compensation for Takagi-Sugeno fuzzy models: New stability conditions and dynamic feedback designs. *IFAC Proceedings Volumes*, 32(2):5374–5379, 1999. ISSN 1474-6670. doi: 10.1016/S1474-6670(17)56915-3 URL https://www.sciencedirect

.com/science/article/pii/S1474667017569153. 14th IFAC World Congress 1999, Beijing, Chia, 5–9 July.

NN. Grid Forming Technology, Bulk Power System Reliability Considerations. Technical report, NERC, North American Electric Reliability Corporation, 3353 Peachtree Road NE, Suite 600, North Tower, Atlanta, GA 30326, 2021.

OAT. *Technical data sheet - Pitch System of OAT*. OAT Osterholz Antriebstechnik GmbH, August-Thyssen-Str. 7, 32278 Kirchlengern, Germany, 2021. URL https://www.oat-gmbh.de/en/produkte/pitch-system/.

Daniel E. Ochoa, Felipe Galarza-Jimenez, Felipe Wilches-Bernal, David A. Schoenwald, and Jorge I. Poveda. Control systems for low-inertia power grids: A survey on virtual power plants. *IEEE Access*, 11:20560–20581, 2023. doi: 10.1109/ACCESS.2023.3249151.

Peter Fogh Odgaard, Jakob Stoustrup, and Michel Kinnaert. Fault tolerant control of wind turbines –a benchmark model. *IFAC Proceedings Volumes*, 42(8):155–160, 2009. ISSN 1474-6670. doi: 10.3182/20090630-4-ES-2003.00026 URL https://www.sciencedirect.com/science/article/pii/S147466701635769X. 7th IFAC Symposium on Fault Detection, Supervision and Safety of Technical Processes.

Hiroshi Ohtake, Kazuo Tanaka, and Hua O. Wang. Fuzzy modeling via sector nonlinearity concept. In *Joint 9th IFSA World Congress and 20th NAFIPS International Conference*, pages 127–132, Vancouver, Canada, 2001.

K. Z. Ostergaard, P. Brath, and J. Stoustrup. Estimation of effective wind speed. Journal of Physics: Conference Series. In *The Science of Making Torque from Wind, IOP Publishing Journal of Physics: Conference Series 75*, number 012082, pages 1–9, 2007. doi: 10.1088/1742-6596/75/1/012082

Kasper Zinck Ostergaard, Jakob Stoustrup, and Per Brath. Linear parameter varying control of wind turbines covering both partial load and full load conditions. *International Journal of Robust and Nonlinear Control*, 19(1):92–116, 2009. doi: 10.1002/rnc.1340

Donghua Pan, Xiongfei Wang, Fangcheng Liu, and Rongliang Shi. Transient stability of voltage-source converters with grid-forming control: A design-oriented study. *IEEE Journal of Emerging and Selected Topics in Power Electronics*, 8(2):1019–1033, 2020. doi: 10.1109/JESTPE.2019.2946310

Qiao Peng, Yongheng Yang, Huai Wang, and Frede Blaabjerg. On power electronized power systems: Challenges and solutions. In *2018 IEEE Industry Applications Society Annual Meeting (IAS)*, pages 1–9, 2018a. doi: 10.1109/IAS.2018.8544488

Zishun Peng, Jun Wang, Daqiang Bi, Yuxing Dai, and Yeting Wen. The application of microgrids based on droop control with coupling compensation and inertia. *IEEE Transactions on Sustainable Energy*, 9(3):1157–1168, 2018b. doi: 10.1109/TSTE.2017.2773512

Florian Pöschke, Jens Fortmann, and Horst Schulte. Nonlinear wind turbine controller for variable power generation in full load region. In *American Control Conference*, pages 1395–1400, Sheraton Hotel, Seattle, USA, 2017.

Florian Pöschke, Eckhard Gauterin, Martin Kühn, Jens Fortmann, and Horst Schulte. Load mitigation and power tracking capability for wind turbines using linear matrix inequality-based control design. *Wind Energy*, 23(9):1792–1809, 2020. doi: 10.1002/we.2516.

Florian Pöschke, Vlaho Petrović, Frederik Berger, Lars Neuhaus Michael Hölling, Martin Kühn, and Horst Schulte. Model-based wind turbine control design with power tracking capability: A wind-tunnel validation. *Control Engineering Practice*, 120(105014):1–13, March 2022. doi: 10.1016/j.conengprac.2021.105014

E. S. Pyatnitskii and V. I. Skorodinskii. Numerical methods of Lyapunov function construction and their application to the absolute stability problem. *System & Control Letters*, 2(2):130–135, August 1982.

Taoufik Qoria, Ebrahim Rokrok, Antoine Bruyere, Bruno François, and Xavier Guillaud. A PLL-free grid-forming control with decoupled functionalities for high-power transmission system applications. *IEEE Access*, 8:197363–197378, 2020. doi: 10.1109/ACCESS.2020.3034149.

R. Ramakumar and J. E. Bigger. Photovoltaic systems. *Proceedings of the IEEE*, 81(3):365–377, March 1993. ISSN 1558-2256.

John William Strutt Rayleigh. *The Theory of Sound*, volume 1. Macmillan and Co., New York. USA, 1877.

Jennifer M. Rinker and Henri P. Gavin. Subspace identification of a 5 MW reference wind turbine. In Michael Mains, editor, *Topics in Modal Analysis, Volume* 10, pages 23–29, Cham, 2015. Springer International Publishing. ISBN 978-3-319-15251-6

Joan Rocabert, Alvaro Luna, Frede Blaabjerg, and Pedro Rodríguez. Control of power converters in ac microgrids. *IEEE Transactions on Power Electronics*, 27(11):4734–4749, 2012. doi: 10.1109/TPEL.2012.2199334

Roberto Rosso, Xiongfei Wang, Marco Liserre, Xiaonan Lu, and Soenke Engelken. Grid-forming converters: An overview of control approaches and future trends. In *2020 IEEE Energy Conversion Congress and Exposition (ECCE)*, pages 4292–4299, 2020. doi: 10.1109/ECCE44975.2020.9236211

Roberto Rosso, Xiongfei Wang, Marco Liserre, Xiaonan Lu, and Soenke Engelken. Grid-forming converters: Control approaches, grid-synchronization, and future trends— A review. *IEEE Open Journal of Industry Applications*, 2:93–109, 2021. doi: 10.1109/OJIA.2021.3074028

Ruben Salazar-Chiralt, Marc Cheah-Mane, Elia Mateu-Barriendos, Eduard Bullich-Massague, Eduardo Prieto-Araujo, Hasan Mehrjerdi, and Oriol Gomis-Bellmunt. Dynamic interactions in large scale photovoltaic power plants with frequency and voltage support. *Electric Power Systems Research*, 207:107848

2022. ISSN 0378-7796. doi: 10.1016/j.epsr.2022.107848 URL https://www.sciencedirect.com/science/article/pii/S0378779622000785.

Peter W. Sauer, M. A. Pai, and Joe H. Chow. *Power System Dynamics and Stability*. John Wiley & Sons, 2nd edition, 2018.

J. Schoukens and R. Pintelon. *System Identification - A Frequency Domain Approach*. IEEE Press, New York, 2001.

Horst Schulte. *Approximative Modellierung, Systemidentifikation und Reglerentwurf mittels gewichteter Kombination lokaler Zustandsraummodelle am Beispiel fluidischer Antriebe: [Dissertation]*. Kassel Univ. Press, Kassel, 2005a. ISBN 3-89958-175-x

Horst Schulte. Approximate modeling of a class of nonlinear oscillators using Takagi-Sugeno fuzzy systems and its application to control design. In *Proceedings of the 44th IEEE Conference on Decision and Control*, pages 3387–3392, 2005b. doi: 10.1109/CDC.2005.1582685

Horst Schulte. Control-oriented description of large scale wind turbines with hydrostatic transmission using Takagi-Sugeno models. In *2014 IEEE Conference on Control Applications (CCA)*, pages 664–668. IEEE, 2014.

Horst Schulte. On nonlinear observers in Takagi-Sugeno form based on fixed and variable structure. In *IEEE International Conference on Fuzzy Systems (FUZZ-IEEE)*, pages 2364–2371, July 2016. doi: 10.1109/FUZZ-IEEE.2016.7737989

Horst Schulte and Eckhard Gauterin. Input-to-state stability analysis of small wind turbines using LMI conditions (in German). *at - Automatisierungstechnik*, 62(10):698–707, 2014.

Horst Schulte and Eckhard Gauterin. Input-to-State Stability (ISS) for nonlinear systems in Takagi-Sugeno's form using quadratic and non-quadratic Lyapunov Functions. In *IEEE International Conference on Fuzzy Systems*, Istanbul, Turkey, August 2015.

Horst Schulte and Hubert Hahn. Fuzzy state feedback gain scheduling control of servo-pneumatic actuators. *Control Engineering Practice*, 12(5):639–650, 2004. ISSN 0967-0661. doi: 10.1016/S0967-0661(03)00148-5 URL https://www.sciencedirect.com/science/article/pii/S0967066103001485. Fuzzy System Applications in Control.

Schulte H. and Pöschke F. (2016). *Sliding Mode Observer for Fault Diagnosis: LPV and Takagi–Sugeno Model Approaches*. In: Rauh, A., Senkel, L. (eds). Variable-Structure Approaches. Mathematical Engineering. Springer, Cham. https://doi.org/10.1007/978-3-319-31539-3_8.

Martin Shan, Weiwei Shan, Friedrich Welck, and Daniel Duckwitz. Design and laboratory test of black-start control mode for wind turbines wind energy. *Wind Energy*, 23(3):763–778, 2020. doi: 10.1002/we.2457

Sakshi Sharma, Vibhu Jately, Piyush Kuchhal, Peeyush Kala, and Brian Azzopardi. A comprehensive review of flexible power-point-tracking algorithms for grid-connected photovoltaic systems. *Energies*, 16(15), 2023. ISSN 1996-1073. doi: 10.3390/en16155679 URL https://www.mdpi.com/1996-1073/16/15/5679.

Md. Nahid Haque Shazon, Nahid-Al-Masood, and Atik Jawad. Frequency control challenges and potential countermeasures in future low-inertia power systems: A review. *Energy Reports*, 8:6191–6219, 2022. ISSN 2352-4847. doi: 10.1016/j.egyr.2022.04.063 URL https://www.sciencedirect.com/science/article/pii/S2352484722008289.

Eric Simley, Lucy Y. Pao, Rod Frehlich, Bonnie Jonkman, and Neil Kelley. Analysis of light detection and ranging wind speed measurements for wind turbine control. *Wind Energy*, 17(3):413–433, March 2014.

A. Singh and B. Pal. IEEE PES Task Force on Benchmark Systems Forstability Controls - Report on the 68-Bus, 16-Machine, 5-Area System. Technical report, IEEE General Meeting Power & Energy Society, 2013.

M. Singh, E. Muljadi, J. Jonkman, V. Gevorgian, I. Girsang, and J. Dhupia. Simulation for Wind Turbine Generators — With FAST and Matlab®-Simulink Modules. Technical Report NREL/TP-5D00-59195April 2014, National Renewable Energy Laboratory / Nanyang Technological University, 2014.

Eduardo D. Sontag and Yuan Wang. On characterizations of the input-to-state stability property. *Systems and Control Letters*, 24:351–359, 1995.

J. F. Sturm. Using SeDuMi 1.02, a MATLAB® toolbox for optimization over symmetric cones. *Optimization Methods and Software*, 11-12:625–653, 1999.

Irina Subotić and Dominic Groß. Power-balancing dual-port grid-forming power converter control for renewable integration and hybrid AC/DC power systems. *IEEE Transactions on Control of Network Systems*, 9(4):1949–1961, 2022. doi: 10.1109/TCNS.2022.3181551

Tomohiro Takagi and Micheo Sugeno. Fuzzy identification of systems and its application to modeling and control. *IEEE Transactions on Systems, Man, and Cybernetics*, 15(1):116–132, 1985.

Kazuo Tanaka and Hua O. Wang. *Fuzzy Control Systems Design and Analysis: A Linear Matrix Inequality Approach*. John Wiley & Sons, Inc., 2001.

K. Tanaka, T. Ikeda, and H. Wang. Design of fuzzy control systems based on relaxed LMI stability conditions. In *Proceedings of 35th IEEE Conference on Decision and Control*, volume 1, pages 598–603, 1996. doi: 10.1109/CDC.1996.574389

Kazuo Tanaka, T. Ikeda, and H. Wang. Fuzzy regulators and fuzzy observers: Relaxed stability conditions and LMI-based design. *IEEE Transactions on Fuzzy Systems*, 6:250–265, 1998.

Remus Teodorescu, Marco Liserre, and Pedro Rodriguez. *Grid Converters for Photovoltaic and Wind Power Systems*. Wiley-IEEE Press, 2007.

Remus Teodorescu, Marco Liserre, and Pedro Rodríguez. *Grid Converters for Photovoltaic and Wind Power Systems*. John Wiley & Sons, Ltd, 2011.

Gijs van der Veen, Jan-Willem van Wingerden, and Michel Verhaegen. Data-driven modelling of wind turbines. In *Proceedings of the 2011 American Control Conference*, pages 72–77, 2011. doi: 10.1109/ACC.2011.5990944

Matthias Wächter, Hendrik Heißelmann, Michael Hölling, Allan Morales, Patrick Milan, Tanja Mücke, Joachim Peinke, Nico Reinke, and Philip Rinn. The turbulent nature of the atmospheric boundary layer and its impact on the wind energy conversion process. *Journal of Turbulence*, 13:N26, January 2012. doi: 10.1080/14685248.2012.696118

Hua O. Wang, Kazuo Tanaka, and M. F. Griffin. An approach to fuzzy control of nonlinear systems: Stability and design issues. *IEEE Transactions on Fuzzy Systems*, 4(1):14–23, 1996.

O. Wasynczuk, D. T. Man, and J. P. Sullivan. Dynamic behavior of a class of wind turbine generators during random wind fluctuations. *IEEE Transactions on Power Apparatus and Systems*, PAS-100(6):2837–2845, 1981.

Bernd Weise, Abdul Korai, and Adrian Constantin. Comparison of selected grid-forming converter control strategies for use in power electronic dominated power systems. In *Proceedings of the 18th Wind Integration Workshop*, Dublin. Ireland, October 2019.

Weidong Xiao. *Photovoltaic Power System*. John Wiley & Sons, 2017.

W. Xiao, M. G. J. Lind, W. G. Dunford, and A. Capel. Real-time identification of optimal operating points in photovoltaic power systems. *IEEE Transactions on Industrial Electronics*, 53(4):1017–1026, 2006. doi: 10.1109/TIE.2006.878355

Amirnaser Yazdani and Reza Iravani. *Voltage-Sourced Converters in Power Systems: Modeling, Control, and Applications*. John Wiley & Sons, Inc., 2010. doi: 10.1002/9780470551578

P. P. Zarina, S. Mishra, and P. C. Sekhar. Exploring frequency control capability of a PV system in a hybrid PV-rotating machine-without storage system. *International Journal of Electrical Power & Energy Systems*, 60:258–267, 2014. ISSN 0142-0615. doi: 10.1016/j.ijepes.2014.02.033 URL https://www.sciencedirect.com/science/article/pii/S0142061514000969.

Qing-Chang Zhong, Wen-Long Ming, and Yu Zeng. Self-synchronized universal droop controller. *IEEE Access*, 4:7145–7153, 2016. doi: 10.1109/ACCESS.2016.2616115

Index

Advanced Control of Grid-Integrated Renewable Energy Power Plants: LMI-based Design in the Takagi-Sugeno Framework, First Edition. Horst Schulte.

www.ingramcontent.com/pod-product-compliance
Lightning Source LLC
LaVergne TN
LVHW011524200926
848484LV00001B/138
* 9 7 8 1 1 1 9 7 0 1 3 9 2 *